T0203569

BALKEMA – Proceedings and Monographs
in Engineering, Water and Earth Sciences

# Water Crisis: Myth or Reality?
## Marcelino Botin Water Forum 2004

*Edited by*

Peter P. Rogers
*Harvard University, Cambridge, Massachusetts, USA*

M. Ramón Llamas
*Royal Academy of Sciences, Madrid, Spain*

Luis Martínez-Cortina
*Spanish Geological Survey, IGME, Spain*

LONDON/LEIDEN/NEW YORK/PHILADELPHIA/SINGAPORE

Published by: Taylor & Francis/Balkema
            P.O. Box 447, 2300 AK Leiden, The Netherlands
            e-mail: Pub.NL@tandf.co.uk
            www.balkema.nl, www.tandf.co.uk, www.crcpress.com

ISBN 10: 0-415-36438-8
ISBN 13: 9-78-0-415-36438-6

Printed in Great Britain

# TABLE OF CONTENTS

PRESENTATION

Media headlines around the world often voice an impending water crisis that is likely to affect a large share of the world population. A good number of experts consider that this looming crisis might be over-hyped, and probably the consequence of a preventive strategy. However, *crying wolf* seems to be backfiring: pessimistic predictions done in a good number of international fora have not occurred and the credibility of a part of the scientific community is in jeopardy.

In an attempt to contribute to dimension the problem in a realistic way the Marcelino Botin Foundation organized and sponsored the first Santander Workshop, entitled **Water Crisis: Myth or Reality?** Topic and expert selection were entrusted to Prof. M. Ramon Llamas, from the Complutense University of Madrid, Spain; and to Prof. Peter Rogers, from Harvard University, USA.

A three-day workshop was held in the Marcelino Botin Foundation headquarters in Santander from June 14 to 16, 2004. Two experts were assigned to discuss each of the topics. The first one to present his/her views and the second to critique those opinions or to present a different perspective on the same topic. A debate by the whole group followed. Experts were required to circulate their manuscripts in advance and those wishing to do so were given time to modify them for publication after the workshop.

This book includes a foreword and nineteen chapters corresponding to the topics selected. Four of the manuscripts previously envisioned could not be finally included in the book for different reasons. The task of collecting the final versions of the manuscripts from the authors and to send them to the publishers has been painstakingly carried out by Dr. Martinez-Cortina.

The reader will find different opinions on the same topics, depending on the views of each author. This is a consequence of the multifaceted character of most water-related issues. However, all experts seem to have reached a general agreement on something: *a potential*

Figure 1.    Headquarters of the Marcelino Botin Foundation in Santander (Spain).

Figure 2.   Participants in the Santander Workshop[1].

*water crisis would not be due to physical water scarcity but rather to water resources misman-*
*agement, or in other words, to poor water resources governance.*

The Marcelino Botin Foundation has had prior involvement in dealing with water resources issues, beginning in 1999, with a four and a half year research project on groundwater resources: the *Proyecto Aguas Subterraneas* (PAS) also led by Prof. Llamas. The results of this activity are summarized in thirteen monographs directly published by the M. Botin Foundation (they can be downloaded from the Foundation Web: *www.fundacionmbotin.org*), six books in Spanish published jointly with Mundi-Prensa, and two books in English published by Balkema Publishers, as a common effort with the Geological Survey of Spain (IGME). A basic Hydrogeology textbook for secondary school students has also been produced jointly with IGME, and distributed by Spain's Ministry of Education to all the country's Secondary Schools. This book has been translated into English for its inclusion among UNESCO's publications. A final result of the PAS project was the organization of an International Symposium on Groundwater Intensive Use in Valencia (December 10–14, 2002). Selected papers of this Symposium gave rise to the second book published with Balkema.

Finally, we are glad to announce that the Marcelino Botin Foundation is already working on the second Santander Workshop, to be held in June 2007, under the title of ***Ethical Issues in Water Resources Management***. Prof. Llamas will also be in charge of organizing this event. UNESCO has already expressed a willingness to be involved.

Santander, September 2005
*Marcelino Botin Foundation*

---

[1]From left to right; (standing): E. Custodio, T. Asano, P. Arrojo, D. Jiménez-Beltrán, J.A. Allan, D. Del Porto, P. Rogers, M.R. Llamas, F. García-Novo, J. Ramírez-Vallejo, P. Martínez-Santos, W.M. Hanemann, L. Martínez-Cortina, and W.J. Cosgrove; (seated): A. Mukherji, E. Schlager, C.A. Sullivan, M.S. Aguirre, and E. López-Gunn.

# FOREWORD

Everyday we are bombarded by the media with tales of gloom and doom – we are running out of water, petroleum, open space, clean air, arable land, etc. While there is good reason to be wary of some of the potential outcomes of these prognostications, the reader should be encouraged to remember that *forecasts* are not *destiny*. The human race has an uncanny knack of proving the doomsayers wrong. This certainly goes back at least as far as the 18th century of Malthus and his gloomy predications of widespread famine unless there were wars and epidemics to curb the natural growth of population. To be sure, there have been many wars and epidemics since Malthus's time, too many, nevertheless the population of the globe continued its giddying increase. The outcome is that 200 years after Malthus, the population has increased several-fold and enjoys greater longevity and health than it enjoyed in his times. Of course, there are more poor people on the globe today than the total population at Malthus's time; that is bad, but there are also many more people who would be considered to be *living like kings* in Malthus's time. Is this good? Both poverty and affluence stress the environment in ways that Malthus could never have envisaged. This book is about fresh water and asks the question: is the impending global shortage of water reality or myth?

Water is a classic *renewable resource*, that is one that will cycle continuously regardless of ice ages and global warming. It is pretty much in a fixed amount on our small planet. What is at stake is the allocation of this amount of water among potential users. For example, before *homo sapiens* roamed the globe, water use was 100% for the ecosystem. As mankind grew and prospered we started to make inroads in this exclusive use of water by the ecosystem, but here was plenty of water to go around for millennia. It took maybe a hundred thousand years before *homo sapiens* demanded, and took, a significant amount of water from nature. This was when agriculture became settled and we began to develop irrigated agriculture in the Fertile Crescent and in China. Settled agriculture allowed a rapid increase in human populations which continually demanded more and more water from nature's account. Seven thousand years brings us up to the present when the largest human demand for water comes from irrigation systems worldwide. As much as 70–90% of all the water consumed on an annual basis is used by irrigated agriculture. This is why so much of the *running-out-of-water* stories are based upon the facts that in many parts of the world the rivers are running dry and the groundwater is over-pumped to provide irrigation for feeding the world's population. Surely, this is a sign of a Global Water Crisis? Because if we are unable to feed ourselves, then Malthusian constraints will take over with disastrous consequences for *homo sapiens* and for the planet.

In addition, the rate at which water sources are becoming contaminated with waste from humans, industry, and agriculture is truly alarming. Human ingenuity is busily creating tens of thousands of new chemicals that ultimately find their way into our drinking water supplies. More than one million three hundred thousand children under the age of four die each year due to diarrhoeal diseases and another one million due to malaria, both largely caused by water supply and management deficiencies. Global climate change is predicted by some to have major impacts upon the availability, spatial distribution, and the variability of water supplies during the next century.

This litany of disasters can make even the most stouthearted falter. This volume attempts to examine these issues and tries to unravel from them what is permanent and unchangeable and what is remediable and changeable. On close examination of the problems we find that there is much to be hopeful about the global water situation. To be sure, there will be places on the globe that will be condemned to permanent water shortages and crises, but these are only a small part of the problems. We find that if the policy prescriptions outlined in these chapters are taken into account, the Global Water Crisis can be dealt with over a fairly short time scale.

September 2005
The Editors

I

# Water Policy and Management

# CHAPTER 1

# Water governance, water security and water sustainability

P. Rogers

*Harvard University, Cambridge, Massachusetts, USA*

ABSTRACT:  This chapter attempts to bring together a set of disparate concepts that are fundamental to examining water as a resource and establishing the seriousness of the current and future water scarcity. As is well known, there is a plentiful supply of water considered at a global scale. However, as we examine scales much closer to individual humans, a pattern of great heterogeneity emerges. Some parts of the world have plentiful supplies of water, others have severe droughts; some plenty of high quality water, others with badly polluted waters; in some the rivers flow full, in others they are devoid of water for many days of the year. It would be simple if these differences were due only to the physical climate, but careful examination shows that there are large differences within the same climate zones that cannot be explained purely by climate and topography. In these cases, one sees the hands of human interference in terms of governance, property rights, and sheer population size. The situation may become much more serious in the badly impacted areas as the great climate change experiment unfolds. There is huge uncertainly associated with the predictions of climate for the 21st century. The chapter is able to be optimistic in the face of such uncertainty by pointing to several technical, economic, and social developments that can reduce the human footprint on the scarce supplies of easily accessible water. By relying more on rainfed agriculture and agricultural trade to meet food needs scarce irrigation water can focus upon higher value and less water using crops or can be diverted to high value municipal and industrial uses; improving the efficiency of current irrigation technologies will free up large quantities of water for other uses; relying upon new ecological sanitation techniques can greatly reduce the impact on water quality; and low cost breakthroughs of desalination cost which are now economically competitive with alternative sources of fresh water to meet the needs of urban populations anywhere in the world. In order for these solutions to the emerging crisis to be adopted much more attention will have to be paid as to how we as individuals and communities approach and the world community approaches the governance of water. A successful shift to effective governance will enable us to have sustainable water supplies for all well through the 21st century.

Keywords:  *Governance, policy, politics, water security, global resource availability, promising options*

> "Of all the social and natural crises we humans face, the water crisis is the one that lies at the heart of our survival and that of planet Earth . . . No region will be spared from the impact of this crisis which touches every facet of life, from the health of children to the ability of nations to secure food for their citizens . . . Water supplies are falling while the demand is dramatically growing at an unsustainable rate. Over the next twenty years the average supply of water worldwide per person is expected to drop by one third" (UN/WWAP, 2003).

## 1 INTRODUCTION

Writing about myths, reality, and crises, in the context of water governance, security, and sustainability provides the writer with a huge number of studies on resource use by humans and nature. When considering sustainability, one has to contend with the 18th century Malthusian hypotheses of fixed resources facing inexorable geometric human population growth. This has to be reconciled with the 21st century's actual water resource picture and the needs of nature and ecosystems. Security of supply is a major issue, where one is faced with highly variable resources having to meet the needs of increasing and highly diverse populations in cities and countryside, both rich and poor, and the demands of nature. Finally, governance has to face up to reconciliation of all these demands placed upon a fixed, or maybe even a shrinking resource base. How can we ever unscramble the myths and realities from these highly interrelated issues?

## 2   SUSTAINABILITY

The problem with the concept of *sustainability* is that everyone seems to understand what is meant by it in its promiscuous usage, however, everybody understands it differently from all others. As long ago as 1992, Pezzey (1992) compiled a list of over 50 definitions in widespread use at that time. No two were identical, yet the users seemed to feel comfortable with their use. Since then, a widespread outpouring of sustainable development literature triggered by the *UN Commission on Sustainable Development* (UNCSD) has led to the addition of several hundreds of indicators of sustainable development.

There are some deep philosophical issues involved in the concept of sustainability. First, we need to understand sustainability for whom – the planet, the ecosystem and nature, the human populations, the standard of living of the industrialized countries, the poor, the rich, for cities, for suburbs, for particular species of fauna and flora, and deposits of metals and minerals, etc. Second, sustainability is not a static concept; whatever we do now will influence, positively or negatively, how sustainable the future course of human development will be. Hence, the time horizon is of great importance. The *Bruntland Commission* (World Commission on Environment and Development, 1987) claimed that sustainability meant a process that "meets the needs of the present without compromising the ability of future generations to meet their own needs". Repetto (1986) was more restrictive than the Bruntland Commission when he defined it thus: "sustainable development as a goal rejects policies and practices that support current living standards by depleting the resource base, including natural resources, and that leaves future generations with poorer prospects and greater risks than our own". But do we mean the next generation or all future generations? The choice might change our conclusions about the outcome. In addition, human populations and the ecosystem itself are quite adaptable to changes in their resource base over time. We only have to examine the geological history of the recent past when *homo sapiens* first showed up on the planet to see how adaptable our species and the planet itself have been. Solow (1991) used a gentler definition: "leave future generations the option or the capacity to be as well off as we are". Pezzey (1992) makes the definition more complex by raising the distinction between *survivability* "which requires welfare to be above a particular threshold in all time periods" and *sustainability* "which requires welfare to be non-decreasing in all time periods". So without any agreed upon operational definition it is difficult to say what in particular is sustainable – but we can unequivocally state what is not sustainable! Our current flagrant over exploitation of energy and water resources is not sustainable.

Our concern in this book is water as a resource and hence we are spared some of the anguish of persons working in broader ecosystems. Nevertheless, without water there are no ecosystems. It is worth noting that sustainability also implies *survivability*, but that the concepts are markedly different. Survivability implies that we will be able to survive particular episodes of extreme stress whereas; sustainability has the implications of being able to sustain a life-style at least at present levels well into the future. In the same vein, Pezzey (2002) further confuses the debate as he points out how *sustainability policy* and *environmental policy*, usually considered one and the same thing, can be theoretically quite distinct.

> "Environmental policy is dynamic, government intervention to maximize inter-temporal social welfare based on the individual's own discount rate path, by internalizing the social values of stocks and flows that agents ignore (externalize) when they privately maximize welfare ... By contrast, sustainability policy aims to achieve some improvement of intergenerational equity, whether a general shift to a lower path of the utility discount rate over time, or a specific goal such as making utility forever constant, or non-declining or sustainable" (Pezzey, 2002: 26–27).

In this chapter, then, the reader will notice the concern for managing to survive the next two generations until 2050. In this stance we are more akin to the supporters of *survivability* than of true *sustainability*. There are, of course, some ambiguities: should effects of climate change be considered as sustainability issues or security issues?

## 2.1 *Basic laws of ecology*

There are three basic laws of ecology that must be obeyed if we are to achieve a semblance of *sustainability* to our human existence on the planet Earth. These can be characterized in plain language as:

(1) Everything must go somewhere;
(2) There is no such thing as a free lunch;
(3) *Plus ça change plus la même chose*.

The first law of ecology is recognized as the *first law of thermodynamics*; in other words the conservation of matter and energy. First law violations are fairly obvious and direct. They confront you immediately, such as when the solid waste that you throw out of the front door soon clogs up the street in front of you. Either, you have to take remedial action, or move your household fairly soon. It also reminds us that when one takes something from one phase of the environment for use in another phase the first phase suffers a loss. For example, diverting water from a river for domestic water supply removes water supply from aquatic species.

The second law of ecology, which holds both in economics and in the natural sciences, is recognized as the *second law of thermodynamics*. The second law is subtler. In the natural sciences it applies where resources are used inefficiently; for example every energy trans-formation increases the entropy of the globe. It speaks to species extinction in ecology, and in economics it reminds us that every transaction has some external non-priced effects that impose a cost upon third parties in addition to the costs borne by the direct participants in the transaction.

The third law of ecology reflects human and social decision making in as much as things change in the environment nations and states react pretty much the same way but with a time lag. Diamond (2005) shows how culturally determined responses to environmental stress tie societies to outmoded practices which can lead to rapid social and environmental collapse. The third law is based upon the nature of human societies and their perceptions of risks and the costs of action, and tells us that developed societies and less developed societies behave in a similar manner when faced with environmental and resource sustainability issues. So for instance, the wealthy countries tolerated massive air and water pollution for decades before the perception of the health and ecosystem damages became too intolerable and the major risks associated with polluted drinking water were dealt with first, followed by clean-up of particulates in urban air, then concerns for the ambient aquatic system were experienced and massive wastewater clean-ups of the ambient water were started, shortly thereafter, the oxides of sulfur in the air became of serious concern leading to the introduction of sulfur scrubbers on smokestacks, and more recently, the issue of automobile emissions and other volatile compounds in the air have become the targets of increasingly stringent controls. Finally, the transboundary issues of long-range sulfur transport, ozone depletion, and carbon dioxide build-up in the stratosphere have become a major concern of all countries, both rich and poor alike.

This sequence is exactly what we observe in the major developing countries: first a concern for drinking water quality, then for particulates in urban air, followed by concerns for ambient water quality protecting fish and marine species, the removal of sulfur from the air, the whole urban mobile air quality improvement, and finally a concern for the greenhouse gas problems caused by oxides of carbon and other gases. What is most interesting is the rapid rate that the developing world is transitioning through this sequence in comparison to how long it took the industrialized countries to make the same transitions. All too often we are impatient with the rates at which change happens in the Third World, but we should bear in mind the distance traveled in a span of decades rather than centuries that the original industrialized countries took. The evidence (Shafik & Bandyopadhyay, 1992) seems to indicate an inverse U-curve, the so-called *Kuznets' Curve*, relating environmental insult to per capita income. In many cases the argument implies that we can spend our way out of environmental degradation as we become richer.

These three laws of ecology basically constrain what is achievable on planet Earth: we have to keep track of all material flows; we have to avoid unpleasant irreversible effects; and we can expect that most human societies, given time and sufficient wealth, will choose environmentally sustainable paths – if it is not too late. The first two laws are immutable and not easily subverted by social and economic tinkering – in the long run the second law of thermodynamics guarantees us a *heat death*, although the speed with which we approach that final extinction is certainly a social choice. The rate at which the third law guarantees a safer path towards sustainability is entirely dependent upon social, political, and economic forces. In comparing several societies' radically different responses to environmental stresses (particularly increasing aridity) Diamond (2005) clearly demonstrates that the outcomes can be radically different depending upon the nature of the environmental threats and socio-cultural rigidities.

In this chapter we choose a pragmatic stance and use operational definitions when possible. Therefore, we work on the principle (attributed to Solow, 1991) that if we can manage to be sustainable for the next couple of generations, say until 2050, we will be able to provide our descendents with the tools to be able to extend sustainability a further couple of generations, and so on. It is important to remind ourselves that no species comes with a written guarantee against extinction, least of all a dominating species like *homo sapiens*. A 10,000-year-in-the-future look back to the present would be helpful, but we will never have such a luxury. We have only our recent history and our brains to guess at such a future.

## 3   GLOBAL WATER: SHORTAGE, SCARCITY, AND STRESS

The reason for this book is to explore the notion that there is a global water crisis and to examine whether it is more of a myth than a reality, or if a reality to suggest paths to reduce the magnitude of the crisis. To address these questions one needs to understand precisely where we stand at present with respect to water availability and use, in order to make any speculation about the future. As mentioned earlier, there is no universally acknowledged reliable global database for water availability and use. The works of Russian geographers, L'Vovich (1979) and Shiklomanov (2000) are considered to be fundamental, and are the usual starting point for the discussions concerning global water use and availability. They have attempted the Herculean task of estimating water availability and use in countries and regions around the globe. These data have been the source of much speculation as to the capacity of the earth's water supplies to sustain future populations, for example, L'Vovich's work lead directly to Falkenmark's (Falkenmark & Widerstrand, 1992) concern with regional water stress, now and in the future.

In order to discuss sustainability of the resource in the future, even if we have a good estimate of current conditions, we still need to rely upon some sort of model to predict the future conditions. At this point we run into the classical Malthusian problem; a growing population meets a fixed resource. Obviously if we carry on using a particular resource the way we have been without changing, then an increased population leads inexorably to a crisis in the amounts of the resource available for consumption. When that will occur will depend upon many social, technical, and economic choices. Recall that Malthus' ideas were first propounded in 1783 and, despite a six-fold increase in the population since then, none of the predicted population-based resource crises have occurred. The reasons for the failure of prognoses based upon such an eminently reasonable proposal, and supported by unassailable logic, are many and lead us far beyond the scope of this chapter. Suffice it to say that as scarcity begins to be felt, the economic relationships of the resource use to other substitutable resources changes in such a way as to reduce the demand for the embattled resource. For example, there is no reason to expect the global population to keep on growing after it plateaus out at around 9000 millions in 2050. In fact there is every reason to expect it to start to decrease, as currently in Europe, with increasing standards of living. And if we remove the population growth Malthus no longer holds sway! In addition, rapid technological developments have improved the efficiency of water use and recycling so much that pushing Malthus into the future is much easier. However, there will be a long lag time between the declining population and declining water use per capita. The reason for this is that the levels of water use in many parts of the world (the majority of the world's population in fact) are currently so low that any improvement in standard of living will necessitate large increases in per capita water use, despite improved technological fixes.

The recent United Nations/World Water Assessment Programme (UN/WWAP, 2003), *Water for People, Water for Life*, gives the current best estimates of water and its uses and concludes that we are...

> "...facing a serious water crisis. All of the signs suggest that it is getting worse and will continue to do so, unless corrective action is taken. This crisis is one of water governance, essentially caused by the ways in which we mismanage water" (UN/WWAP, 2003: 4).

It is important to note that the UN/WWAP study concludes that the crisis faced is a *governance crisis* not a *resource crisis*. They are concerned about a water crisis, but one that can be remedied by management and governance; not the classic Malthusian end point!

As mentioned above, we have chosen to use the water availability data from Shiklomanov (2000) in the sketches of global water stocks and flows. In Figure 1 we sketch the total water cycle showing both stocks (shown in square boxes) and annual flows (shown by arrows). All of the numbers are in thousand km$^3$. As expected there is typically a huge difference between the stocks and annual flows with one major and surprising exception – the atmosphere. The stock of water in the atmosphere at any one time is only about 12,900 km$^3$. Given that the annual precipitation is on the order of 577,000 km$^3$, the retention time of moisture in the atmosphere must be only in the order of 8 to 9 days! Glaciers and groundwater are the major terrestrial reservoirs of fresh water on the globe. Of the 187,900 km$^3$ in surface storages only about 8000 km$^3$ are due to man-made storage reservoirs. Rivers also have very little storage in comparison to annual runoff. In Figure 2 we rely upon a slightly different database used by Postel *et al.* (1996), which estimates the fate of the renewable fresh water (RFWS in the sketch) falling on land surfaces.

This diagram shows the division of the available fresh water into Falkenmark's blue (surface runoff; shown here in the medium grey shade on the right hand side of the figure), green (water that falls to the surface and does not contribute to runoff; shown in light shading on the left side), and brown (water diverted for human use and which is contaminated before being

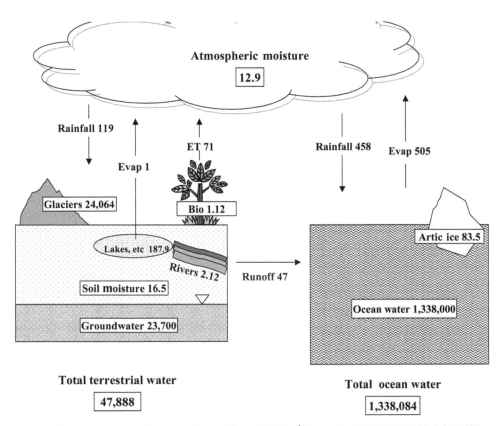

Boxes indicate stocks; arrows indicates annual fluxes. All units in '000 km³'. Source data: Shiklomanov and Sokolov (1983).

Figure 1.    Schematic global water balance.

returned; darker shade on the lower right). The important points of this figure are that more water evaporates from the land surface than flows over or under it. Of this green water humans are able to capture 18,200 km$^3$ for rainfed crops, forest products, etc. Following the blue water side of the diagram we see that the amounts of water appropriated for human use is less than that appropriated from the green water. Of the total brown water (6780 km$^3$) indicated in the figure, at least 3000 km$^3$ should be removed, as this is the estimated amount of additional evaporation due to surface water irrigation. Hence, the total amount of brown water to be dealt with as human pollutants is on the order of 3700 km$^3$. According to this figure, the human appropriation of accessible renewable fresh water is on the order of 30% of the total.

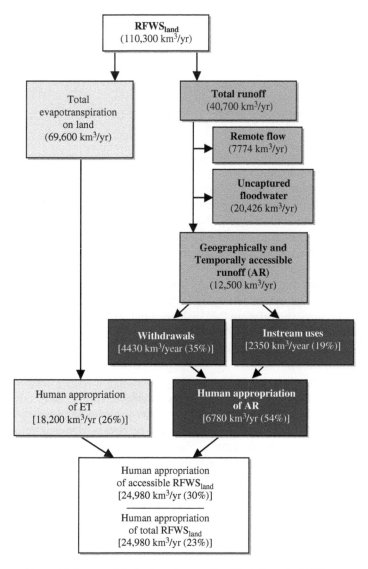

Figure 2.    Annual terrestrial renewable fresh water supplies (Postel *et al.*, 1996).

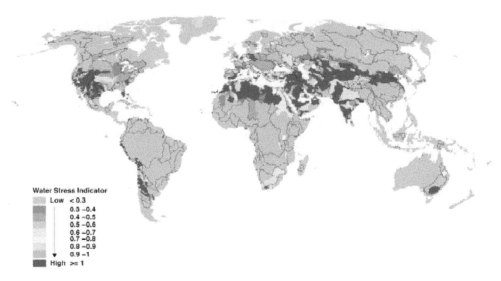

Figure 3.   Environmental water scarcity index by basin (World Resources Institute, 2003).

The questions that arise are: whether this is a large amount now, how large will it get in the future, and will it be sustainable?

If we assume that we would need a 17% increase in the water diverted to irrigated agriculture, based upon IWMI (2000), then the numbers in Figure 2 indicate that globally on the average there is still a large margin of water available. There are, however, large regional variations in water availability for agriculture. But if we look towards rainfed agriculture there is a huge amount of green water evaporation that could be additionally appropriated for agriculture in several productive agricultural regions. This may imply, however, the necessity of large transshipments of grains and other foodstuffs from the rainfed regions of the globe to the less well-watered regions. The option of relying on *virtual water* will always be available, as a backstopping technology should more local solutions fail. To achieve this, a transformation in many arid country's policy on *food self-sufficiency* to *food security* would be needed. Figure 3 shows the estimates of current levels of scarcity by major river basins worldwide (World Resources Institute, 2003).

The recycling of irrigation water evapotranspiration is an interesting question. Water vapor has only a residence time of 8 to 9 days in the atmosphere. This may not seem to be a long time, but given the circulation in the troposphere the additional water vapor typically has moved hundreds of kilometers from its origin and reappears as precipitation in those new locations mostly over the oceans. In terms of water balances carried out on an annual basis, the evapotranspiration is already accounted for in the other locations. It should be remembered that virtually no water is lost (or gained) annually on planet Earth. What we see is a reapportionment of the location of the water (and water vapor) between fresh surface water, groundwater, and the oceans with the atmosphere being one of the temporary channels to do this. When we talk about consumptive use of water, we are, hence, referring to it as measured from the perspective of an annual water balance. So irrigation return flows from crop drainage are available almost immediately whereas the evapotranspiration component goes out of the immediate system and is already accounted in the precipitation in other regions.

## 3.1   *Security*

One of the basic needs for humans is security. Before anything else, humans need security of food and shelter, and security of personal freedom. Security of possessions, and life-style come a distant third in this ranking. When we talk about water, however, it is not just the cup of water today that humans' need, it is the ensured supply and access well into the future, measured in terms of the life expectation of the individuals concerned. Security of supply then, concerns access and property rights that allow different groups to satisfy their needs. One additional wrinkle with water security is that not only is a shortage of water (drought) a security risk, but also a surplus of water (flood) is a security risk. In many parts of the world floods follow droughts leading to large-scale disruptions of human activities and loss of life. So, humans need security against droughts and floods. Moreover, with the widespread pollution of the surface and ground waters, humans need security from contamination of the supply. Since the terrorist attacks of September 11, 2001, security in the USA has focused more on dealing with willful terrorist acts rather than broader acts of man and nature.

There are many concepts floating around water security that are often used synonymously with *water security*. Water *stress* and water *scarcity*, water *vulnerability*, water *shortage*, and *basic human needs* appear in the literature when various aspects of water security are under discussion. Winpenny (pers. comm.) expanded on some of these concepts and gave examples of their use in managing water security:

> "In popular usage, *scarcity* is a situation where there is insufficient water to satisfy normal requirements. However, this commonsense definition is of little use to policy makers and planners. There are degrees of scarcity – absolute, life-threatening, seasonal, temporary, cyclical, etc. Populations with normally high levels of consumption may experience temporary *scarcity* more keenly than other societies, who are accustomed to using much less water. Scarcity often arises because of socio-economic trends having little to do with basic needs. Defining scarcity for policy-making purposes is very difficult.
>
> (1) *Water shortage*: a dearth, or absolute shortage; low levels of water supply relative to minimum levels necessary for basic needs. Can be measured by annual renewable flows (in $m^3$) per head of population, or its reciprocal, *viz*. the number of people dependent on each unit of water (e.g. millions of people per $km^3$).
>
> (2) *Water scarcity*: an imbalance of supply and demand under prevailing institutional arrangements and/or prices; an excess of demand over available supply; a high rate of utilization compared to available supply, especially if the remaining supply potentials is too difficult or costly to tap. Because this is a relative concept, it is difficult to capture in single indices. However, current utilization as a percentage of total available resources can illustrate the scale of the problem and the latitude for policy makers.
>
> (3) *Water stress*: the symptoms of water scarcity or shortage, e.g. growing conflict between users and competition for water, declining standards of reliability and service, harvest failures and food insecurity. Difficult to capture in numbers, though a checklist approach is possible" (Winpenny, pers. comm.).

For absolute levels of these definitions Population Action International (2003) gave the following:

> "A country whose renewable fresh water availability, on an annual per capita basis, exceeds about $1700 \, m^3$ will suffer only occasional or local water problems. Below this threshold countries begin to experience periodic or regular water stress. When fresh water availability falls below $1000 \, m^3/yr$ per person, countries experience chronic water scarcity, in which the lack

of water begins to hamper economic development and human health and well-being. When renewable fresh water supplies fall below 500 m$^3$ per person, countries experience absolute scarcity" (Population Action International, 2003).

A problem with these definitions is that they do not distinguish between security concerns due to the actions of humankind or of nature. The issue of potential climate change is a real test for the need to make sharp distinctions. Many climate change effects appear to be purely natural events unfolding on a geological time frame; others appear to be human induced. How do we separate these effects? Should we? If it is an issue of control we have little choice; we can only influence the human-induced effects and work out strategies to mitigate the natural effects. This dichotomy is referred by the *Intergovernmental Panel on Climate Change, Third Assessment Report* (IPCC, 2001) as:

> (1) "Climate change will lead to an intensification of the global hydrological cycle and can have major impacts on regional water resources, affecting both ground and surface water supply for domestic and industrial uses, irrigation, hydropower generation, navigation, in-stream ecosystems and water-based recreation".
> (2) "The impacts of climate change will depend on the baseline condition of the water supply system and the ability of water resource managers to respond not only to climate change but also to population growth and changes in demands, technology, and economic, social and legislative conditions" (IPCC, 2001).

The first quote deals with climate change from all sources, the second addresses the ability to control human causes and adjust to the natural causes. This is essentially what water professionals do in the normal course of their professional careers. From the climate data on the recent past it is very difficult to discern trends in water availability and the model results reported in the IPCC (2001) are even more difficult to interpret. For example in Figure 4 based upon the IPCC, which compares two of the IPCC's scenarios: A2 which refers to a scenario with a slow decrease in fertility rates and regional patterns, and B2 that has more rapid fertility declines and with local rather than regional economic growth rates. These are both middle of the road scenarios, neither is at the extremes of the projection ranges. Nonetheless, Figure 4 shows wide and confusing differences between the forecast made by the IPCC models for the middle of the century. Agreement as to sign and magnitude of changes in precipitation mainly occur at the poles, with mostly inconsistent results in the mid-latitudes. It is hard to draw strong conclusions from these simulations. Alcamo *et al.* (2000) using the IPCC models and data on river basins under severe stress in 1995, predicted in Figure 5 critical regions in the world for 2025. The results are not much different from what we already know about regions currently under water stress.

Given the conflicting views on the effects of climate change on water availability and use, and given our knowledge of the rapidity of demographic, technological, economic, and social impacts on water use, leads us to endorse the IPCC positions quoted above and focus upon management of the resource in all of its dimensions. In other words, we believe that effective governance of water is the one fundamental approach that we know that will enable us to achieve our *survivability* goal by 2050.

## 3.2   *Solving water security issues*

### 3.2.1   *Use more green water*
Based upon Falkenmark's conceptual breakthrough, now in book form (Falkenmark & Rockstrom, 2004), of partitioning water into blue and green water, we now have a radical new look

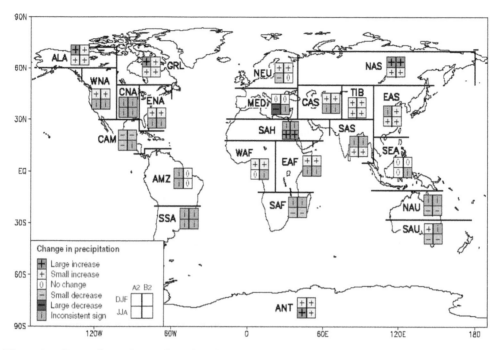

Figure 4.    Comparison of scenarios A2 and B2 of TAR (IPCC, 2001).

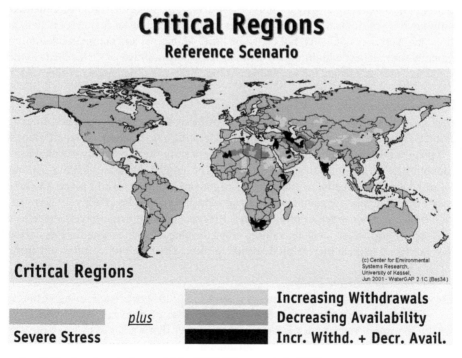

Figure 5.    Critical regions in the world for 2025 (Alcamo *et al.*, 2000).

at the hydrological cycle. Traditionally water balances kept track only of the blue water. By this is meant all of the water that appears as runoff to the streams and as groundwater recharge, but neither the green water involved in supporting the non-irrigated lands of the globe, nor the used water (brown water), appear directly in the balances. Using this new concept, we have redrawn the Postel *et al.* (1996) water balances in Figure 2 to emphasize a global water balance that truly reflects the roles of blue, green, and brown water for supporting humanity and the ecosystem. Of the 110,300 km$^3$/yr *Reusable Fresh Water Supply* (RFWS in Figure 2) available from precipitation on the land surface of the globe, only 40,700 km$^3$/yr goes to the blue water account, and 69,600 to the green water account. Figure 2 shows the brown water; at most this amounts to about 25% of the human withdrawals, or 1100 km$^3$/yr. The Figure also shows that human appropriations of the green water were almost three times as large as those of the blue water. The possibility of apportioning more of the green water for crop production and reducing the demands on the blue water are active questions that could lead to dealing more easily with feeding the future populations.

Paying more attention to rainfed agriculture research and to trade of agricultural commodities between the rainfed areas and the more arid areas are two policies that immediately suggest themselves in this context as potent solutions to feeding the world.

### 3.2.2    *Exploit the asymmetries in water use*
There are two ways of characterizing water use: withdrawal use, and consumption use. Withdrawals reflect the amounts of water removed from the water source for use and are the amounts of water actually consumed or evaporated in the production process. For most uses small amounts of water are actually consumed with the bulk returning, albeit in many cases polluted, to the water source. Irrigation uses, however, typically evaporate more than 70% of the withdrawn water. Industry, particularly heavy industry and fossil fuel electric generation, withdraw huge amounts of water, but consume little. Municipal and commercial withdrawals are typically the smallest withdrawals with about 20% consumptive use. This leads to the role of irrigation water as a balancing mechanism for meeting water demands in other sectors.

### 3.2.3    *The irrigation flywheel*
In many countries of the world as much as 60–95% of all water is consumed by irrigated agriculture. This fact is extremely important when looking toward future potential water shortages. Typically, a 10% reduction in irrigation water consumption would make available a more than doubling of the amounts of water withdrawn for industrial and municipal uses. In other words, provided that effective wastewater management programs are in place, the amounts of water made available for municipalities and industries could be effectively doubled. The question then hinges on whether or not such a 10% reduction in irrigation efficiencies could be achieved (or equivalent areas for rainfed crops developed) at the same time as agriculture is being expanded to meet increasing demands for food, fiber, and other industrial crops.

### 3.2.4    *Virtual water escape hatch*
Closely related to the irrigation flywheel is what Allan (2001) characterized as virtual water: the use of imports of food crops as a substitute for use of domestic water for irrigation. Essentially the importing country is importing the water that was used to grow the crops in the exporting country. This virtual water can amount to as much as 1–5000 tons of water per ton of crop imported. Hence, an import of 2 million tons of food from a rainfed source will

save the importer 2000–10,000 million tons (or 2000–10,000 Mm$^3$; the annual average flow of the Nile is approximately 60,000 Mm$^3$) of domestic water. Virtual water, through the global agricultural trade is already helping overcome the wide disparities between the distributions of water resources among countries, and can be expanded to help future needs. Based upon estimates of the world agricultural trade, the total virtual water already amounts to as much as 700,000–800,000 Mm$^3$/yr (Hoekstra & Hung, 2002; Ramirez-Vallejo & Rogers, 2004).

### 3.2.5 *Urban water: low cost desalination breakthrough*
Recent developments in membrane filtration technologies combined with reverse osmosis have revolutionized the potential for widespread application of desalination. Costs as low as US$ 0.50/m$^3$ (about US$ 1.80 per 1000 gallons) are expected for large water treatment plants. At this cost, all urban areas in the world with access to saline water could have a plentiful water supply comparable to current typical prices for urban water supply. While this does not give an instant solution to the problem in many poor cities in the developing world, it does indicate that reasonable economically achievable technology can solve the problems as they develop.

### 3.2.6 *The ecological revolution*
The major reason for widespread dissatisfaction about water issues in developing countries stems from the vast numbers who have no access to adequate drinking water or sanitation. The numbers often used are 1000 million without adequate water supply and 2000 million without access to adequate sanitation. Whatever the actual numbers are, they represent a massive failure by national governments and the world community to deal with these two issues, which involve human dignity much more than economics. It is economics, however, which drives the lack of action in meeting the water needs. It is simply beyond the financial capabilities of existing governmental and international agencies to cover the capital cost of providing conventional water and sanitation to these persons and their increasing numbers. Either large new funding sources will be needed, or the conventional views of water-borne sanitation will have to be changed toward well-known ecological sanitation practices that use little or no water. Low cost ecological engineering breakthroughs are likely to revolutionize the supply of water and sanitation in the near future.

### 3.2.7 *Resolving transboundary conflicts*
Two or more riparian countries share more than 261 river basins worldwide, and 40% of the global population lives within these basins (Wolf *et al.*, 2003). Often the basins are shared by water-rich and water-poor countries or by heavily polluting industries in some countries and not in others. Since there is no strong international law governing the resolution of transboundary water quality and quantity disputes, the potential for physical and armed conflict remains high. This issue needs careful consideration by riparians and the regional family of nations to head off these conflicts from becoming worse. If these issues could be resolved, for example, 500 million inhabitants in the Ganges-Brahmaputra basin could improve their sense of security about future water supplies and their access to them. Similar cases are exemplified in the Nile, Mekong, and the Tigris-Euphrates basins.

### 3.2.8 *Reform of idiosyncratic water institutions*
Unfortunately, one water fact-of-life makes it very difficult to implement the types of technical and social breakthroughs already conceptually available: the hurdle is that property rights,

institutions, and laws, vary widely from country to country even within countries. The plethora of institutions and legal regimes in all countries tends to get in the way of rational water management. In many cases it is impossible to reallocate or trade between conflicting water uses. The confusion allows for the entrenchment of conventional institutional views and blocks consideration of newer approaches. Approaches such as public/private partnerships are often difficult to implement under these Balkanized conditions. Any progress in this dimension will make the potential for implementing the above measures that will greatly enhance water security for all.

## 4   WATER GOVERNANCE

At the start of the new millennium the concept *governance* has become part of the lexicon of all development planners. The World Bank, the Asian Development Bank, and most UN Agencies have embraced the concept. It also features large in the discussions of the bilateral agencies and the NGO community. On closer examination, however, we can see that in most cases it is merely a renaming of concern for institutions and institution building that goes a long way back in time in most agencies. The concern for institutions, and now governance, has grown out of the slow realization that the problems of development are not simply shortages of capital. Indeed there were so many cases where development happened without large capital transfers that some other, X-factor, was needed to explain them. The fads and fashions swung from education to health, from poverty alleviation to globalization, etc. We now stand at a period of time at which the role of governance – the relationship between the state and civil society has been identified as the critical X-factor.

> "Governance is the exercise of economic, political and administrative authority to manage a country's affairs at all levels . . . It comprises the mechanisms, processes and institutions through which citizens and groups articulate their interests, exercise their legal rights, meet their obligations and mediate their differences . . . Water governance refers to the range of political, social, economic and administrative systems that are in place to develop and manage water resources, and the delivery of water services, at different levels of society" (GWP, 2000).

Governance relates to the broad social system of governing, which includes, but is not restricted to, the narrower perspective of *government* as the main decision-making political entity. There is no single definition of governance that effectively includes all of these disparate ideas and different approaches are followed. Some may see governance as essentially preoccupied with questions of financial accountability and administrative efficiency. Others may focus on broader political concerns related to democracy, human rights and participatory processes. There are also those who look at governance with a focus on the match and mismatch between the politico-administrative system and the ecological system or in terms of operation and management of services. Water governance is already practiced in all countries and our aim here is to make it more transparent and relate it directly to sustainable water development and security.

Governance covers the manner in which *allocative and regulatory* politics are exercised in the management of resources (natural, economic, and social) and broadly embraces the formal and informal institutions. Institutions are interpreted here to include both the formal (codified and legally adopted) and the informal (traditionally, locally agreed and non-codified). The new term for discussing this combination of formal and informal institutions is *distributed governance*. There is a profoundly political element to governance, which involves balancing various interests and political realities that must be taken into account. Although politics may set the agenda, the priorities and the vision, people need governance systems that give the political vision credibility and ownership. Finally, management structures must be established to run things and carry out the day-to-day tasks.

The notion of governance for water includes the ability to design public policies and institutional frameworks that are socially accepted and mobilize social resources in support of them. Water policy *and the process for its formulation* must have as its goal the sustainable development of water resources, and to make its implementation effective, the key actors/ stakeholders must be involved in the process. Governance aspects overlap with technical and economic aspects of water, but governance points us to the political and administrative elements of solving a problem or exploiting an opportunity. Governance of water is a subset of the more general issue of the creation of a nation's physical and institutional infrastructure and of the still more general issue of social cooperation.

Water governance is concerned with those political, social and economic organisations and institutions (and their relationships), which are important for water development and management. Given the complexities of water use within society, developing, allocating and managing it *equitably* and *efficiently* and ensuring *environmental sustainability* requires that the disparate voices are heard and respected in decisions over common waters and use of scarce financial and human resources. Water governance is concerned with the functions, balances and structures internal to the water sector (*internal governance*). It includes the framing of social agreements on property rights and the structure to administer and enforce them known as the law. Influences also come from civil society and from the *current* government and these latter are all considered parts of the *external governance* of water, which will be discussed later. Although issues can arise for water governance from the economic and technical spheres, in most countries the driving force is *politics*. Effective governance of water resources and water service delivery will require the combined commitment of government and various groups in civil society, particularly at local/community levels, as well as the private sector.

To achieve more effective water governance it is necessary to create an enabling environment, which facilitates efficient private and public sector initiatives and stakeholder involvement in articulating needs.

## 4.1  *Good governance matters*

Social analysts have shown that there is a strong causal relationship between better governance and better development outcomes such as higher per capita incomes, lower infant mortality and higher literacy (Kaufmann *et al.*, 1999). A stable and just social order founded on clear institutional rules and effective and equitable markets enhances poverty reduction. Effective governance is thus essential to poverty reduction and can help the poor to help themselves. Poor governance is a barrier to development and hurts the poor through both economic and non-economic channels, making them more vulnerable and unable to adapt to changes. As a result, markets will be weak and distorted thus holding back growth and employment opportunities. Structural and institutional reforms are needed to turn poor governance into more effective governance; including measures such as creating accountability in the use of public funds, building national capacity for better policy formulation, implementation, and enforcement mechanisms. It includes making decision-making and implementation more inclusive processes where civil society and the private sector have clear roles to play with shared responsibilities on the basis of public-private partnerships. The division of labour between the different actors and the sharing of responsibilities and balancing power relations are all part of the same process, that of defining the governing system.

## 4.2    *Historical context of water governance*

Since the Dublin conference in 1992, significant international goals have been set that relate to water governance. At the 2000 World Water Forum in The Hague, the GWP Framework for Action (GWP, 2000) stated that "the water crisis is often a crisis of governance", and identified making water governance effective as one of the highest priorities for action. The 2000 Hague Ministerial Declaration reinforced this view and called for "governing water wisely to ensure good governance, so that the involvement of the public and the interests of all stakeholders are included in the management of water resources". At the Bonn 2001 Freshwater Conference the ministers recommended action in three areas, with water governance as the most important. They proposed that "each country should have in place applicable arrangements for the governance of water affairs at all levels and, where appropriate, accelerate water sector reforms". The UN 2000 Millennium Assembly, emphasised conservation and stewardship in protecting our common environment and especially "to stop the unsustainable exploitation of water resources, by developing water management strategies at the regional, national and local levels, which promote both equitable access and adequate supplies". This was endorsed at the World Summit on Sustainable Development in 2002 where Heads of State agreed a specific target "to prepare IWRM (Integrated Water Resources Management) and water efficiency plans by 2005". To be meaningful these plans will need to take cognisance of prevailing governance systems and allow for necessary reforms.

At the beginning of the 21st century, we are searching for coherence and accountability in the maze of organisations within national (and international) political systems. However, many of today's institutions and government systems were developed in the 19th century to supervise states with much more limited functions than today. The developing countries face particular problems as they often have layers of systems – some indigenous and others imported. It is not expected that developing countries can or even should adopt the same systems as industrialised countries but there are basic principles for effective governance that they need to adopt in their own way. With regard to water, the state may need to act quickly to develop the essential infrastructure for development and cannot wait for the *ideal* governance systems to be established. Nevertheless, any water development should be done hand in hand with broader governance reforms that will help to make the development sustainable. In pushing for this care should be taken not to further weaken the weak state.

The traditional bases of political power have been eroded in the last 20 years or so and the institutional strength of the state is being challenged. Some recent changes in society have facilitated this weakening of the central state and making traditional water governance irrelevant. Pierre (2000) claims that some of these changes (there are others) include:

- Fiscal crises within the state (limitations on raising taxes).
- The globalisation process, including deregulation of financial markets and volatility of capital that restricts the state's ability to govern/control the economy.

- Technological advances that facilitate networking and subsidiarity.
- A more assertive sub-national democracy in cities or semi-autonomous regions.
- Excessive workload and responsibilities on smaller government bureaucracies.
- Large concentrations of people and political power in urban areas.

There are several modes of governance that are often being pursued at the same time in within a given state. In the following discussion we rely heavily on Pierre (2000) and Rogers & Hall (2003).

*Hierarchical governance*: Part of modernisation is generally seen as the evolution of political systems from top-down, hierarchical government systems with centralised institutional settings, to more decentralised administrative forms. There is no evidence that more decentralised systems are necessarily more effective than centralised ones. The real test here is, *what works in the particular setting*. There is, however, a perceived ever-widening gap between those countries that have managed to move toward subsidiarity – or the performance of functions at the lowest effective level – and those that remain centralised and stagnant.

*Market-led governance*: With the end of the cold war in the closing decades of the 20th century, many in the western countries proposed the market as the solution to economic growth, social equity and environmental problems. This led to deregulation and more involvement of the private sector and a changed role for the civil service and civil society. This institutional restructuring of the state aimed to reduce government command and control functions with more individualism (fewer collective solutions) and *private enterprise and the market* as the superior resource allocation mechanisms. This *market-led* governance model is the immediate background in which we now examine governance with respect to water resources management and the delivery of water services. Today the honeymoon with the *laissez-faire* market-led model is over, and hard questions are being asked. It is considered by many to be too simplistic (hierarchies may not work well but markets do not necessarily work well either in all situations) and not necessarily representative of wider societal values. More people are examining what new instruments and new forms of exchange between state and society can be developed to ensure political control and societal support. From this examination, propositions for management in partnership, co-management and co-governance, and distributed governance, have developed.

*Distributed governance*: At the beginning of the new millennium the state's role of *directing* or *steering* society, particularly in the areas of environment and water, is being challenged by cohesive local networks (civil society, private sector) and global networks (international organisations and NGOs) with these same entities also supporting the state in its aims to develop society. This gives a dynamic relationship between different social forces. Many politicians (mainly in the West) see the state increasingly as part of the problem rather than the solution. There are more calls for a return to smaller government, reversing the post Second World War ideology of a hierarchical central state caring for its citizens. The state no longer believes it can solve societal problems acting alone; particularly socio-environmental ones and the private sector alone cannot address the problems of the poor and the environment. The *command and control or hierarchical* model and the *market-led* governance models are both thus much weakened. The *Dublin Principles* manifestly reflect this concept of distributed governance.

Clearly modern governance sees formal authority being supplemented by an increasing reliance on informal authority; for example, through genuine public-private co-ordination and co-operation to the benefit of both of these as well as the customer/citizen (organisations such as the GWP (*Global Water Partnership*) and international NGOs such as *Transparency International* are examples of such co-operative networks). The state thus needs to adapt to a new situation and distributed governance is an institutional response to the changed environment. Distributed governance is thus the empirical manifestations of state adaptation to its external environment. It is the conceptual representation of the co-ordination of social systems and specifically the role of the state in that process.

### 4.3    *Water governance principles and legal bases*

The *Dublin Water Principles* (WMO, 1992) bring water resources firmly under the state's function of clarifying and maintaining a system of property rights, and through the principle of participatory management at the lowest appropriate level asserts the relevance of meaningful decentralization. The Dublin Principles that guide the IWRM approach are:

- Fresh water is a finite and vulnerable resource, essential to sustain life, development and the environment.
- Water development and management should be based on a participatory approach, involving users, planners and policy-makers at all levels.
- Women play a central role in the provision, management and safeguarding of water.
- Water has an economic value in all its competing uses and should be recognized as an economic good.

There is increasing pressure to recognise and formalise water rights and this is happening in many countries. Formalising rights raises complex questions about the plurality of claims and the balancing of the distribution of benefits among the social groups. It also imposes responsibilities including that of pollution prevention and financial sustainability. The process of formalisation is often biased in favour of the rich and powerful who may abuse the system and capture rights. Informal *rights*, as defined locally with their historical rules and principles, are equally important and improper formalisation may lead to conflict between the formal and traditional. The formalisation of rights may therefore be neither necessary nor sufficient to secure access to water resources. The capacity to defend rights against competing claimants is essential for the rights to be meaningful, whether they are formal or informal. An important matter is to what extent the processes of devolving water rights serve segments of a population, or its entirety.

### 4.4    *Water law is about property rights*

The state has an important role to play through its core function of defining property and use rights and responsibilities. In modern pluralistic democratic societies, the foundation of the state rests upon the *publicisation* (the term for the shift from the private to the public sphere, also called nationalization in some countries) of the costly monitoring and policing needed to protect productive assets from being redistributed to intruding claimants. Without this policing, called the *law*, systems of property would never have advanced beyond appropriative behaviour backed by force. Discussions of water rights usually focus upon the rights of the property right holder and ignore the contingent responsibilities that holder has with regard to others in society who do not share the rights. These obligations need to be stressed in any discussion of governance.

The state has an important role to play in as much as it defines property rights. Some examples of the main types of property rights and civil society's responsibilities toward different property rights with their associated rights and obligations:

- *Open access property*
  There is no defined group of users or owners and the benefits are available to anyone. Individuals have both privilege (the ability to act without regard to the interests of others) and no right (the incapacity to affect the actions of others) with respect to usage and maintenance of the asset.
- *Common pool property*
  The management group (the owners) has a right to exclude non-members and non-members have a duty to abide by the exclusion. Individual members of the management group have both rights and duties with respect to usage and maintenance of the property.
- *Private property*
  Individuals have the right to undertake socially acceptable uses and a duty to refrain from socially unacceptable uses. Others (non-owners) have a duty to allow socially acceptable uses and a right to expect that only socially acceptable uses will occur.
- *State property*
  Individuals have a duty to observe use and access rules determined by the controlling agency of the state.

In most countries water is state property, with lesser amounts owned privately. However, water property rights usually start out as an *open access property*, which is initially appropriated by a group and becomes a *common pool property* resource. Ultimately the state tends to appropriate these rights from the common-pool resource ownership group to create *state property*. The state is then faced with the problem of how to deploy the resource to the national advantage. Water rights and land use are closely linked, sometimes formally through riparian rights, and landowners can affect water through land use changes such as deforestation. The key to water governance at the beginning of the 21st century is how, through politics, states can achieve this fairly and equitably, without reducing incentives for efficient use of the resource.

### 4.5   *Theories of collective action*

The theoretical basis of governance lies in theories of collective behavior. Unfortunately, while intellectually stimulating and of great historical interest going back to the Greek philosophers and earlier, there is no one simple theory to explain every situation. Starting from an analysis of property rights and experimentation with these rights over time has led the USA to a flexible approach to water governance. This approach allows for shifts when the economic and social conditions change without having to build institutions that cover all possible eventualities.

Many questions can be posed about the viability of any of the property rights regimes based upon the collectivity of the individual players and their initial resource endowments. Robert Wade (1987) cites Mancur Olson (1965) as the source of pessimism about the viability of collective action in *common pool* or *common property resources*:

> "One of the theories that have generated pessimism about the viability of collective action is Mancur Olson's *logic of collective action* (which might better be called the illogic of collective action, or the logic of collective inaction). His core proposition is this: unless there

is coercion or some other special device to make individuals act in their common interest, *rational, self-interested individuals will not act to achieve their common or group interests*" (Wade, 1987: 221).

Based upon his experiences with water management in Indian villages, Wade concludes that this is not the case. Others, such as Ostrom (1990), again based upon Asian villagers' collaboration on water allocation and use, back up Wade's arguments. Wade and Ostrom claim that groups can build mutually binding agreements in spite of substantial individual differences, and that heterogeneity could facilitate cooperation when some members of the privileged group value a collective good enough that they are willing to provide it in spite of the actions or inaction of the remaining group members.

Tendler (1997) noted that we know a lot more about what constitutes *bad governance* than we do about achieving *good governance*. Her case studies tend to question some conventional nostrums and preconceptions of how governance should be and drive us back to a close functional analysis of each individual case. Maass & Anderson (1978) provide in-depth analyses of the development of the governance of irrigation since the 15th century in Valencia, Murcia, and Alicante, in Spain. In all of these empirical studies the authors found strong evidence to support the notion that, despite a wide range of property rights regimes, user groups could develop into sustainable institutions over many years (centuries in the case of the Spanish irrigation property rights sharing systems). Essentially, there is a possibility of identifying a level of centralisation and decentralisation and regulation to produce effective water governance. Whilst empirical evidence suggests there can be no dogmatic solutions it would be helpful to establish some universal attributes that make water governance effective in practice.

## 5    THE POLITICS OF WATER GOVERNANCE

The driving force in any area of governance is *politics*. The conventional view of the relationship of politics to governance portrays an orderly world where politicians act as rational legislators in formulating laws for the general welfare, which in turn are implemented by institutions, which carry out the legislated water policies through rules and regulations. Real politics, however, is not so neat and in 1936 Harold Lasswell said it best merely in the title of his book: *Politics: Who gets What, When, and How?* (Lasswell, 1936). Governance is now not seen as a simple linear process, but as discursive and a highly complex set of interactions between laws and institutions, and personal and group interests as well as the general interest. Stone (2000: 208) says: " . . . policy is more like an endless game of Monopoly than a bicycle repair".

For a water enterprise, politics is certainly part of the governance domain. Contestatory, often very personalized maneuvering, aimed at the building of consensus and support for policies and persons, is certainly part of the water politics concept. The application of politics to water problems has been called *hydropolitics* by John Waterbury in his cautionary tale in the *Hydropolitics of the Nile Valley* (Waterbury, 1979), which demonstrated the powerful combination of local and international politics. *Hydropolitics* can occur inside or outside a water agency or enterprise, but the politics of water governance are primarily the play of the sociological factors (structures, institutions, even leaders' personalities) that lie outside the water enterprise itself. These reflect more general political-sociology, that is to say, of the water institution's setting. Hence, framing the political decision is a fundamental question.

An agency's own governance is nested within these factors and the boundary between the agency and its environment is permeable, and social capital, which can take the form of political power, can also move in both directions across it.

## 6  APPROACHES TO POLITICAL DECISION-MAKING

The literature on politics and political theory and their implications for effective water governance leads to many different pathways for explaining political decision-making. A brief introduction to them is given below.

### 6.1  *Economic theory of politics*

The literature on politics and political theory and their implications for effective water governance in the USA leads to at least three different pathways for explaining political decision-making. First, and *a priori* the most attractive to professional planners, is in the direction of the *Economic Theory of Politics* (ETP), which is nicely summarized in a review article by Jan-Peter Olters (2001). This theory aims at a synthesis of the ethics of Rawls with the political theories of Lasswell, Dahl and Lindblom, and the welfare economics of Pareto, Hicks, and Bergson. The ETP in essence attempts to replace the maximization of the Hicksian welfare function with the maximization of a social welfare function. Downs (1957) led the attack on the use of the classical economic social welfare function in ETP. Downs hypothesized that political parties act in order to obtain income, prestige, and power, and politicians were motivated by the vote maximization objective rather than by altruistic or ideological objectives, and concluded that "parties formulate policies in order to be elected rather than win elections to formulate policies". Downs' model is pure Adam Smith with selfish behavior of individual politicians resulting in economic and other policies that guarantee a *social optimum*. In other words, "governments continue spending until the marginal vote-gain from expenditure equals the marginal vote-loss from financing" (Downs, 1957: 73). Despite Downs's efforts, the ETP, however, appears to be less relevant than other approaches to informing practical real political decisions. An attempt to model water quality decisions using the ETP based upon *Paretian Environmental Analysis* was made by Dorfman *et al.* (1972). Their rational approach provided the tools for political bargaining, but did little to enlighten the bargaining process itself. Political solutions do not necessarily rely upon Pareto feasibility to still claim to be good solutions.

### 6.2  *Institutional approach*

The second approach is an *institutional approach* exemplified by Maass *et al.* (1962) and more recently by Crenson & Ginsberg (2002), which examines the evolution, development, and erosion of popular democratic institutions and sees subtle shifts away from *popular democracy* towards *personal democracy*. Popular democracy depends upon group political mobilizations such as labor unions, fraternal associations, etc., whereas personal democracy has developed as American's become increasingly atomized to small groups who exert their political influence through political contributions leaving the political aspects to paid political operatives. This pathway is very much in keeping with the *Bowling Alone* view of institutions of Putnam (2000). The institutional approach assesses the institutions and sees how well they are meeting their stated goals. *Muddy Waters* (Maass, 1951) used this approach in analyses of the relationships between the Congress and the Corps of Engineers. More recently, Crenson & Ginsberg (2002) are particularly pessimistic about the general health of democratic institutions and democracy in the USA. They see citizens becoming customers of the government, not the controllers of the government. They give many reasons including the size of the population, the rise of multitudinous NGOs, the lack of direct taxation specifically to fund government's war-making ability, and the role played by TV and media in converting citizens to passive consumers.

To give a sense of how much has changed in citizen participation in environmental and water governance over the past decades consider the plethora of new citizen organizations that have sprung up worldwide to help address these and other problems. Bornstein (2004) states:

> "Consider that 20 years ago Indonesia had only one independent environmental organization. Today it has more that 2000. In Bangladesh, most of the country's development work is handled by 20,000 NGOs: almost all of them were established in the past 25 years. India has well over one million citizen organizations. Between 1988 and 1995, 100,000 citizens groups opened shop in the former communist countries of Central Europe. In Canada, the number of registered citizens groups has grown by more than 50% since 1987, reaching close to 200,000. In Brazil, in the 1990s, the number of registered citizen organizations jumped from 250,000 to 400,000, a 60% increase. In the USA, between 1989 and 1998, the number of public service groups registered with the Internal Revenue Service jumped from 464,000 to 734,000, also a 60% increase" (Bornstein, 2004).

It is difficult to get one's head around the magnitudes involved. Even if they had only 50 active members in each group this would amount to more than 35 million active members of these groups in the USA alone. Such participation levels would seem to contradict Crenson & Ginsberg's comments about the decline in participatory democracy in the USA.

### 6.3    *The Polis model*

The third, and most intriguing approach replaces the highly rational models of policy analysis based upon ETP by the *Polis* model of Stone (2000). Stone argues that rational behavior models (usually the ETP) miss the point of politics. She characterizes the rational behavior model taught in schools of Public Policy as follows:

- Identify objectives.
- Identify alternative courses of action for achieving the objectives.
- Predict the possible consequences of each alternative.
- Evaluate the possible consequences of each alternative.
- Select the objective that maximizes the attainment of objects.

We can all recognize these steps as constituting most of what we do, or attempt to do, in our professional and intellectual lives. They are the foundation of documents like the *US Army's Corps of Engineers Principles and Guidelines* (US Water Resources Council, 1983), which would be unthinkable apart from the above criteria. So what could possibly be wrong with such an approach? Stone's response is:

> "...a model of political reasoning ought to account for the possibilities of changing one's objectives, of pursuing contradictory objects simultaneously, of winning by appearing to lose and turning loss into victory, and most unusual, of attaining objectives by portraying oneself of having attained them... Political reasoning is reasoning by metaphor and analogy" (Stone, 2000: 9).

In a series of tables she compares and contrasts the *market model* of policy analysis, which is essentially based upon the ETP model, with her *political (polis) model*. For example in Table 1, the comparison considers: what are the units of analysis, sources of conflicts, sources of ideas, and what is the nature of collective action. From this table we see two radically different perceptions emerging of the nature of society. The unit of analysis shifts from the individual to the community. Self-interest is extended also to include public interest. The conflicts now change from conflicts between self-interests to between self-interests and the public interest. The nature of collective activity moves from competition to a mixture of cooperation and competition. Most importantly, the criteria for decision-making change from maximizing self-interest to the promotion of the public interest and loyalty to people and places. The nature of the information used tends to move from objective to ambiguous, interpretive, incomplete, and strategically manipulated. Ultimately, what matters under the Polis model are ideas, the pursuit of power, and the maintenance of alliances. Polis approaches to politics and policy are quite different from those implied by the *rational* model of politics.

## 6.4   *The bureaucratic politics and process model*

This model is based on political-bureaucratic bargaining in a federal system. Its focus is typically the executive branch, with the elected legislature hardly in the picture. Classic cases are drawn from USA foreign policy problems (for example, the Cuban missile crisis) where Congress was not a major player, but this is the opposite from the situation in water, where the executive branch until recently was largely excluded by Congress.

Table 1.   Comparison of *market model* with *polis model*.

|  | Concepts of society | |
|---|---|---|
|  | *Market model* | *Polis model* |
| 1. Unit of analysis | Individual | Community |
| 2. Motivations | Self-interest | Public interest (as well as self-interest) |
| 3. Chief conflict | Self-interest *vs.* self-interest | Self-interest *vs.* public interest (commons problems) |
| 4. Source of people's ideas and preferences | Self-generation within the individual | Influences from outside |
| 5. Nature of collective activity | Competition | Cooperation and competition |
| 6. Criteria for individual decision-making | Maximizing self-interest, minimizing cost | Loyalty (to people, places, organizations, products), maximize self-interest, promote public interest |
| 7. Building blocks of social action | Individuals | Groups and organizations |
| 8. Nature of information | Accurate, complete, fully available | Ambiguous, interpretive, incomplete, strategically manipulated |
| 9. How things work | Laws of matter (e.g. material resources are finite and diminish with use) | Laws of passion (e.g. human resources are renewable and expand with use) |
| 10. Sources of change | Material exchange. Quest to maximize own welfare. | Ideas, persuasion, alliances. Pursuit of power, pursuit of own welfare, pursuit of public interest. |

### 6.5   The congressional behavior model

A second federal model concentrates on the elected congress, with the view that to under-stand congressional behavior is to understand that congressmen are *single-minded seekers of reelection*. It follows from this that congressmen's goals are to improve the welfare of their constituents in the shortest possible time frame. The realities of information processing are also important in describing congressional behavior. With humanly limited capacities to absorb and judge, legislators are so overloaded with information that they have to be extremely selective in committing their attention. Legislators deal with this by specializing in a particular and limited area; in other domains they take their cues from other sources (colleagues, outside groups, committee reports) that they have learned to trust.

### 6.6   The interest group model

When a national legislator thinks about the constituency that elected him, he or she rarely, if ever, sees an undifferentiated mass of individual voters. They see categories of interests. In some cases, they see only a few dominant interests. In the USA there are literally thou-sands of active interest groups – environmental groups, water resources groups, professional associations, and industry associations – involved with water policy (a total of 734,000 public interest groups are registered with the Internal Revenue Service for all purposes including water and environment). These interest groups often have overlapping concerns and overlap-ping memberships. They constitute vital channels for particular publics to participate in the federal governmental process. Pork barrel projects are the fodder for the well known *iron triangles* of legislators, bureaucrats, and active interest groups that develop in specific issue fields (the term *pork barrel* was first used to describe the acceptances of non-economic water projects by congressmen in other districts in return for equally dubious projects in their own districts in the bi-annual *US Rivers and Harbors Acts*).

   Useful developments of interest group theory are found in Dahl's (1961) *regime theory* and in Buchanan & Tullock's (1962) *public choice theory*. They attempt to predict which pattern of decision-making will prevail based upon the concentration or diffusion of costs and benefits of public choices. If we examine the distribution of effects of a particular action by the state, depending upon whether the benefits (and the costs) are concentrated on a few persons or widely dispersed throughout the economy, we can predict what type of political system will dominate. Table 2 shows schematically how this can work. For example, it is often said that the problem with water governance in the USA is the "tendency to privatize the benefits and socialize the costs". These are situations such as federally financed irrigation, where the benefits are concentrated in the hands of a few farmers and the costs are widely dispersed over society. Such behavior would be in the upper right hand box in Table 1, and one would expect *Client Politics* to prevail. If so, this may be an acceptable situation, since regimes which distribute benefits and costs widely lead to *Majoritarian Politics*, and hence, to inertia and under-investment. There are obviously large overlaps among the ETP, the Polis model, and the interest group model in their analysis of any particular situation.

### 6.7   Governance failures

An underlying theme of social science literature is that all governing structures *fail* and all markets and hierarchies have their limitations and also *fail*. More effective governance regimes or systems need to be designed/created to overcome *government failure*, *market failure* and

Table 2.   Framing regimes through *public choice theory.*

| Distribution of benefits of state intervention | Distribution of costs of state intervention | |
| --- | --- | --- |
| | Concentrated | Diffused |
| Concentrated | *Interest group politics* Organized lobby activity: high but contradictory. | *Client politics* Organized lobby activity: high but one-sided. |
| | Expected outcome: deadlock, compromise, policy see-saw. | Expected outcome: stable capture. |
| Diffused | *Entrepreneurial Politics* Organized lobby activity: low unless *policy entrepreneur* intervenes. | *Majoritarian Politics* Organized lobby activity: low. |
| | Expected outcome: Inertia bias, may be offset by entrepreneur activity. | Expected outcome: Inertia bias except after calamity. |

*Source*: Judith Rees (pers. comm., 2001).

*system failure* or a combination of these. For example, water is not a simple economic good; it is sometimes a public good, sometimes a private good and often lies somewhere in between. Its development can lead to natural monopolies, and it presents major economic and physical externalities, etc.

These failures are listed in Table 3. They are inherent in most countries and have to be addressed. Institutional and communication gaps are likely to be the most difficult. An empirical examination of how to overcome the problems caused by market, government and system failures is essential for each specific setting if effective water governance is to be achieved. There are failures that cannot be easily addressed by water sector professionals as they lay outside the water domain: for example, national institutional structures that impede political vision, poor mechanisms for inter-sectoral dialogue, coping with unpriced assets and public goods such as flood control and drought management. The water community nevertheless needs to understand such external governance constraints and engage with non-water organisations to seek solutions.


## 7   PRINCIPLES AND PRACTICE FOR EFFECTIVE WATER GOVERNANCE

### 7.1   *New forms of water governance*

When proposing changes to water governance systems, it is important to understand and distinguish between the different functional levels in water management: *operational*, *organisational* and *constitutional*. The first focuses on the use or control of water for specific purposes to fulfil specific needs. There are always a plethora of operational enterprises covering municipal and industrial water supply, wastewater treatment, hydropower, irrigation, environmental management, tourism, etc., and they can be in public or private hands.

Table 3.    Sources of market, government, and system failure.

***Sources of market failure:***
Externalities (environmental, economic, and social).
Unpriced assets and missing markets.
Public goods.
Economies-of-scale.
Transaction costs.
Property rights.
Ignorance and uncertainty.
Short-sightedness.
Irreversibility.
Existence of monopolistic situations.

***Sources of government failure:***
Failure to correct market distortions.
Price regulation.
Subsidies to resource users and polluters.
Inappropriate tax incentives and credits.
Over-regulation or under-regulation.
Bureaucratic obstacles or inertia.
Conflicting regulatory regimes.
Short-sightedness.
Voter ignorance and imperfect information.
Special-interest effect.
Little entrepreneurial incentive for internal efficiency.
Imprecise reflection of consumer preferences.
The inability of the government to control and regulate sustainable use.
The lack of payment to other social and environmental services linked
    to water.
The independence and impartiality of the regulatory agencies.
The lack of effective knowledge of the resource, the demands imposed
    on the resource, and the current uses that are made of it.

***Sources of system failure:***
Gaps in the institutional structure that impedes the positive use of
    politics.
Absence of mechanisms for coordination, decision, and conflict
    resolution.
Lack of effective mechanisms for intersectoral dialogue.
Lack of mechanism for participation of the community and interested
    parties.

The organisational level co-ordinates and reduces conflict between these competing enterprises, administers the rules and polices for water use and the users in a water system. This function resides within the public sector – and includes for example river basin authorities and regulatory bodies – the latter should be autonomous (within constitutional boundaries) if they are to act impartially. Finally, the constitutional function creates the enabling environment within which the other functions operate. It sets the policies and legislation, taking into account external governance and political imperatives. In many countries such functions are unclear and often governments may be unable or unwilling to exercise their responsibilities. In this case *ad hoc* arrangements at local government or community level are often established.

These are vulnerable as they may lack any formal basis and can be adversely affected by vested interests or by central government policies and laws. A participatory and consultative approach when reforming water governance systems can help to strengthen local government and bring the positive aspects of such arrangements into the formal system and reduce vulnerability.

Hydro-geographical boundaries – the river basin – often provide opportunities for modern governance networks. A basin is a closed region where there are incentives for people to come to an agreement on governance systems with water as the focus. Although basins cut across formal jurisdictional boundaries and thus local government and other government entities that do not necessarily work together, the *basin society* (a river basin agency or commission) could require them to do so. The basin society may thus have specific governing capacities and needs. National governments acting alone cannot easily allocate and regulate water in a basin, as they are unlikely to appreciate local interests or priorities. Government should, however, provide the rules and regulations and establish a framework for local people to meet (for example, the basin community has a spatial footprint such as in the *Catchment Management Agencies* in South Africa and the *River Basin Agencies* in France). Regulation within a basin must address issues of quality as well as allocate quantity to users. Regulation of sectoral users such as agriculture and industry is very weak. Preventing pollution from agricultural water use (salinity, nitrates in groundwater) and from industries such as tanneries and mining is becoming increasingly important, and Pakistan has recently recognised the need to regulate irrigated agriculture. Catchment planning and management, combining land and water use, is a means to regulate at the basin level but hitherto the tools have not been readily available to make this practical. New approaches, for example as in the European Union (EU) *Water Framework Directive* and the *Streamflow Reduction Strategies* in South Africa, are now starting to incorporate this into governance systems.

## 7.2 *Lower water use, lower conflict levels*

Demand for water can be reduced voluntarily by using many different technical, social, and economic tools. Essentially, this means that the consumer will change his or her consumption preferences. Regulatory instruments involving permits, restrictions, and allocations to various users and uses can also reduce water demand.

It is obvious that the water crises are due to an increase in demand and reducing that demand would help greatly even though there would still be problems of existing levels of resource conflicts and environmental degradation. For example, total water demand in the USA has declined from a high in 1980, despite large increases in wealth and population (Hutson *et al.*, 2004). This means that maintaining aquatic environmental quality is getting progressively easier. In this case direct water pricing policies have not brought about this decline. It appears to be largely due to external factors such as higher energy costs and mandated energy efficiency improvements to domestic and commercial water appliances and decline in the value of irrigated crops. Specific water policy measures such as effluent limitations on wastewater discharges and enforcement of federal in-stream water requirements for ecosystem maintenance have also had a significant impact. It is worth noting how well-informed public pressure acted as a driver for policy change and technological innovation to achieve water savings. Each person reduced his or her water use, and overall, this has made a big difference in water availability in the USA.

### 7.3    *Governance external to the water sector*

Water governance can draw strength from the governance structures obtaining in other sectors in the country, for example through the stabilisation of property rights, broad rules and laws. Certain more general Californian state laws for example aided the creation of Californian groundwater basins. The end of apartheid in South Africa facilitated significant changes to water laws, and the accession of Eastern European countries to the European Union has acted as a spur to improved water governance. Conversely, if the service provider succeeds, it can also validate and strengthen the politics that made it possible. There are several examples of water governance influencing external governance. The best known of these is perhaps the co-operative water development in the Netherlands in the early part of the 20th century, which was an important part of nation building for the modern Dutch welfare state.

Water governance traditionally begins from the social and economic policies set by government. However, with the growing liberalisation of trade, water services are becoming increasingly affected by international trade agreements. Often Trade Ministry officials who know little about water and may not necessarily consult with water officials negotiate such trade agreements. Recent concern has been expressed by some NGOs about the inclusion of water services in the *General Agreement on Trade and Services* (GATS). Whilst liberalisation of such services may be beneficial in raising foreign direct investment, countries need to take care in negotiating the rules under the GATS. Government negotiators can place limitations on the commitments it makes in a specific service sector thus restricting the application of GATS rules but this is a complex issue and often developing country negotiators are in a weak position in such negotiations.

Of particular concern is the conflict between promoting trade and protecting the regulatory rights of national government. It is accepted by all that the ability of government to regulate water services providers is essential for effective private or public sector provision of water services but the government's right to regulate may be restricted under GATS. Apart from GATS other trade agreements such as *North American Free Trade Agreement* (NAFTA), can affect water. For example, the negotiations recently started on the *Doha Round* of talks on agricultural trade liberalisation could impact on water use for food production. Similarly, debt repayments and HIPC (*Heavily Indebted Poor Countries*) agreements may skew a government's ability to allocate budgetary provisions for water services.

Approaches to water governance should be:

- *Open and Transparent*: Institutions should work in an open manner. They should use language that is accessible and understandable for the general public to increase confidence in complex institutions.
- *Inclusive and Communicative*: The quality, relevance and effectiveness of government policies depend on ensuring wide participation throughout the policy chain – from conception to implementation.
- *Coherent and Integrative*: Policies and action must be coherent. The need for harmony and coherence in governance is increasing as the range of tasks has grown and become more diverse.
- *Equitable and Ethical*: All men and women should have opportunities to improve or maintain their well-being. Equity between and among the various interest groups, stakeholders, and consumer-voters needs to be carefully monitored throughout the process of policy development and implementation.

Performance and operation of water governance should be:

- *Accountable*: Roles in the legislative and executive processes need to be clear. Each of the institutions must explain and take responsibility for what it does. The *rules of the game* need to be clearly spelled out, as should the consequences for violation of the rules, and have built-in arbitration enforcing mechanisms to ensure that satisfactory solutions can still be reached when seemingly irreconcilable conflicts arise among the stakeholders.
- *Efficient*: Classical economic theory demands efficiency in terms of economic efficiency, but there are also concepts of political, social, and environmental efficiency which need to be balanced against simple economic efficiency.
- *Responsive and Sustainable*: Policies must deliver what is needed on the basis of demand, clear objectives, an evaluation of future impact, and, where available, of past experience. Water governance must serve future as well as present users of water services.

## 8   CHALLENGING THE WATER CRISIS

At any time in the history of the development of resources management there is the suspicion that, somehow or other, things are not working out as originally planned. At the outset of the 21st century this is true of many areas such as energy, agriculture, and climate, and is particularly true for water resources management. It is tempting to say that everything that could be said about water management has already been said. A couple of centuries of diligent research and development have provided scientific concepts, technology, laws, and management institutions which, by and large, have served to meet most of society's material needs. So, why challenge this wisdom, and why now? It turns out, however, that the application of conventional approaches has been shown to be ineffective in protecting the resource even for the current situation and to be widely viewed as not being able to address future needs and demands, under a highly uncertain future. The ever increasing need to spend large sums to correct mistakes of excess pollution, or inappropriate allocation of the resource, made in the recent past tells us that even under the best conditions (in the developed countries) our approaches are lagging behind needs, and under the worst conditions (the developing countries) have produced major problems which cannot be easily solved. All of these concerns are magnified by the suspicion that we may be entering a period of rapid human-induced climate change with very uncertain implications for water resource management. Despite what may occur in the future, the overhang of existing problems in the developed and developing countries seems to necessitate fairly radical reevaluation of the prevailing approaches that has placed us in this situation.

### 8.1   *In defense of conventional wisdom*

In spite of the indictment of conventional wisdom above, I believe that it has within its purview all the tools needed to help humankind survive the water resources challenges posed. It is often that the conventional wisdom, if properly examined, already contains the seeds for challenging the way the water community currently does business. I believe that conventional wisdom and conventionally accepted concepts and ideas about water could, if understood in a holistic way, significantly improve the sustainability of water as a resource and as a pillar of the ecosystem within which we all live in the face of the real possibilities of a water crisis. Part of the problem is that conventional wisdom is not so widely understood as it appears and not so widely applied; in other words, it is not conventionally accepted conventional wisdom!

## 8.2    Nine important facts of life for water

The nine very important facts-of-life concerning water discussed above, are conventionally well known, but generally not understood by all of the different water constituencies.
They are certainly conventional wisdom, but never seem to be quoted together as the fabric
or the framework under which to view water problems. These important facts cover a multitude of issues, but by themselves they constitute a sort of *Nine Commandments* of water
sustainability!

- *Blue/Green/Brown water*: It is estimated that human appropriations of the *green* water were
  almost three times as large as those of the *blue* water. The possibility of using more of the
  *green* water and reducing the demands on the *blue* water are active questions that could lead
  to dealing more easily with feeding the future populations.
- *Asymmetries in water use*: For most uses small amounts of water are actually consumed
  with the bulk returning, albeit in many cases polluted, to the water source. Irrigation users,
  however, typically evaporate more than 70% of the withdrawn water. Industry, particularly
  heavy industry and fossil fuel electric generation, withdraw huge amounts of water, but consume little. Municipal and commercial withdrawals are typically the smallest withdrawals
  with about 20% consumptive use. This leads to the role of irrigation water as a balancing
  mechanism for meeting water demands in other sectors. A 10% reduction in water use by
  irrigation would more than double the amounts needed for municipal and industrial uses.
- *Irrigation flywheel*: In many countries of the world as much as 60–95% of all water is
  consumed by irrigated agriculture. This fact is extremely important when looking toward
  future potential water shortages. The question then hinges on whether or not such a 10%
  reduction in irrigation water use could be achieved by efficiency improvements alone, or
  the equivalent areas for rainfed crops and additional trade need to be developed, in order to
  meet increasing demands for food, fiber, and other industrial crops.
- *Virtual water escape hatch*: Directly related to the irrigation flywheel is what Allan (2001)
  characterized as *virtual water*: the use of imports of food crops as a substitute for use of
  domestic water for irrigation. Many arid countries in the world already rely upon the trade
  in *virtual water* to meet their needs for food. The question remains: as to what extent can
  the trade be expanded by developing the use of more *green water*?
- *Low cost desalination breakthrough*: Recent developments in membrane filtration technologies combined with reverse osmosis have revolutionized the potential for widespread
  application of desalination. While this does not give an instant solution to the problem
  in many poor cities in the developing world, it does indicate that reasonable economically
  achievable technology is available to solve the urban water supply problems as they develop.
- *The ecological revolution*: The major reason for widespread dissatisfaction about water
  issues in developing countries stems from the vast numbers who have no access to adequate
  drinking water or sanitation. It is simply beyond the financial capabilities of existing governmental and international agencies to cover the cost of providing conventional water and
  sanitation to these persons and their increasing numbers. Recent developments and cost
  breakthroughs in ecological sanitation practices that use little or no water, point the way to
  resolving this major issue.
- *Transboundary conflicts*: Since there is no strong international law governing the resolution
  of transboundary water quality and quantity disputes, the potential for conflict remains
  high. This issue needs careful consideration by riparians and the regional family of nations to

head off these conflicts from becoming worse. New approaches through regional governance organizations such as the EU, SAARC (South Asian Association for Regional Cooperation), NAFTA, SADAC (Southern African Development Community), seem to hold the best chance of resolving the issues.

− *Uncertainty of water availability*: By the end of the 20th century, hydrologists were searching for evidence of long-term anthropogenic climate change caused by greenhouse gases. Many conflicting data sets and theories have been reviewed with contradictory results and while at present there is no general agreement as to the change in variability of precipitation over the next 5 decades, there seems to be a growing consensus that precipitation will increase over large areas of the globe and that sea level may rise between 0.1 and 0.3 meters. Most of the discussion about the uncertain future of water has focused upon the water supply side of the equation; much larger uncertainties can be expected on the water demand side influenced by demographic parameters, economic growth, technology change, and life styles. There is a pressing need for water planners and agencies to improve their approaches to resolving resource issues with such high levels of uncertainty.

− *Idiosyncrasy of water institutions*: Finally, one water fact-of-life makes it very difficult to implement the types of technical and social breakthroughs already available; the hurdle is that property rights, institutions, and laws, vary widely from country to country, even within countries. The legal and political institutions tend to get in the way of rational water management. In many cases it is impossible to reallocate or trade between conflicting water uses. The confusion allows for the entrenchment of bad conventional institutional views and blocks consideration of newer approaches.

## 9 CONCLUSIONS

This chapter has provided an introduction to the issues of water security and scarcity in the context of governance of the resource. The discussion points to many issues which are not yet resolved in analyzing the question of *crisis or reality*? It does, however, make a strong plea for the consideration of the political and social dimensions of the question. Later chapters in the book deal with the physical, technical, and economic aspects of the *water crisis*, but the reader is urged not to ignore the cautionary parts of this first chapter. The water crisis is real if the players involved in development and management of water ignore the powerful influence of politics on how the problems are structured and that the outcomes depend upon the political structuring of the problem.

The commentator on this chapter at the workshop in Santander (Domingo Jiménez Beltrán), summarized the main points of the chapter as follows:

• Sustainability is now a well-established concept that contains both the more restricted concepts of security and survivability, and makes redundant any approaches that are solely aimed at them.
• By establishing sustainable development you can improve both the environment and the resource availability and at a lower cost.
• Sustainability and governance is one and the same thing. If the principles of good governance (efficacy, efficiency, equity, coherence, transparency, accountability, and public participation) are pursued, then sustainability must be a consequence.

- There is no *water crisis*; there is a clear *water management crisis*. This is demonstrated by the failure to bring water use onto a sustainable path.
- The principles and practices for future water governance are well known and tested, but it seems to be difficult to escape from the weight of past practices. New forms of governance (*Nueva Cultura del Agua* in Spain) using a different logic from the ones that created the current situation are desparately needed.
- Most, if not all, of the needed technical, economic, political, and social tools needed to effect this transition are already well known. What is needed is some way to synthesize all of these components into a coherent whole. This should be the role of *Integrated Water Resources Management* (IWRM).

## REFERENCES

Alcamo, J.; Heinrichs, T. & Rösch, T. (2000). *World Water in 2025*. World Water Series, Report 2. Centre for Environmental Systems Research, University of Kassel, Germany.

Allan, J.A. (2001). *The Middle East Water Question: Hydropolitics and the Global Economy*. I.B. Tauris, London, UK.

Bornstein, D. (2004). *How to Change the World: Social Entrepreneurs and the Power of New Ideas*. Oxford University Press.

Buchanan, J. & Tullock, G. (1962). *Calculus of Consent: Logical Foundations of a Constitutional Democracy*. University of Michigan Press, Michigan, USA.

Crenson, M.A. & Ginsberg, B. (2002). *Downsizing Democracy: How America Sidelined its Citizens and Privatized its Public*. Johns Hopkins.

Dahl, R.A. (1961). *Who Governs*. Yale University Press.

Diamond, J. (2005). *Collapse: how societies choose to fail or succeed*. Viking, New York, USA.

Dorfman, R.; Jacoby, H.D. & Thomas, H.A. Jr. (eds.) (1972). *Models for Managing Regional Water Quality*. Harvard Press.

Downs, A. (1957). *An Economic Theory of Democracy*. Harper and Row.

Falkenmark, M. & Rockstrom, J. (2004). *Balancing Water for Humans and Nature*. EARTHSCAN, London, UK.

Falkenmark, M. & Widerstrand, C. (1992). *Population and Water Resources: A delicate Balance*. Population Bulletin, 47(3). Population Reference Bureau, Washington, D.C., USA.

GWP (Global Water Partnership) (2000). *Towards Water Security: A Framework for Action*. GWP, Stockholm, Sweden.

Hoekstra, A.Y. & Hung, P.Q. (2002). Virtual Water Trade, a quantification of virtual water flows between nations in relation to international crop trade. Technical Report. *Virtual Water Research Report Series*, No. 11. IHE, Delft, the Netherlands.

Hutson, S.S.; Barber, N.L.; Kenny, J.F.; Linsey, K.S.; Lumia, D.S. & Maupin, M.A. (2004). *Estimated Use of Water in the United States in 2000*. USGS (United States Geological Survey), Circular 1268.

IPCC (Intergovernmental Panel on Climate Change) (2001). *Climate Change 2001: Third Assessment Report*. Cambridge University Press.

IWMI (2000). *World Water Supply and Demand*. International Water Management Institute, Colombo, Sri Lanka.

Kaufmann, D.; Kraay, A. & Zoido-Lobatón, P. (1999). *Governance Matters*. Policy Research Working Paper 2196. World Bank Institute, October, 64 pp.

Lasswell, H.D. (1936). *Politics: Who gets What, When and How?* McGraw-Hill.

L'vovich, M.I. (1979). *World Water Resources and their Future*. Translation by the American Geophysical Union. LithoCrafters Inc., Chealsea, Michigan, USA.

Maass, A. (1951). *Muddy Waters: The Army Engineers and the Nations Rivers*. Da Capo Press.

Maass, A. & Anderson, R.L. (1978). *... and the desert shall rejoice: conflict, growth, and justice in arid environments*. MIT Press.

Maass, A.; Hufschmidt, M.M.; Dorfman, R.; Thomas. H.A. Jr.; Marglin, S.A. & Fair, G.M. (1962). *Design of Water Resources Systems: New Techniques for Relating Economic Objectives and Engineering*. Harvard University Press.

Olson, M. (1965). *The Logic of Collective Action: Public Goods and the Theory of Groups*. Harvard University Press.

Olters, J.P. (2001). *Modeling Politics with Economic Tools: A Critical Survey of the Literature*. Working Paper 01/00. International Monetary Fund.

Ostrom, E. (1990). *Governing the Commons: The Evolution of Institutions for Collective Action*. Cambridge University Press.

Pezzey, J.C.V. (1992). *Sustainability Development Concepts: An Economic Analysis*. World Bank, Environment Paper no. 2. Washington, D.C., USA.

Pezzey, J.C.V. (2002). *Sustainability Policy and Environmental Policy*. Economics and Environment Network, Australian National University. Working Paper, EEN0211. Canberra, Australia: 43 pp.

Pierre, J. (ed.) (2000). *Debating Governance, Authority, Steering and Democracy*. Oxford University Press.

Population Action International (2003). *Sustaining Water: Population and the Future of Renewable Water Supplies*. Population Action International, Washington, D.C., USA.

Postel, S.L.; Daily, G.C. & Erhlich, P.R. (1996). *Human Appropriation of Renewable Fresh Water*. Science, 271: 785–788.

Putnam, R.D. (2000). *Bowling Alone: The Collapse and Revival of American Community*. Simon and Schuster.

Ramirez-Vallejo, J. & Rogers, P. (2004). Virtual Water Flows and Trade Liberalization. *Journal of Water Science and Technology*, 49(7): 25–32.

Repetto, R. (ed.) (1986). *World Enough and Time*. Yale University Press.

Rogers, P. & Hall, A. (2003). *Effective Water Governance*. Global Water Partnership. TEC Background Paper No. 7. Stockholm, Sweden.

Shafik, N. & Bandyopadhyay, S. (1992). *Economic Growth and Environmental Quality: Time-Series and Cross-Country Evidence*. Background paper for the World Development Report. WPS 904. World Bank, June.

Shiklomanov, I.A. (2000). Appraisal and Assessment of World Water Resources. *Water International*, 25(1): 11–32.

Shiklomanov, I.A. & Sokolov, A.A. (1983). Methodological basis of world water balance investigation and computation. In: *New Approaches in Water Balance Computations*. Proceedings of the Hamburg Symposium. International Association for Hydrological Sciences. Publication No. 148.

Solow, R.M. (1991). *Sustainability: An Economist's Perspective*. The eighteenth J. Seward Johnson Lecture. Woods Hole Oceanographic Institution, Massachusetts, USA.

Stone, D.A. (2000). *Policy Paradox: The Art of Political Decision Making*. W.W. Norton & Company, New York, USA.

Tendler, J. (1997). *Good Governance in the Tropics*. Johns Hopkins Press.

UN/WWAP (United Nations/World Water Assessment Programme) (2003). *Water for People, Water for Life*. UN World Water Development Report. UNESCO (United Nations Educational, Scientific and Cultural Organization) and Berghahn Books.

U.S. Water Resources Council (1983). *Principles and Guidelines*. Executive Order 11747, February 3.

Wade, R. (1987). The Management of Common Property Resources: Finding a Cooperative Solution. *World Bank Research Observer*, 2(2): 219–234.

Waterbury, J. (1979). *Hydropolitics of the Nile Valley*. Syracuse University Press, Syracuse, New York, USA.

WMO (World Meteorological Organization) (1992). *The Dublin Principles Keynote Papers*. International Conference on Water and Environment: Development Issues for the 21st Century (Dublin, Ireland). WMO, Geneva, Switzerland.

Wolf, A.; Shira, T.; Yoffe, B. & Giordano, M. (2003). International waters: identifying basins at risk. *Water Policy*, 5(1): 29–60.

World Commission on Environment and Development (1987). *Our Common Future*. Oxford University Press, New York, USA.

World Resources Institute (2003). *Water Resources and Freshwater Ecosystems*. Earth Trends (World Resources Institute environmental information portal).

CHAPTER 2

# Water for growth and security

W.J. Cosgrove
*Bureau d'Audiences Publiques sur l'Environnement, Québec, Canada*

ABSTRACT: The lack of action to address the need for better management of the world's scarce water resources, and in particular about the lack of investment in water services and infrastructure that are essential to economic and social development should be of deep concern to all. We could say that the efforts of the water community in raising awareness of the critical state of the world's water resources and of the need for water for health and development have succeeded. Yet the reality is that picture remains bleak. Progress can be made with an approach having three basic elements: (1) Clear goals that target the poor and use the right means; (2) Partnerships; and (3) Using processes underway.

Keywords: *Water resources, infrastructure gap, economic development, targeting poor, benefits*

## 1 INTRODUCTION

I am deeply concerned about the lack of action to address the need for better management of the world's scarce water resources, and in particular about the lack of investment in water services and infrastructure that are essential to economic and social development. I have for many years seen the connection between economic development and the presence of water infrastructure.

## 2 TO-DAY'S SITUATION

On the planet to-day over 1000 million live in absolute poverty, with incomes of less than one US\$/day. Thousands of millions of people lack access to safe water supply and sanitation. Rural communities lack access to water for subsistence farming. Droughts and floods wreak havoc and destroy years of hard work and savings. Scarce resources are polluted and overused. The natural sources of environmental services and of life to the species with which we share this planet are being destroyed.

The international community is not unaware of this situation. For example, the Millennium General Assembly of the United Nations (UN) adopted a number of Millennium Development Goals (MDGs). Among them was a goal to reduce by 2015 half the proportion of people without sustainable access to safe drinking water. The World Summit on Sustainable Development (WSSD) in Johannesburg two years later added a goal to reduce by half the population without access to basic sanitation by 2015. Leaders gathered at WSSD also called upon all countries to prepare country plans for Integrated Water Resource Management (IWRM) by 2005. In Kyoto in 2003 at the 9th Meeting of the Committee of the Parties (COP9) those responsible for

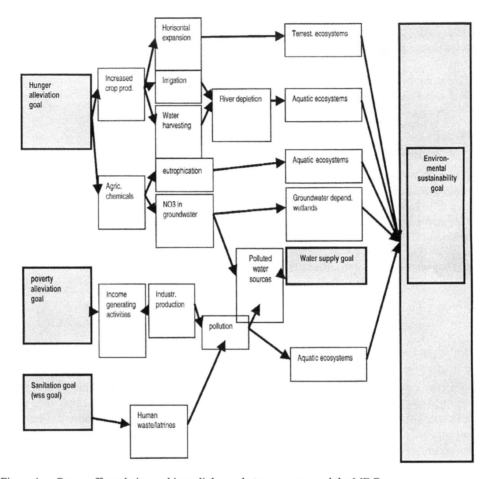

Figure 1.   Cause-effect chains and inter-linkages between water and the MDGs.

implementing the *Kyoto Protocol* recognized that adaptation to climate variability and change is an emerging issue for water managers.

The UN Task Force on Water for the Millennium Development Goals has recognized the links between water and achievement of nearly all of the MDGs (Figure 1: UN Task Force on Water and Sanitation, 2004).

Thus there is no doubt that water has been recognized as a critical factor on the Sustainable Development agenda. We could say that the efforts of the water community in raising awareness of the critical state of the world's water resources and of the need for water for health and development have succeeded.

Yet the reality is that the bleak picture remains that I described in the opening of this chapter and there are few signs that it is changing. Former President Jimmy Carter of the USA in his speech accepting the Nobel Peace Prize (Carter, 2002) said:

> "Citizens of the 10 wealthiest countries are now 75 times richer than those who live in the 10 poorest ones. The separation is increasing every year. Not only between nations, but within

them. The results of this disparity are the root causes of most of the world's unresolved problems, including starvation, illiteracy, environmental degradation, violent conflict, and unnecessary illnesses that range from guinea worm to HIV and AIDS".

He might well have mentioned lack of access to water for life and livelihoods.

## 3 ADDRESSING THE GAP

Water management (supply, demand and availability) is a universal concern and response requires local and global cooperation in partnerships between governments at all levels, civil society, private sector and knowledge institutes. Progress can be made with an approach with three basic elements:

- Clear Goals
  - Targeting the poor
  - Using the right means
- Partnerships
- Using processes underway.

### 3.1 Clear goals

#### 3.1.1 Targeting the poor

The UN Convention states that *special regard* should be given to *the requirements of vital human needs*. In this context world community thinks of Africa. In Africa only 175 million out of a rural population of 410 million have access to an improved water supply. The total number to be served will have increased through population growth to 517 million by 2015. The situation in urban areas appears to be better. Reports show that 175 million out of a total population of 215 million currently have access to an improved source. However through population increase and rural-urban migration, the number requiring service by 2015 will have increased to 389 million. Moreover, the definition used of those with/without service is technological. Those in urban areas connected to a distribution system are considered to be served. Yet recent reports demonstrate that in many cases these distribution systems are not operated reliably, with no water in them much of the time and the water that is there often polluted. The figures for access to basic sanitation services and hygienic living conditions are much worse. The absolute numbers of those without access are higher in Asia. However it is the appalling high percentages that are seen in Africa.

#### 3.1.2 Targeting means: the infrastructure gap

Statistics show that as a level of country's income grows the value of its infrastructure increases. For example Chad which in 1990 had a per capita income (1985 PPP[1] terms) of roughly US$ 200, had invested only US$ 10 per capita in infrastructure through its history. Australia, with a per capita income about 600 times as high in the same year had invested close to US$ 8000 per capita in infrastructure. The poor countries of Africa are grouped together with low investments in infrastructure. The industrialised countries with high incomes have invested

---

[1] Purchasing power parity.

large amounts in infrastructure. But have these countries invested more as their economies grew, or did the countries' economies grow when they invested in infrastructure? Experience in many places has shown that it is the latter. Consider, for example, the Tennessee Valley in the USA where the investments in infrastructure by the Tennessee Valley Authority clearly lifted the residents of the valley out of poverty as well as contributing to development beyond the valley through power generation.

I share the hypothesis put forward by David Grey, Senior Water Adviser of the World Bank, that water security is essential for sustained economic growth and poverty eradication and that there is a minimum platform of water resources infrastructure and institutions to achieve *water security*. Developed countries have invested public funds heavily to provide this infrastructure that is a public good.

### 3.1.3   *Needs and benefits*

This infrastructure stores water to reduce flooding and provide sources of supply during dry seasons. The need for such infrastructure is especially great in Africa. In Kenya, for example, fluctuations in rainfall may vary plus and minus 40% year on year. Subsequent flood and drought years between 1998 and 2000 have been estimated to have resulted in losses to the Kenyan economy of 22% of the average Gross Domestic Product (GDP) of the country over those two years. A correlation between rainfall and GDP has been clearly demonstrated in Zimbabwe, and should be expected in any country in which agriculture counts for a significant percentage of total GDP.

The provision of water supply and sanitations services is perhaps a greater priority given the reduction in premature deaths, sickness and other hardships and indignities that will be eliminated, and the benefits from a healthier population at study and at work. Often overlooked is the fact that investment in infrastructure and services is not just an expense. Enterprises, whether public or private, are necessary to produce the materials required, to construct the systems and to operate and maintain them. Their establishment and strengthening add to a country's institutional framework and benefit further development.

### 3.2   *Partnerships*

Partnerships can provide a major means of accelerating investment in water services and infrastructure. These may be at the national, regional (continental) or international level.

Many governments have begun the process of decentralizing responsibility for the construction and management of services to local governments, including water and sanitation services. This approach has great potential for increasing the rate at which improved access to these services is achieved and maintained. However, it requires that the capacity be created at the local level, and it further requires that mechanisms be found to bring the financial resources to this level. The international and multinational banks are looking at ways to do this through sub-sovereign lending. However I believe that the solutions to these questions will best be found within countries, with national governments searching for appropriate solutions in partnership with local elected officials. The local governments, in turn, are working more and more in partnership with community-based organizations and non-governmental organizations.

Moreover, the solution is not to be found in the traditional approach that requires that the capacity be created first before investments are made. Rather investments must be forthcoming that are in keeping with available capacity but stretch it, so that capacity to build and manage services grows as the size of the systems grow.

The most prominent example at the regional level is without a doubt the New Partnership for Africa's Development (NEPAD). African heads of state have committed themselves to a participatory development approach and to being held accountable for their progress by their peers. African governments and civil society are partners in this process. African water ministers are working together in the African Ministerial Conference on Water (AMCOW). They are supported by UN Water for Africa and encouraged by other members of an independent group known as the African Water Task Force. The African Development Bank has responded to requests from NEPAD and AMCOW by the establishment of the African Water Facility to accelerate investments and innovation in the water sector, and the Rural Water Supply and Sanitation Initiative. Like-minded donors and investors are committed to responding with assistance to this commitment by Africans to help themselves.

Internationally a major development was the creation of *Type II* Partnerships at the World Summit on Sustainable Development in Johannesburg. One of these, tentatively titled the North-North Partnership, aims at developing water partnerships within developed countries, linking these to share their practices and to make links to countries of the South. Many of the latter have already developed national water partnerships under the aegis of the Global Water Partnership. A similar partnership has been at work through the Cooperative Program on Water and Climate (formerly the Dialogue on Water and Climate). This network brings together climatologists, meteorologists, hydrologists and water managers from national and basin level from both developed and developing countries to address issues of water management in the face of climate variability and long-term climate change.

## 3.3 *Using processes underway*

Water management (supply, demand and availability) is a universal concern. As described above successful action requires local and global cooperation in partnerships between governments at all levels, civil society, private sector and knowledge institutes. Progress can be made using processes underway if the goals are clear, appropriate means are employed and all work in partnership.

Many governments have not yet included investments in water among their priorities. One interpretation of this can be that they have not yet understood the importance of water to achieving all of their economic and social development goals. We know that awareness of the needs and benefits are well understood by ministers responsible for the sector. However we must all work at impressing on the ministers of finance and economic development the critical role that water plays in all sectors. Clear targets for access to water services must be supported by short- and long-term budgets, policies, and investment strategies shared by all members of government.

At the intergovernmental level one of the important existing processes is that of the Commission on Sustainable Development (CSD). The current cycle of the CSD deals with water supply and sanitation and housing. The meeting of CSD 13 in April 2005 provides an opportunity for countries to demonstrate that they have already begun to fulfill their responsibilities using the national processes described above. CSD 13 should not be a place where ministers of the environment from around the world put forward proposals for policy approaches to be followed by developing countries in managing their water, but rather as an occasion for developing countries to put forward the policies that they know will work and to ask the international community to support them in their implementation.

The United Nations General Assembly in September 2005 marking the fifth anniversary of the adoption of the Millennium Development Goals (MDGs) provides another such occasion. However, it will be even more important as it will permit the heads of state of developing countries to refer to the policies that they have presented at CSD 13 and to ask their partners of the developed world to fulfill theirs.

In March 2006 the World Water Council will sponsor the Fourth World Water Forum hosted by the Government of Mexico in Mexico City. Its overall theme will be *Local Actions to Address Global Challenges*. This reflects the recognition that it is local actions that unleash local social energy. The Forum will support the institutional networks that support local actors and processes, hopefully developing synergies and increasing collaboration between different stakeholder groups. National and local governments and their partners in civil society will have an opportunity to present cases that demonstrate how their new policy approaches are being translated into actions on the ground.

## REFERENCES

Carter, J. (2002). *The Nobel Peace Prize Lecture*. December 10, Oslo, Norway.
United Nations Task Force on Water and Sanitation (2004). *Achieving the Millennium Develop-ment Goals for Water and Sanitation: What Will It Take?* Interim Full Report. Task Force on Water and Sanitation. Millennium Project. February. [www.unmillenniumproject.org/documents/tf7interim.pdf].

CHAPTER 3

# Collective systems for water management: is the *Tragedy of the Commons* a myth?

E. Schlager
*School of Public Administration and Policy, University of Arizona, Tucson, USA*

E. López-Gunn
*London School of Economics and Political Science, UK*

ABSTRACT: The tragedy of the commons is a myth in water management. Hundreds of case studies of local water users devising institutional arrangements to successfully govern their use of shared water resources have been documented. Failures in water governance that have occurred are rarely due to the tragedy of the commons, that is, to the lack of institutional arrangements. Failures are more often due to the challenges of devising, adapting, and maintaining institutional arrangements in dynamic water settings. This chapter explores two institutional challenges that repeatedly confront water users – devising institutional arrangements that are well matched to the physical and social features of the water setting; and devising complementary and supportive relations among organizations and governments operating at different scales. The chapter begins with a review of the literature on common pool resources and the creation and maintenance of local, self-governing institutional arrangements illustrated through case studies of water governance. It then turns to an examination of cross-scale linkages – relations among organizations and governments at different scales, from the local to the international. The chapter provides an in-depth examination of cross-scale linkages, what appear to be the most important features of such linkages, and, in particular, important features of higher level governments that promote support for the self-governing efforts of local resource users. Cases of water governance from the USA and Spain are used to illustrate cross-scale linkages and the role they play in common pool resource management.

Keywords: *Institutions, self-governance, cross-scale linkages, common pool resources, multi-level governance, water governance*

## 1 INTRODUCTION

The model of the tragedy of the commons predicts the failure of appropriators to sustainably use a common pool resource. Appropriators race to harvest what they can from a resource, helpless to prevent the race from occurring. Appropriators are trapped in a Hobbesian state of nature, incapable of communicating and, consequently, unable to devise agreements or institutional arrangements that would allow them to sustainably govern a common pool resource.

While all commons tragedies result in failures, not all failures are due to commons tragedies. In devising long enduring, self-governing arrangements for managing water, appropriators must not only pay close attention to carefully matching institutional arrangements with the physical and social setting, but also to the interactions among different levels of organization and government. In other words, multiple institutional hurdles must be safely traversed for long enduring, self-governing arrangements to emerge and to be sustained.

Failures in water governance are rarely due to the tragedy of the commons, that is, to the lack of institutional arrangements. Failures are more often due to the challenges of devising, adapting, and maintaining institutional arrangements in dynamic water settings. Increasingly in this context, sustainability is perceived as a dynamically maintained equilibrium, not a static one (Agrawal, 2002). This chapter explores two institutional challenges that repeatedly confront water appropriators – devising institutional arrangements that are well matched to the physical and social features of the water setting; and devising complementary and supportive relations among organizations and governments operating at different scales.

Well-matched institutional arrangements require that local water users, with their wealth of local time and place information, participate in the creation, maintenance, and revision of water institutions. For example, questions over whether institutions are created top down or bottom up can have implications at the local scale on the ownership and empowerment local people might feel towards these institutions. Local level participation is necessary, but often not sufficient, for sustainable governance.

The spatial extent of water resources often means that water is not just a local affair, indeed in some cases it is a highly conflictive transboundary international issue. Many local communities, regional governments, and national governments participate in water governance. In a globalized world, institutions are increasingly nested within one another and "cross-scale institutions and vertical linkages become even more important" (Berkes, 2002: 294). Issues like whether higher level authorities crowd in or, on the contrary, facilitate better local governance of water resources is progressively more important, i.e. whether interdependencies are symmetrical or benign or unidirectional (and malign) (Young, 2002). The quality of the cross-scale linkages among organizations and governments directly affects the ability of water appropriators to govern and use water resources in a sustainable fashion. As Lowndes & Wilson (2003: 279) state: "partnerships and networks are as important as hierarchical intra-organizational relationships".

In the following two sections the literature on common pool resources and the creation and maintenance of local, self-governing institutional arrangements is surveyed and illustrated through case studies of water governance. These two sections provide the foundation for the heart of the paper – cross-scale linkages. Cross-scale linkages refer to relations among organizations and governments at different scales, from the local to the international. Now that considerable progress has been made in recognizing and explaining local, self-governing efforts, increasing attention is being devoted to the context in which local institutional arrangements operate, or cross-scale linkages. The last section of the chapter provides an in-depth examination of cross-scale linkages, what appear to be the most important features of such linkages, and, in particular, important features of higher level governments that promote support for the self-governing efforts of local resource users. Cases of water governance from the USA and Spain are used to illustrate cross-scale linkages and the role they play in common pool resource management.

## 2    ROBUST LOCAL SELF-GOVERNING INSTITUTIONAL ARRANGEMENTS

As the title of this chapter implies, for decades the most widely accepted explanation for natural resource degradation and destruction was Garrett Hardin's (1968) model of the tragedy of the commons. It dovetailed nicely with other widely accepted models of individual decision-making in the social sciences. For instance, Olson's (1965) theory of collective action predicted that only under very special circumstances would people be able to overcome free riding problems and engage in collective action. From game theory, the prisoners' dilemma game made equally grim predictions: people will pursue their own self-interest, even if in doing so they make themselves worse off than if they had cooperated and pursued the group's interest. Natural resource economics models also highlighted conflicts between individual self-interest and the efficient use of natural resources (Gordon, 1954; Scott, 1955; Anderson, 1986). All models pointed in a single direction – self-interested individuals will make choices and take actions that will lead to the economic ruin, if not the actual destruction, of natural resources.

This simplistic view of individual rationality as self-interested is now much more nuanced, as "*homo reciprocans*, agents are neither completely self-interested not blatantly altruistic … agents as homo reciprocans in the sense that people typically exhibit altruistic tendencies only towards those who cooperate with them and reciprocate their altruistic gestures" (Onyeiwu & Jones, 2003: 236). This means that while in some cases grim predictions were borne out, in others cases quite the opposite occurred. Users of shared resources cooperated to define and administer self-governing institutional arrangements that allowed them to coordinate their actions and use resources in sustainable ways (McCay & Acheson, 1987; Berkes, 1989). Individuals are complex, and the challenge is to increase the numbers of *conditional cooperators* (Ostrom, 2000a) through strong institutional arrangements.

Based on in-depth and carefully constructed analyses of numerous cases of both successful and failed efforts at local, self-governance of natural resources, Ostrom (1990) posited a set of institutional design principles that appeared to be common among all cases of successful, long-lasting, efforts at local self-governance (see Table 1). Design principle one, exclusion, is critical if appropriators are to commit to following a set of institutional arrangements over time and investing in modifying them as circumstances warrant. Exclusion, while critical, is insufficient to ensure long-term commitment to the rules. The rules themselves must make sense, they must be crafted to the exigencies of the situation, and as the situation changes, the appropriators must have the ability to modify the rules. Accountable monitors and graduated

Table 1. Ostrom's Institutional Design Principles for long enduring Common Pool Resource (CPR) arrangements (Ostrom, 1990: 90).

---

1 Individuals or households who have rights to withdraw resource units from the CPR must be clearly defined, as must the boundaries of the CPR itself.

2 Appropriation rules restricting time, place, technology, and/or quantity of resource units are related to local conditions and to provision rules requiring labor, material, and/or money.

3 Most individuals affected by the operational rules can participate in modifying the operational rules.

4 Monitors, who actively audit CPR conditions and appropriator behavior, are accountable to the appropriators or are the appropriators.

5 Appropriators who violate operational rules are likely to be assessed graduated sanctions (depending on the seriousness and context of the offense) by other appropriators, by officials accountable to these appropriators, or by both.

6 Appropriators and their officials have rapid access to low-cost local arenas to resolve conflicts among appropriators or between appropriators and officials.

7 The rights of appropriators to devise their own institutions are not challenged by external governmental authorities.

8 Appropriation, provision, monitoring, enforcement, conflict resolution, and governance activities are organized in multiple layers of nested enterprises.

---

sanctioning maintain appropriators' commitment to institutional arrangements. Knowing that most appropriators are following the rules most of the time sustains rule following behavior. In instances of rule violations, graduated sanctions act to bring appropriators' actions in line with the rules. Finally, conflict resolution mechanisms and at least a minimal recognition of the right to organize prevent these institutional arrangements from unraveling due to internal strife or invasion from external governmental authorities.

The principles have received considerable support from in-depth research conducted across several different types of common pool resources. In relation to water, studies of irrigation systems and groundwater basins both support the conclusion that resource users can devise and manage robust self-governing arrangements. Irrigation studies have confirmed the importance of exclusion and carefully crafted rules (design principles one and two), and the importance of appropriators participating in devising and modifying rules (design principle three) (Tang, 1989, 1992, 1994; Lam, 1998). Tang studied 43 irrigation systems, examining the differences between high performing systems, those that were well-maintained and in which farmers followed the water allocation rules, and low performing systems, those that either were not well maintained and/or farmers failed to follow the rules consistently. Tang (1994) argues that among irrigation systems that perform well, rules that govern water allocation and maintenance activities are better crafted to the specific conditions of each irrigation system. High performing systems were associated with multiple rules that adequately limited access to the system and that fairly allocated water among the irrigators. Poorly performing irrigation systems were characterized by a single simple rule set or by no rules at all. Access to the irrigation systems was not adequately regulated and water allocation rules often did not work well.

While low performing irrigation systems were similar in the rules of access and allocation that were used, such systems were much more likely to be government owned rather than farmer owned. Government officials, since they are not directly subject to the irrigation rules that they devise, face few incentives to design rules that ensure the effective operation of irrigation systems. Instead, they face incentives to devise rules that increase their political support and that lighten their administrative burdens. The rationalist institutionalist approach takes a dim view of civil servants acting as self-interested individuals. Conversely, because farmers directly experience the consequences of their rule-making decisions, they confront incentives to carefully craft the rules to the particular situation that they face (Tang, 1992).

Tang's research was replicated and extended by Lam (1998). Extensive data on 150 Nepalese irrigation systems was analyzed by Lam (Lam *et al.*, 1994; Lam, 1998). Like Tang (1992, 1994), Lam (1998) found that farmer managed irrigation systems performed significantly better than did government managed irrigation systems. Irrigators in farmer managed systems exhibited significantly higher levels of: (1) entrepreneurial activities in attempting to coordinate irrigation activities; (2) information and understanding of the irrigation system; and (3) mutual trust (Lam, 1998: 26–133). Irrigators in farmer managed systems also used more varied and complex sets of rules for governing their activities.

Equally in Spain in the context of groundwater management, a classic example of a common pool resource, water users associations in the same region have had very different experiences in mutually beneficial collective action. The water users associations that were created top down (in Western Mancha and Campo de Montiel) have not fared well compared to the one created bottom up (in Eastern Mancha). In Eastern Mancha where farmers themselves have designed and created their own institutions, with the support of higher-level authorities, empowerment has led to farmers slowly devising their own rules. Thus in these three cases,

Table 2.  Comparative application of Ostrom (1992) design principles in three spanish aquifers.

| Design principles (Ostrom, 1992) | | Western Mancha aquifer | Campo de Montiel | Eastern Mancha |
|---|---|---|---|---|
| 1. Clearly defined boundaries. Both the boundaries of the service area and the individuals or household with rights to use water from an irrigation system area clearly defined. | Geographic (aquifer boundary) | ✓ | ✓ | ✓ |
| | Water rights | ✗ | ✗ | ✗ |
| 2. Collective choice agreements. Most individuals affected by operational rules are included in the group that can modify these rules. | | ✗ | ✗ | ✗ |
| 3. Monitoring. Monitors, who actively audit physical conditions and irrigator behavior, are accountable to the users and/or are the users themselves. | | ✗ | ✗ | ✓ |
| 4. Graduated sanctions. Users who violate operational rules are likely to receive graduated sanctions. | | ✗ | ✗ | ✓ |
| 5. Conflict resolution mechanisms. Users and their officials have rapid access to low-cost local arenas to resolve conflict between users or between users and officials. | | ✗ | ✗ | ✓ |
| 6. Minimal recognition of rights to organize. The rights of users to devise their own institutions are challenged by external government authorities. | | ✓ | ✓ | ✓ |
| 7. Nested Enterprises. Appropriation, provision, monitoring, enforcement, conflict resolution, and governance activities are organized in multiple layers of nested enterprises. | | ✗ | ✗ | ✓ |

the design principles identified by Ostrom (1992) have been a good predictor of the success or failure of collective action and self-governance (see Table 2). In the absence of these design principles (i.e. farmer self-management) higher level authorities have attempted to fill the gap of farmer initiatives through perverse subsidies to change behavior, effectively crowding out farmers' initiatives, and generating and re-enforcing a rent-seeking behavior which then creates a path dependency (Pierson, 2000) difficult to change (López-Gunn, 2003a, 2003b).

The empirical evidence is overwhelming. Resource users are capable of devising rules that carefully guide and constrain their harvesting activities in sustainable ways. Resource users are not inevitably trapped in a tragedy of the commons. Furthermore, the most resilient institutional arrangements devised by local level resource users are those with well-defined boundaries with resource users as the primary decision-makers. Such arrangements are well matched to the problems and challenges that the resource users face and they function in a manner that supports continued interaction and cooperation. Monitoring supports commitment to the rules by providing assurances that most users are following the rules most of the time. When rule violations are detected, graduated sanctions are used as a means of correcting the actions of the rule breaker and not as a means of punishment. Also, there is a recognition that conflict and disagreements are inevitable and that some means of peaceably and legitimately settling differences is necessary.

It is important to stress that although research and analysis of internal collective arrangements are plentiful, the relevance of context and external factors impacting on these internal arrangements was taken as given or not explored in depth. Yet due to the nature of complex networks which now dominate institutional arrangements, the focus is now shifting towards a deeper understanding of the two design principles identified by Ostrom (1990), centered on recognition of the right to organize and nested enterprises. For example, the acknowledgement by higher-level authorities of the right of appropriators to devise their own institutional arrangements is in itself an indicator of a very basic (but crucial) trust cementing cross-scale linkages. In the context of theories on institutional learning, some basic level of social capital that can be built upon, harnesses local knowledge and creativity (Lowndes & Wilson, 2003: 287).

## 3   THE EMERGENCE OF LOCAL SELF-GOVERNING ARRANGEMENTS

Identifying critical features of long enduring self-governing institutional arrangements differs from identifying the conditions under which resource users may be willing to invest in developing or revising such arrangements. For institutional arrangements to persist and to perform well over time, they must be revised to address new challenges and changing circumstances. After considerable additional research, Ostrom (2000b) has suggested a set of conditions that support the emergence of cooperation to devise or revise institutional arrangements. These conditions consist of attributes of common pool resources and attributes of appropriators (see Table 3). The attributes of common pool resources focus on the ability of appropriators to readily understand the resource and their actions in relation to it. For instance, indicators, predictability and spatial extent are supportive of resource users learning the critical features and underlying dynamics of a resource. In this line there are recent developments on communities developing their own set of community indicators to help with local water management particularly grounded on that community. Such information is vital if resource users are to develop rules well matched to the physical setting. Feasibility of improvement, however, is the critical attribute. Whether it is worthwhile investing in knowledge of the resource and in institutional arrangements governing the resource depends on the returns to be realized. If changes in behavior will have little effect because the resource is too degraded, resource users are unlikely to take such steps.

Even if resource users know and understand the resource well, they are only likely to engage in institutional change, according to Ostrom (2000b) if they exhibit certain attributes. If they are heavily dependent on the resource and if they envision themselves, their children, grandchildren and community remaining heavily dependent on the resource well into the future, then they are more likely to engage in rule development, particularly if they share a common understanding of the resource. Conversely, if they are only moderately dependent on the resource for income, or if they envision their children pursuing alternative income generating activities, then they may be less willing to invest in new rules. Furthermore, the resource users must trust one another and they must have leaders or organizational skills to draw upon to engage in rule development. If they do not trust one another, if they perceive that other resource users will not consistently follow the rules, or will act in ways that undermine cooperative relationships, then they are unlikely to invest in institutional arrangements. Engaging in institutional change is no easy matter. Resource users must be committed to the resource, they must have a relatively well-developed understanding of it, they must trust one another, and they must have sufficient leadership and autonomy to engage in the task of rule change.

Table 3.   Ostrom's attributes that support institutional change (Ostrom, 2000b: 40).

---

*Attributes of the resource*

1  *Feasible improvement*: Resource conditions are not at a point of deterioration such at it is useless to organize or so underutilized that little advantage results from organizing.
2  *Indicators*: Reliable and valid indicators of the condition of the resource system are frequently available at a relatively low cost.
3  *Predictability*: The flow of resource units is relatively predictable.
4  *Spatial extent*: The resource system is sufficiently small, given the transportation and communication technology in use, that appropriators can develop accurate knowledge of external boundaries and internal microenvironments.

*Attributes of appropriators*

1  *Salience*: Appropriators are dependent on the resource system for a major portion of their livelihood or other important activity.
2  *Common understanding*: Appropriators have a shared image of how the resource system operates … and how their actions affect each other and the resource system.
3  *Low discount rate*: Appropriators use a sufficiently low discount rate in relation to future benefits to be achieved from the resource.
4  *Trust and reciprocity*: Appropriators trust one another to keep promises and relate to one another with reciprocity.
5  *Autonomy*: Appropriators are able to determine access and harvesting rules without external authorities countermanding them.
6  *Prior organizational experience and local leadership*: Appropriators have learned at least minimal skills of organization and leadership through participation in other local associations or studying ways that neighboring groups have organized.

---

Whether these attributes of the resource and of appropriators are key to explaining institutional change remains to be determined. Increasingly, however, there is more and more evidence of the crucial role social capital (both bonding and bridging) can play in cementing different institutional arrangements to foster collaborative behavior in water management. As Onyeiwu & Jones (2003: 243) state: "social capital appears to be a built in mechanism, albeit a subtle one for ensuring cooperative behavior". The same authors identify a series of features in social capital: (1) fostering mutually beneficial exchanges through the development of norms of civic behavior, trust and cooperation; (2) establishing a consultative and collective decision-making process that reduces externalities and promotes the reduction of public goods; and (3) reducing the transaction costs that also facilitate the flow of information and innovation. That is, cooperative behavior is underpinned by social capital, and there is a direct proportional relationship between social capital (both bonding and bridging) and collaborative behavior to encourage rational egoists to behave like (conditional) cooperators.

While considerable research has investigated robust institutional arrangements, little research has focused on institutional change and stability in relation to common pool resources. Research is needed on questions related to institutional learning like studies centered on triple loop learning (Maarveland & Dangbegnon, 1999). In single loop learning the outcomes of decision-making are evaluated in the context of goals and expectations, in double loop learning, feedback starts to generate change in current practices, however, in triple loop learning institutions learn to learn, or institutional reflexiveness. This is compared to its opposite, learned helplessness when individuals or communities are unable to influence their contexts through their behavior. Once again in institutional learning cross-scale linkages are crucial to foster triple loop learning and avoid learned helplessness.

The example of common pool resource management demonstrates why it is crucial that water users themselves are involved in management leaving room for experimentation and trial and error. If allowed opportunities to experiment, water users can progress to double loop learning and the development of adaptive management. As Maarveland & Dangbegnon (1999) state: "space needs to be created to diversify learning patterns". In this context, the key relevance of higher-level authorities' involvement to strengthen or rejuvenate local management becomes apparent. The role of higher level authorities cannot and should not be underestimated; in providing support, granting legitimacy to local institutions, creating enabling legislation and backing capacity building (Berkes, 2002). A delicate balance needs to be established between *push and pull*, to create space for social learning in common pool resource management.

## 4   NEXT STEPS IN COMMON POOL RESOURCE RESEARCH

The most critical tasks presently facing researchers are theory development and testing (Agrawal, 2002; Stern *et al.*, 2002). Agrawal argues that too much attention is still being devoted to the identification of factors that support self-governance and sustainable use of common pool resources, so much so that researchers face an almost embarrassing wealth of factors. Agrawal (2002: 62–63) has identified 33 distinct factors that a variety of scholars argue account for success. So many possible factors present a variety of problems. First, lists of factors alone do not explain success or failure in common pool resource management. Rather, the factors are configural, they interact, and through that interaction outcomes are realized. To date, however, little attention has been devoted to identifying or exploring the interactions among the factors – that is, to use the factors as a means to develop theory. Second, the factors rarely apply uniformly across all common pool resource settings. The values that the factors take and their effects on an outcome are influenced by the larger context within which the common pool resource is situated. The factors are contingent on a variety of contextual variables, such as ties to regional, national and global markets, or the linkages and relations among different levels of government. A key factor that keeps coming to the fore is social capital cementing cross-scale institutional arrangements, leading to the *sustainability of the commons* – the durability of institutions that frame the governance of common pool resources (Agrawal, 2002: 44).

### 4.1   *Cross-scale linkages*

One of the more important tasks facing water policy researchers is to develop theoretically grounded explanations of the relations among organizations and governments that use, manage, and govern water. The spatial extent of water resources often means that water is not just a local affair. Many local communities, regional governments, and national governments participate in water governance. The quality of the cross-scale linkages among organizations and governments directly affects the ability of water appropriators to govern and use water resources in a sustainable fashion.

The importance of cross-scale linkages is demonstrated in research by Blomquist *et al.*, (2004) that examined and compared the conjunctive water management activities of local jurisdictions among three USA states – California, Arizona, and Colorado. Conjunctive water management refers to the coordinated use of surface water and groundwater as a means of developing additional water supplies and limiting the waste of existing supplies. Typically, the coordination of the two water sources occurs through recharge projects. Surplus surface water is stored, or recharged, in an underground basin for use at a later time.

Local water appropriators contemplating investing in recharge projects must have assurances that the water they store underground will be available to them at a later time. In other words, appropriators must be able to exclude those who did not contribute to their project from accessing their stored water. The spatial extent of a basin must be such that water appropriators can gain control over the basin and exercise exclusion if they are to cooperate to engage in conjunctive water management.

In California, conjunctive water management activities occur *only* in basins in which appropriators have organized themselves and have developed a set of institutional arrangements whereby they control access to and use of the basins (Heikkila, 2001; Blomquist *et al.*, 2004). Surprisingly, however, even though appropriators in Arizona and Colorado do not control access to and use of groundwater basins, they too actively engage in conjunctive water management. Why? The larger institutional setting – specifically cross-scale linkages. The governments of Arizona and Colorado, unlike the government of California, have devised and allocated private property rights in surface water, groundwater, and in stored, or recharged, groundwater. Appropriators in Arizona and Colorado, because of their state granted private property rights in water, are assured that if they store water underground they will be able to retrieve it at a later date because of their property rights in that water. They do not first have to gain control over a groundwater basin. Appropriators in California, on the other hand, must first gain control over their basin and define property rights in it before they are sufficiently secure to invest in conjunctive water management projects. Thus, the necessity of appropriators controlling access to a common pool resource depends on the larger institutional setting–context matters.

Research in Spain highlights that enduring institutional arrangements are dependent on the positive synergy between the regional government, water authorities and the water users themselves. Debates that center on property rights and the trade off between flexibility and security, are mirrored on current debates on institutional arrangements and the trade off between robustness and flexibility. The comparison of three Spanish aquifers mentioned above and their institutional arrangements to solve aquifer over-use are an interesting example. In the case of Western Mancha, institutional arrangements are in a state of flux and therefore are flexible, however, institutional arrangements are not robust, which is translated into free-riding, corruption and aquifer over-use. By contrast, the institutional arrangements in the neighboring aquifer, Campo de Montiel, are robust, yet not flexible enough to acknowledge the legitimate concerns of non-farming water uses. The third case is more promising because the institutional arrangements being developed are robust, yet also flexible. Why the difference? The answer lies in a case of *earned autonomy* on the part of the farmers, as a successful story of co-management of water resources, and thus rides on the strength of cross-scale linkages. The concept of earned autonomy refers to the performance of local government and how – if it delivers in terms of policy outcomes – higher level authorities will grant further autonomy.

Greater attention to cross-scale linkages potentially signals a major change in the social sciences. As Lowndes & Wilson (2003) comment: "the *postmodern paradox* is based on the idea that institutional fragmentation and multilevel interdependencies is now the norm in governance. In this context eclecticism is an institutional design principle whose time has come and which aims at encouraging reflexivity and learning and signals the end of modernist grand narratives about the right way to govern". Or as Rhodes (1997: xii) states: "messy problems require messy solutions and there is no right institutional fix, it's the mix that matters in the end governing code".

Young (2002) argues that greater attention must be paid to the quality and complementarity of the relationships among organizations. According to Young (2002), the issue is not so much getting the tasks divided up appropriately among different levels of government; rather the issue is more one of developing supportive and cooperative relations, or cross-scale linkages, among organizations and governments at different levels of society. The quality of relations among organizations and governments appears to depend on a variety of factors such as features of an organization, authority relations among organizations, and interactions among organizations.

### 4.1.1   *Features of organizations*

Young (2002) argues that the ability of an organization, such as a national government, to make commitments to other governments and to follow through with those commitments depends on three different features. Capacity refers to material resources and social capital necessary to realize the activities the government has committed itself to (Young, 2002: 278). If the government does not possess the resources or does not allocate the resources to meet its commitment then, according to Young, it is lacking in capacity. Compatibility refers to the degree of congruence between the commitments made by a government and the internal structure and functioning of the government and society (Young, 2002: 278). If a government that oversees a centrally planned economy agrees to engage in market activities, the compatibility between how it operates and how it has committed to act is relatively low. Finally, competence refers to the legal and political authority needed to make commitments to others (Young, 2002: 277). As Young notes, competence depends on the structure of the government. The competence of federal systems is relatively low because agreement must be gained from different branches of government or from sub-national governments. Young (2002) focuses on the characteristics of organizations as critically defining cross-scale linkages. Capacity, compatibility, and competence together determine the ability of an organization to make and follow through with its commitments to other organizations. However, Young's notion of competence suggests that authority relations among organizations also play an important role.

### 4.1.2   *Authority relations*

The ability of a government to follow through with its commitments depends on its relations with sub-governments and with citizens. Conversely, the ability of citizens to engage in governing common pool resources depends on the authority and autonomy granted them by a government. If resources users have constitutionally protected property rights in a resource they are more able to protect their interests than if they do not have such protections. Their protected right to be in the resource provides them with considerable leverage in determining who else can and cannot be in the resource. Furthermore, the extent to which higher levels of government can influence the collective choice processes of local appropriators depends on the decision-making authority that local appropriators have been granted and the extent to which that authority is protected. If the authority is embedded in legislation or in a constitution, higher levels of government will be more limited in their ability to directly affect local level choices. In order to override local decisions, external government officials will have to change legislation and/or change constitutional provisions.

The quality of cross-scale linkages is also determined by features of higher-level governments. Local, self-governing efforts are supported by higher-level governments that provide technical expertise, funding, and independent and dispassionate monitoring and conflict resolution mechanisms (Ostrom, 1990; Blomquist, 1992; Landre & Travis, 1998). Local self-governing efforts are much more likely to be successful if appropriators have considerable autonomy to devise their own governing arrangements, secure from arbitrary interventions from higher levels of government and if higher levels of government provide supportive contexts that encourage self-governing activity.

Institutional features that support local level self-governing activity, however, present real challenges for higher-level governments engaged in addressing larger scale common pool resource problems that local communities alone cannot readily address. Higher-level governments will find it much more difficult to command compliance from relatively autonomous local resource users. Rather than commanding compliance, higher-level governments may have to rely more heavily on other mechanisms to bring the activities of local resource users in line with their interests and goals. Enabling legislation, grants, funding for projects are mechanisms that higher-level governments can use to try to align local interests with their own.

### 4.1.3 *Interactions among organizations*

Scharpf (1997) suggests that relations and interactions are likely to differ considerably among organizations depending on whether they are attempting to coordinate their actions to reach a better collective outcome or whether they are locked in a zero-sum game. Relations among organizations are much more likely to be complementary and cooperative if they are attempting to realize a shared benefit than if they view one another as strict competitors. Even if actors realize that they need to work together to create a shared benefit, relations may be tense because of what Scharpf labels the negotiators' dilemma. In a situation in which all actors' contributions are important for realizing a desired outcome, actors often must address two closely related issues: production – what the final project or outcome will be, and distribution – how the benefits and costs will be allocated among the participants. Conflict may be intense as the parties address production and distribution issues, or the creation and sharing of value (Scharpf, 1997: 120). According to Scharpf, conflict is likely to be manageable if organizations are bargaining over the production of a desired outcome and not over the distribution of the costs and benefits or realizing the outcome. If, however, organizations are bargaining over the distribution of benefits and costs, conflict may become so intense that agreements are not reached and cooperation and coordination do not occur.

The literature on cross-scale linkages is relatively sparse. Scholars who have attended to such issues suggest that attention be paid to features of the organizations, authority relations among them, and the dynamics of their interaction in order to explain the role of cross-scale linkages in supporting or hindering local level self-governance.

### 4.1.4 *Case studies: attributes of higher-level authorities*

According to Young (2002: 266) "the extent to which environmental or resource regimes yield outcomes that are sustainable is a function not only of the allocation of tasks between and among institutions operating at different levels of social organizations, but also of cross-scale interactions among distinct institutional arrangements ... it seldom makes sense to focus exclusively on finding the right level or scale at which to address specific problems ... the key to success lies in allocating specific tasks to the appropriate level".

This section will argue that a key to sustainability lies in strengthening cross-scale linkages, and in particular, paying careful attention to encouraging certain characteristics in higher level authorities. In Spain, the key role of higher level authorities can be seen in the three neighboring aquifers already discussed, with quite distinct and complex internal institutional arrangements yet very similar higher level authorities. This chapter will not analyze the internal factors described by Ostrom (1990, 1992), which are described in a previous paper (López-Gunn, 2003a). Rather the analysis focuses on the key role played by external, higher level authorities in strengthening or weakening cross-scale linkages.

The case study concentrates on three aquifers located in the Mancha region of Spain: the Western Mancha aquifer, the Eastern Mancha aquifer, and the Campo de Montiel aquifer. The three aquifers are relatively large and are heavily utilized mainly for irrigation, which boomed from the 1970s leading to cases of severe aquifer drawdown. However, the institutional response has been very different both internally in terms of the activities of their respective water user associations and externally in relation to their interaction with higher level authorities and cross-linkages with state and non-state actors. The three case studies exemplify the complex set of networks that now characterizes water (Rogers & Hall, 2003). This section analyses attributes of higher level authorities that can favor or hinder mutually beneficial collective action to prevent or halt aquifer over-use. It identifies what is called *attributes of higher level authorities*, which are listed below:

(1) *Clear boundaries in water rights*: It is essential that as far as practicable higher-level authorities facilitate the clear definition of property rights in water. Ideally, this should be undertaken in joint collaboration with water users, who have the benefit of local knowledge. In the Spanish case studies the situation in relation to clear definition of water rights (and therefore boundaries) is very different. In the case of Western Mancha the situation is chaotic with estimated thousands of illegal abstractors. By contrast, in the case of both the Eastern Mancha and the Campo de Montiel slowly boundaries are being established for property rights. The reason for this success compared to the Western Mancha by and large lies in the collaboration between farmers and the water authorities to determine these water rights. However, it is in the case of the Eastern Mancha where boundaries are actually being drawn jointly in a clear case of a partnership drawing on the synergy between localized knowledge and regulatory power. By contrast, in the case of the Campo de Montiel, although boundaries are clear, the foundations of the relationship between higher level authorities and farmers is not necessarily very positive and boundaries have been defined because farmers themselves saw it in their self-interest to do so in order to exclude other users. Therefore, this brings us to the next attribute of higher level authorities.

(2) *Legitimization of appropriators*: Ostrom (1992) already identified as a factor the key importance of the recognition of the *right to organize*. Local organization cannot operate in a vacuum, and in the long term it is essential that legitimization is granted to local organizations from higher-level authorities. Local organization can be legitimate yet have no supports from higher-level authorities. This in the longer term will harm the prospects of these institutions. In the case of the Campo de Montiel and the Western Mancha water user associations have not been acknowledged fully by the water authority and in both cases, despite the strong internal institutional arrangements of these water user associations, this has led to conflict, which has undermined the activities of these water user

associations, either by leading to litigation or through time spent over intractable water issues.

(3) *Facilitates and support initiatives of appropriators*: The example above has to be contrasted with the experience in the Eastern Mancha, where both the regional government and the water authority have been very supportive and facilitated initiatives from the water user association leading, in the first instance to partnership agreements, at a later stage to co-management and eventually hopefully to self-governance.

(4) *Builds trust with other cross-scale linkages*: Yet the question remains why have in some cases higher level authorities, like Eastern Mancha facilitated and acknowledged the legitimacy of water users and supported local institutions, whilst in other cases, particularly in the Western Mancha it has been the opposite (a latent conflict that spills over from time to time). It is argued that the key difference, as stated in previous sections in the chapter has been bridging trust cementing cross-scale linkages. It is important to stress the dynamic nature of social capital, i.e. it can be eroded and it can be created. In the Eastern Mancha, it is clear that third party trust was created between the water user associations and higher-level authorities and a key ingredient was the leadership role played by individuals at different scales.

(5) *Clear division of responsibilities*: Equally, it is increasingly important to provide, as far as possible, the subsidiarity principle, decisions should be taken at the appropriate level. However, echoing Young (2002), it does not necessarily mean just finding the right scale but rather clarity in the interplay of increasing functional dependencies. For example, in many occasions, more so in the context of common pool resources, it might be relevant to "combine the strength of government level and local level resource management, to mitigate the weaknesses sin of each" (Berkes, 2002: 301). For example, favoring from the outset a co-management approach as a partnership between local users and agencies focuses not only on outcomes but also on the process itself as social learning.

(6) *If relevant; from supervisory control, to co-management to self-regulation (from extrinsic reward to intrinsic motivation)*: Increasingly in the policy making arena where the shift has left government behind, to be overtaken by governance it is more and more pertinent for higher level authorities to be more explicit (and upfront?) on asymmetrical power relations and institutional bargaining. For example, acknowledge more directly issues related to access, use, management, exclusion and transferability of water resources, and the rules that might (or might not) be up for negotiation. A networked society calls for a new type of institution imbued with a particular culture.

(7) *Institutional culture*: The right institutional culture when faced with the increasing pattern of decentralization prevailing in much of the developing world in natural resource management (including water) would "seek to harness rather than override the local knowledge and creativity of a multiplicity of designers" (Lowndes & Wilson, 2003: 287). Institutional culture should as far as possible, be receptive, flexible and robust. For instance, the dominant institutional culture in higher level authorities should support positive patterns of behavior (such as community leadership) and eradicate negative traditions (such as departmentalism, paternalism, and social exclusion) (Lowndes & Wilson, 2003). Most recent research in common pool resource management pinpoints to a renaissance in local level governance and new pushes for decentralization. This chapter attempts to contextualize this renaissance, since local institutions do not operate in a vacuum, and

in particular to cross-scale linkages and the role of higher level authorities and external events.

## 5   CONCLUSION

The challenge facing the tragedy of the water commons is to design robust yet flexible institutional arrangements. This chapter has argued that increasingly the focus of attention is shifting to questions on the strength of cross-scale linkages. This is not surprising, since the 21st century the questions posed by water turn around *water governance* not government and ever more complex set of networks surrounding mutually beneficial collective water management.

In the context of institutional cross-scale linkages, the challenge lies in developing cross-scale linkages that are capable of delivering adaptive management – uncertainty is integrated into the decision-making process and institutions that are capable of learning from successes and failures.

For cross-scale linkages it might be safer to opt for a more post-modernist view of policy, or of historical institutionalism (as compared to rational institutionalism) where institutions are less abstracted and more contextualized; i.e. "where actions are not the product of calculated decisions, rather they are embedded in institutional structures of rules, norms, expectations and traditions that severely constrain the behavior of social actors" (Mule, 1999: 148). Strong institutions and strong cross-scale linkages that can cope with critical junctures and unanticipated consequences (Eaton, 2004). In this context, more research should be undertaken on questions centered on institutional change and stability, and by this one has to distinguish, as Lowndes & Wilson (2003: 280) state: "between organizational change and institutional change, while the former may involve no more than structural re-organization, the latter requires effective rules of behavior are altered through specifying and embedding new norms, incentives and sanctions". In the coming decades one will need to assume change and therefore stability will have to be explained (Berkes, 2002).

However, questions still remain on external and endogenous changes from critical junctures, and on the speed of change. For example, local institutions as Berkes (2002) comments are more likely to adapt over a period of decades, rather than months, yet higher-level authorities increasingly operate on election cycles.

In a globalized world, and almost as a counter-reaction, attention is turning towards local level institutions rooted in local circumstance and heavily contextualized. It is important to distinguish local governance from local government and that "top down and bottom up institutional influences interact in important ways. The extent of local distinctiveness in governance arrangements is related to the degree of autonomy and diversity that higher levels of government will tolerate" (Lowndes & Wilson, 2003: 280). Equally one has to distinguish between decentralization of political authority and decentralization of governing capacity. In this second case it is a question of institutional change. In more and more countries of the world, local level common property institutions are the new *institutional messiah*; local institutions can be robust, flexible, reflexive, varied; they can be quick to learn, innovate and adapt compared to centralized institutions (Berkes, 2002). However, this chapter has argued that the Achilles heel of these institutions – potentially crucial for integrated water management – are cross-scale linkages and the role of higher level authorities.

# REFERENCES

Agrawal, A. (2002). Common Resources and Institutional Sustainability. In: E. Ostrom, T. Dietz, N. Dolsak, P.C. Stern, S. Stonich & E.U. Weber (eds.), *The Drama of the Commons*. National Academy Press, Washington, D.C., USA: 41–86.

Anderson, L. (1986). *The Economics of Fisheries Management*. Rev. ed. Baltimore, Maryland, USA. Johns Hopkins University Press.

Berkes, F. (ed.) (1989). *Common Property Resources: Ecology and Community-based Sustainable Development*. Belhaven, London, UK.

Berkes, F. (2002). Cross-Scale Institutional Linkages: Perspectives from the Bottom Up. In: E. Ostrom, T. Dietz, N. Dolsak, P.C. Stern, S. Stonich & E.U. Weber (eds.), *The Drama of the Commons*. National Academy Press, Washington, D.C., USA: 293–322.

Blomquist, W. (1992). *Dividing the Waters: Governing Groundwater in Southern California*. ICS Press. San Francisco, California, USA.

Blomquist, W.; Schlager, E. & Heikkila, T. (2004). *Common Waters, Diverging Streams*. Resources for the Future. Washington, D.C., USA.

Eaton, K. (2004). Designing sub-national institutions: Regional and municipal reforms in post authoritarian Chile. *Comparative political Studies*, 37(2): 218–244.

Gordon, H.S. (1954). The Economic Theory of a Common Property Resource: the Fishery. *Journal of Political Economy*, 62: 124–142.

Hardin, G. (1968). The Tragedy of the Commons. *Science*, 162: 1243–1248.

Heikkila, T. (2001). *Managing Common Pool Resources in a Public Service Industry: The Case of Conjunctive Water Management*. Ph.D. Dissertation, University of Arizona, USA.

Lam, W.F. (1998). *Governing Irrigation Systems in Nepal: Institutions, Infrastructure and Collective Action*. ICS Press. San Francisco, California, USA.

Lam, W.F.; Lee, M. & Ostrom, E. (1994). An Institutional Analysis Approach: Findings from the NIIS on Irrigation Performance. In: J. Sowerwine, G. Shivakoti, U. Pradhan, A. Shukla & E. Ostrom (eds.), *From Farmers' Fields to Data Fields and Back, a Synthesis of Participatory Information Systems for Irrigation and Other Resources*. International Irrigation Management Institute, Colombo, Sri Lanka; and Institute of Agriculture and Animal Science, Rampur, Nepal: 69–93.

Landre, P. & Travis, L. (1998). *Collaborative Watershed Management in the Finger Lakes Region, New York*. [http://www.indiana.edu/~iascp/Final/landre.pdf].

López-Gunn, E. (2003a). The role of collective action in water governance: a comparative study of groundwater user associations in La Mancha aquifers (Spain). *Water International*, 28(3): 367–378.

López-Gunn, E. (2003b). *Policy change and learning in groundwater policy: a comparative analysis of collective action in La Mancha (Spain)*. Ph.D. Thesis. King's College, University of London, UK.

Lowndes, L. & Wilson, D. (2003). Balancing revisability and robustness? A new institutionalist perspective on local government modernization. *Public Administration*, 81(2): 275–298.

Maarveland, M. & Dangbegnon, C. (1999). Managing natural resources: a social learning perspective. *Agriculture and Human Values*, 16: 267–280.

McCay, B.J. & Acheson, J.M. (1987). Human Ecology of the Commons. In: B.J. McCay & J.M. Acheson (eds.), *The Question of the Commons: the Culture and Ecology of Communal Resources*. University of Arizona Press, Tucson, USA: 1–34.

Mule, R. (1999). New Institutionalism: distilling some *hard core* propositions in the works of Williamson and March and Olsen. *Politics*, 19(3): 145–151.

Olson, M. (1965). *The Logic of Collective Action: Public Goods and the Theory of Groups*. Harvard University Press, Cambridge, Massachussets, USA.

Onyeiwu, S. & Jones, R. (2003). An institutional perception of cooperative behavior. *Journal of Socio-economics*, 32: 233–248.

Ostrom, E. (1990). *Governing the Commons: the Evolution of Institutions for Collective Action*. Cambridge University Press, New York, USA.

Ostrom, E. (1992). *Crafting institutions for self-governing irrigation systems*. ICS Press, San Francisco, California, USA.

Ostrom, E. (2000a). Collective action and the evolution of social norms. *Journal of Economic Perspectives*, 14(3): 137–158.

Ostrom, E. (2000b). The danger of self-evident truths. *PS: Political Science and Politics*, 33(1): 33–44.

Pierson, P. (2000). Increasing Returns, Path Dependence, and the Study of Politics. *American Political Science Review*, 94(2): 251–267.

Rhodes, R.A.W. (1997). *Foreword. Managing complex networks*. W. Kickert. Sage, London, UK.

Rogers, P. & Hall, P. (2003). *Effective Water Governance*. Global Water Partnership, Technical Committee.

Scharpf, F. (1997). *Games Real Actors Play*. Westview Press, Boulder, Colorado, USA.

Scott, A. (1955). The Fishery: the objectives of sole ownership. *Journal of Political Economy*, 63: 116–124.

Stern, P.C.; Dietz, T.; Dolsak, N.; Ostrom, E. & Stonich, S. (2002). Knowledge and questions after 15 years of research. In: E. Ostrom, T. Dietz, N. Dolsak, P.C. Stern, S. Stonich & E.U. Weber (eds.), *The Drama of the Commons*. National Academy Press, Washington, D.C., USA: 443–490.

Tang, Y.S. (1989). *Institutions and Collective Action in Irrigation Systems*. Ph.D. Dissertation. Indiana University, USA.

Tang, Y.S. (1992). *Institutions and Collective Action: Self-Governance in Irrigation*. ICS Press, San Francisco, California, USA.

Tang, Y.S. (1994). Institutions and Performance in Irrigation Systems. In: E. Ostrom, R. Gardner & J. Walker (eds.), *Rules, Games and Common-pool Resources*. Ann Arbor, University of Michigan Press, USA: 225–246.

Young, O.R. (2002). Institutional Interplay: the Environmental Consequences of Cross-Scale Interactions. In: E. Ostrom, T. Dietz, N. Dolsak, P.C. Stern, S. Stonich & E.U. Weber (eds.), *The Drama of the Commons*. National Academy Press, Washington, D.C., USA: 263–292.

II

The Economic Value of Water

CHAPTER 4

# The economic conception of water

W. M. Hanemann
*University of California, Berkeley, USA*

ABSTRACT: This chapter explains the economic conception of water – how economists think about water. It consists of two main sections. First, it reviews the economic concept of value, explains how it is measured, and discusses how this has been applied to water in various ways. Then it considers the debate regarding whether or not water can, or should, be treated as an economic commodity, and discusses the ways in which water is the same as, or different than, other commodities from an economic point of view. While there are some distinctive emotive and symbolic features of water, there are also some distinctive economic features that make the demand and supply of water different and more complex than that of most other goods.

Keywords: *Economics, value of water, water demand, water supply, water cost, pricing, allocation*

## 1  INTRODUCTION

There is a widespread perception among water professionals today of a crisis in water resources management. Water resources are poorly managed in many parts of the world, and many people – especially the poor, especially those living in rural areas and in developing countries – lack access to adequate water supply and sanitation. Moreover, this is not a new problem – it has been recognized for a long time, yet the efforts to solve it over the past three or four decades have been disappointing, accomplishing far less than had been expected. In addition, in some circles there is a feeling that economics may be part of the problem. There is a sense that economic concepts are inadequate to the task at hand, a feeling that water has value in ways that economics fails to account for, and a concern that this could impede the formulation of effective approaches for solving the water crisis.

My own personal assessment is that the situation is somewhat more complex than critics suggest. On the one hand, as environmental and resource economics has evolved over the past forty years, it has developed a conceptual toolkit that I think is well suited for dealing with many of the issues of water supply and water resource management. On the other hand, economists sometimes slip into older ways of thinking and characterize economic value in terms that are inadequate or misleading. Moreover, even among economists there is an inadequate appreciation of the complexities of water as an economic commodity; these render it distinctive from other commodities and they contribute to the explanation of the current crisis in water.

This chapter examines the economic conception of water – how economists think about water – at least partly in light of these concerns[1]. It consists of two main sections. Section 2

---

[1] I am well aware that there are other conceptions of water coming from other disciplines. I see those as complements, rather than substitutes, for the economic conception of water. For a fascination account of alternative conceptions of water in the 19th century, see Hamlin (2000).

reviews the economic concept of value, explains how it is measured, and discusses how this has been applied to water in various ways. Section 3 takes on the debate regarding whether or not water can or should be treated as an economic commodity, and discusses the ways in which water is the same or different as other commodities. The chapter ends with a few concluding observations in section 4.

## 2    WHAT IS ECONOMIC VALUE? HOW IS IT MEASURED?

Is economic value measured by market price? If an item has a price of US$ $X$, is this also the amount of its economic value? Most people assume the answer is yes, and economists sometimes also make this statement. For example, the following passage equates the economic value of water with its market price:

> "In a market system, economic values of water, defined by its price, serve as a guide to allocate water among alternative uses, potentially directing water and its complementary resources into uses in which they yield the greatest total economic return" (Ward & Michelsen, 2002).

If it were true that economic value is measured by market price, this would imply that only marketed commodities can have an economic value. Items that are not sold in a market – including the natural environment, and public goods generally – would have no economic value. If this were so, economic value would indeed be a narrow concept and at variance with many people's intuitive sense of what is valuable.

In fact, however, economic value is different than price. Price does not in general measure economic value, and items with no market price can still have a positive economic value.

This was first pointed out by Dupuit (1844) and Marshall (1879). But, as explained below, it took until the 1970s for this to become well accepted within modern economics. It was around this time that operational procedures became available to measure economic value separately from price; and it was around this time that non-market valuation emerged as a field in economics. It so happens that water as a commodity played a role in these developments, both clarifying the economic concept of value and developing operational procedures for measuring it.

### 2.1    *The meaning of economic value*

The distinction between market price and economic value was famously noted by Adam Smith in a passage in the *Wealth of Nations* describing the paradox of water and diamonds:

> "The word *value*, it is to be observed, has two different meanings, and sometimes expresses the utility of some particular object, and sometimes the power of purchasing other goods which the possession of that object conveys. The one may be called *value in use*; the other, *value in exchange*. The things which have the greatest value in use have frequently little or no value in exchange; and, on the contrary, those which have the greatest value in exchange have frequently little or no value in use. Nothing is more useful than water; but it will purchase scarce anything; scarce anything can be had in exchange for it. A diamond, on the contrary, has scarce any value in use; but a very great quantity of other goods may frequently be had in exchange for it" (book I, chapter IV).

Smith was using the comparison between water and diamonds to illustrate a distinction between two different meanings of *value*. In fact, neither the distinction between the definitions of value nor the use of water to illustrate it was original with Smith[2]. Two thousand years before Smith, Plato had observed that: "only what is rare is valuable, and water, which is the best of all things . . . is also the cheapest"[3]. In fact, Plato and Smith were both expressing a thought that had occurred to many other people over the ages, namely that the market price of an item need not reflect its true value. Market price reflects the fluctuating circumstance of daily life, whether the vagaries of supply (sudden scarcity, monopoly, etc.) or demand (temporary needs, changes in taste, fads and fashions), while the true value is something more basic, enduring, and stable.

Just what this true value is has been seen differently at different times. For Plato, the true value was intrinsic to the ideal form underlying the item. For Aristotle, it was intrinsic to the natural end that the item served. Aristotle also originated the distinction between this value – in effect, value in use – and value in exchange: "of everything which we possess there are two uses . . . one is the proper, and the other the improper or secondary use of it. For example, a shoe is used for wear, and is used for exchange; both are uses of the shoe"[4]. For Saint Thomas Aquinas, the true value of an item was determined by its *inner goodness*, an intrinsic quality of the item stemming from its relation to the divine purpose. In the 14th century, some Scholastics propounded a view closer to Aristotle's that the intrinsic value of an item arises from its inherent usefulness and ability to please man according to rules of reason. However, starting with Davanzati in 1588, Italian humanists stressed subjective human preference rather than objective human need as the basis of true value. Men seek happiness, Davanzati wrote, by satisfying all their wants and desires, and they value items as these contribute to this end. While value reflects human preference – not only wants of the body but also what one later writer called "wants of the mind, most of them proceeding from imagination" – price reflects not only demand but also supply, and that is influenced by scarcity. As Barbon wrote in 1690, "things may have great virtues, but be of small value or no price if they are plentiful".

These quotations from the 16th and 17th centuries demonstrate an awareness of three key principles. First, demand is separate from supply. Demand indicates what things are worth to people; supply indicates what things cost. Second, market price reflects the interaction of both demand and supply and, in principle, is separate from each of them. Third, the value that people place on an item (their demand for the item) inevitably reflects their subjective preferences[5].

Returning to Adam Smith, given his distinction between value in use and price (value in exchange), which is fundamentally the more useful measure of value? Here I part company from Smith because, following Hume and Locke before him, he associated the true value of an item largely with its cost of production. This English School held that, while the market price of an item at any particular point in time is determined by demand and supply, in the long run this will tend towards what Hume called a *fundamental price*, and Smith a *natural price*,

---

[2] Moreover, as I shall argue, Smith's analysis of this example is incomplete, and the conclusion he drew from it is largely incorrect.

[3] Plato *Euthydemus*, as cited in Bowley (1973). The discussion that follows draws on Bowley, Schumpeter (1954), Gordon (1975), Pribram (1983) and Niehans (1990).

[4] Aristotle, *Politics*, Book I, 9.

[5] This applies to water, too, even though it is obviously an essential want of the body. Without wishing to demean the importance of water, I will present some evidence below that, compared to other items they could buy, people sometimes place a lower value on improving their access to water than what the public health professionals would recommend.

which is determined by the underlying cost of production. Implicitly, they were assuming a horizontal long-run supply curve, so that consumer demand has no influence on price in the long run. This is now seen as a special case, and the modern economic concept of value focuses essentially on value in use.

The modern concept was first formulated by Dupuit (1844) and Marshall (1879, 1890). Dupuit stated that the "maximum sacrifice expressed in money which each consumer would be willing to make in order to acquire an object" provides "the measure of the object's utility". Marshall used a very similar formulation; he defined the "economic measure" of a satisfaction as "that which a person would be just willing to pay for any satisfaction rather than go without it". These definitions highlight the distinction between demand and supply: the measure of value is what the item is worth to the individual, not what it costs. Thus, an item can be cheap to produce, in the sense that its total cost is low, but highly valuable to the owner, in that its total value to him is large, or conversely.

Generalizing from this, the modern economic concept of value is defined in terms of a trade-off. When an economist states that, for some individual, $X$ has a value of 50 in terms of $Y$, this means no more, and no less, than that the individual would be willing to exchange $X$ for 50 units of $Y$. $Y$ is said to be the numeraire in terms of which value is measured. This numeraire can be money but it need not be; it could, for example, be some specific commodity. The trade-off is in no way limited to market goods; it can be between any two items that the individual values, regardless of whether these have a market price.

Before any further discussion of the relation between value and price, it is necessary to introduce a distinction which was lacking in Smith's analysis but was understood by Dupuit and Marshall, namely the distinction between *marginal*, on the one hand, and *average* or *total*, on the other. Thus, marginal value needs to be distinguished from average or total value, and marginal cost from average or total cost. The marginal quantity measures the change in total value, or total cost, associated with a unit change in quantity, while the average measures total value, or total cost, averaged over the total quantity. Admittedly, there is one case where they are the same: if the marginal value (or marginal cost) is constant as quantity changes, then marginal cost (or value) coincides with average cost (or value). But, in general, marginal value and marginal cost are *not* likely to be constant. In particular, the general presumption is that marginal value (and marginal utility or marginal benefit) decline with quantity.

The notion of declining marginal utility was the cornerstone of Dupuit's analysis. Dupuit recognized that if the consumer is free to vary the quantity of an item purchased, she will choose this quantity so as to equate her marginal value (utility) for the item to its price. In that case, the market price provides an accurate measure of the *marginal* value associated with the last unit of consumption. But, Dupuit stressed, the *total* payment does not accurately reflect the *total* value of all units consumed. This is because of diminishing marginal utility: if the marginal value of the last unit just equals the market price, it follows that the marginal utility associated with the *infra-marginal* units will be *higher* than this market price. In effect, the consumer earns a *profit* on the infra-marginal units because they are worth more to her than the price she pays, which in fact is why she consumes a larger quantity thereby rendering these units infra-marginal. Marshall, who independently formulated a similar argument thirty years later, called this profit the *consumers' rent* in 1879, and the *consumer's surplus* in 1890.

In summary, if there is a market price for the item in question and if the consumer is free to vary the quantity of this item that she purchases, its marginal value to her is reflected in, and can be measured by, the market price; otherwise, not. Even when price reflects marginal value, *total* expenditure does *not* reflect *total* value; instead, total expenditure *understates* total value because of the presumption of diminishing marginal utility. The distinction between marginal and total is the key to the full resolution of the diamond and water paradox: water may have a smaller value than diamonds at the margin, but it undoubtedly has a larger total value.

Although Dupuit and Marshall correctly enunciated the economic concept of value in its modern formulation, it actually dropped out of favor with economists around the turn of the last century. Marshall himself came to be troubled that his use of the demand curve to measure consumer's surplus was inexact and relied on the assumption of a constant marginal utility of income. And, as the ordinal utility revolution took hold in economics, Marshall's analysis based on cardinal utility appeared hopelessly out-dated and irrelevant. It took until the 1970s before these issues were fully resolved and the Dupuit-Marshall concept was recognized as being both fully consistent with modern, ordinal utility theory and susceptible of rigorous empirical measurement. This came about as a result of several important conceptual advances.

First, Hicks rehabilitated the Marshallian concept of consumer's surplus in a series of papers starting with Hicks (1939), which demonstrated that this concept is in fact consistent with ordinal utility theory and that it could be measured exactly if one were given an indifference map. However, this was a pyrrhic victory because the general view was that, while it is a useful theoretical construct, the indifference map is not itself directly observable. Hence, Marshall's measure as re-interpreted by Hicks was not measurable in practice. This view finally changed around 1970 as the result of the development of what is known as duality theory, including the demonstration by Hurwicz & Uzawa (1971) of a theoretically rigorous yet practical numerical procedure for identifying the specific utility function underlying any given system of demand equations that satisfies the formal requirements of modern ordinal utility theory. This now made it possible to start with an econometric estimate of a suitably specified demand equation for a marketed commodity, or a system of demand equations for a set of commodities, and derive a theoretically consistent and rigorous estimate of the Dupuit-Marshall measure of the economic value of these commodities.

Second, building on Hicks (1939), Henderson (1941) discovered an alternative way of characterizing the trade-off that underlies the economic concept of value. When one says that a person is willing to exchange $X$ for 50 units of $Y$, this could mean either: 1) the person would be willing to give up (pay) 50 units of $Y$ to obtain $X$; or 2) the person would accept 50 units of $Y$ to forego $X$. The first uses *maximum willingness to pay* (WTP) as the measure of value, and is the measure mentioned by Dupuit and Marshall and analyzed by Hicks (1939, 1941). The second is the new measure that was suggested by Henderson; it uses *minimum willingness to accept* (WTA) as the measure of value. Together, these exhaust the logically possible ways of expressing a trade-off. Hicks (1942, 1943, 1946) analyzed the relationship between them in the case of a price change and showed that they differ by an income effect[6].

The third development was the extension of the economic concept of value to a broader class of items than market commodities. In fact, nothing limits $X$ in the definition of economic value given above to being a market good; it could actually be *anything* from which people derive satisfaction. This suggests that the same definition of economic value can be applied to

---

[6] An important paper by Willig (1976) showed how one could use Hurwicz and Uzawa's result to develop a tight numerical bound on the possible difference between WTP and WTA in the case of a price change for a marketed commodity. The Hurwicz-Uzawa result implies that both WTP and WTA can be derived from an econometric estimate of suitably specified demand equations.

*non-market* items. For example, one could say that a person values some aspect of his health at 50 units of $Y$ if he would be willing to exchange 50 units of $Y$ to preserve that aspect of his health; that he values a beautiful sunset at 50 units of $Y$ if he would be willing to exchange 50 units of $Y$ to experience it; or that he values an endangered species of animal at 50 units of $Y$ if he would be willing to exchange 50 units of $Y$ to ensure its preservation. In each case, it should be evident that there are two possible ways to formulate the exchange: a WTP formulation and a WTA formulation. This was demonstrated formally by Maler (1971, 1974) who showed that, when $Y$ is money, the Hicksian analysis and its modern formulation in terms of duality theory carry over from the valuation of market goods to non-market items[7]. Maler's analysis thus provides a formal justification for the field of non-market valuation, including the monetary evaluation of the natural environment.

## 2.2   *Non-market valuation and water*

Economic valuation deals with the valuation in monetary terms of items that people might care for. Non-market valuation applies the same notion to items that are not sold in a market. It is important to emphasize that the Dupuit-Marshall concept of economic value carries over to such items. This is because, even for something that is not sold in a market, it is still meaningful to conceptualize the economic measure of the satisfaction from the item as the monetary amount which the person would be just willing to exchange for the item if it were possible to make such an exchange. In effect, this generates a monetary measure of the change in the person's welfare by using the change in the person's monetary income that she would consider equivalent to the item in question in terms the overall impact on her satisfaction[8,9,10].

The history of non-market valuation in the USA is closely intertwined with water projects, since these were an important motivation for the development of cost-benefit analysis. The idea of cost-benefit analysis originated in the USA, in Hammond's (1960) phrase, as "an administrative device owing nothing to economic theory" in the context of managing the activities of the US Army Corps of Engineers around the beginning of the last century. The 1902 *River and Harbor Act* had created a Board of Engineers to review navigation projects;

---

[7] However, there is an important difference. With valuation of non-market items, the difference between the WTP and WTA measures involves not only an income effect but also a substitution effect (Hanemann, 1991).

[8] The equivalence can be conceptualized in two possible ways – the maximum amount that the person would be willing to pay to gain the item, or her minimum WTA to forego it.

[9] It should be noted that this definition provides a unified approach to welfare measurement for both firms and households. In the case of households, whose objective function is defined in terms of utility, the monetary measure is the change in income that is considered equivalent to the change in utility. In the case of firms, whose objective function is defined in terms of profit, the monetary measure is the change in profits itself.

[10] Although the modern economic concept of value is defined in terms of a trade-off, it is possible that some people find themselves unable to make a trade-off because for them the two items being compared are incommensurable. A type of preference that gives rise to such trade-off aversion is where there is a lexical ordering over commodities: certain goods in *any* quantity or quality *always* take precedence over *all* quantities or qualities of other goods, so that no amount of increase in the latter can ever compensate for any reduction in the former. This is known as lexicographic preferences. A modified version of lexicographic preferences is where the lexical ordering applies only below a threshold level of the good (Lockwood, 1996); among other things, this generates a situation where the individual might have an infinite WTA for a reduction in an item.

in conducting a review, the Board was required to consider the commercial benefits from such projects in relation to their costs. The *River and Harbor Act* of 1920 further required the separate reporting of special, or local, benefits as opposed to general, or national, benefits for the purpose of ensuring proper local cost-sharing. In 1934, the National Resources Board appointed a Water Resources Committee to consider "the development of an equitable system of distributing the cost of water resource projects, which should include not only private but also social accounting". Finally, the *Flood Control Act* of 1936 permitted the Army Corps of Engineers to involve itself in flood control provided that, in a famous phrase, "the benefits to whosoever they may accrue are in excess of the estimated costs". In 1946, a Subcommittee on Benefits and Costs of the federal Inter-Agency River Basin Committee was appointed to investigate the practices of the various federal agencies that were engaged in the evaluation of federal water resource projects and to formulate some "mutually acceptable principles and procedures". This led ultimately to the publication in 1950 of what became known as the *Green Book* which attempted to codify the principles of cost-benefit analysis for use by federal agencies. The following decade saw the publication of many academic journal articles and six major academic books dealing with the economic analysis of water projects.

The 1950s were when the field of non-market valuation began to come into existence. The approach that emerged first is what became known as the travel cost method or, more generally, the revealed preference method. It arose initially out of an effort by the National Park Service (NPS) to measure the economic value associated with the national parks. At the time there were no entrance fees at national parks, so the NPS could not use park revenues as a measure of their value. The issue was assigned to a staff economist who wrote to ten distinguished economic experts for their advice. All but one replied that it was impossible to measure recreational values in monetary terms, but the tenth, Harold Hotelling disagreed. He saw that, even though there was no entrance fee for a national park, it still cost visitors something to use the parks because of expenses for travel, lodging and equipment. These expenditures were not captured by the NPS but, they still set a price on the park. Moreover, this price would vary among people coming from different points of origin. By measuring the price and graphing it against visitation rates one could construct a demand schedule for visits to the site, and then determine consumer's surplus in the usual manner as the area under this demand curve.

The NPS report followed the majority view and asserted that it was not possible to set a monetary value on outdoor recreation. However, in 1956 the State of California hired an economic consulting company to estimate recreational benefits associated with the planned *State Water Project*. This company learned of Hotelling's idea and decided to apply it. A survey of visitors was conducted at several lakes in the Sierras and data were collected on how far they had traveled and how much they had spent. Using these data, a rough demand curve was traced out, and an estimate of consumer's surplus was constructed. This analysis appeared in Trice & Wood (1958), the first published application of the travel cost method. At the same time, Marion Clawson (1959) at Resources for the Future had begun collecting data on visits to Yosemite and other major national parks in order to apply Hotelling's method to them, which was the second published application. By 1964, there were at least five more applications in various parts of the USA, and the travel cost method was an established procedure.

The insight behind the travel cost method, and revealed preference generally, is that, while people cannot buy non-market goods such as clean water or an unspoiled environment directly, there sometimes exist market goods that serve as a partial surrogate for the non-market good because the enjoyment of these goods is enhanced by, or depends on, the non-market good. In that case, the demand for the market goods is used as a surrogate for the demand for the non-market good.

The limitation of this approach is that there may not exist a market good that can serve as surrogate for the non-market good of interest. Moreover, even if such a good exists, it may not capture *all* of people's preferences for the complementary non-market good. The conceptual identification of what might be omitted by the revealed preference approach came about as a result of papers by Weisbrod (1964) and Krutilla (1967). Both authors started from the premise that some of people's motives for valuing the natural environment may differ from those for valuing a market good. People may value the natural environment out of considerations *unrelated* to their own immediate and direct use of it. Weisbrod focused on uncertainty and what became known as *option value*: some people who do not now visit a national park, say, may still be willing to pay money to protect it from destruction or irreversible damage because they want to preserve their option of visiting it in the future. Krutilla focused on what became known as *bequest value* and *existence value*[11]. With bequest value, the notion is that some people would be willing to pay because they want to preserve the park for future generations. With non-use value, the notion is that some people would be willing to pay even if they knew that neither they nor their children would ever visit it; in Krutilla's example, people may "obtain satisfaction from mere knowledge that part of the wilderness in North America remains". These are legitimate sources of value, Krutilla and Weisbrod felt, but they would not be respected by private managers of the environmental resource. Nor would they be adequately measured by a conventional revealed preference analysis such as the travel cost method. Consequently, some other method of measurement is needed.

The alternative approach, suggested by Ciriacy-Wantrup (1947), is to interview people and elicit their monetary value; this became known in economics as the contingent valuation (CV) method[12]. Ciriacy-Wantrup was discussing soil conservation and he noted that several of the benefits were non-market goods, such as reduced siltation of rivers or reduced impairment of scenic resources. He characterized the problem as being how to obtain a demand curve for such goods, and suggested the following solution: "[Individuals] may be asked how much money they are willing to pay for successive additional quantities of a collective extra-market good. The choices offered relate to quantities consumed by all members of a social group . . . If every individual of the whole social group is interrogated, all individual values (not quantities) are aggregated". The results correspond to a market-demand schedule. While noting the possible objection that "expectations of the incidence of costs in the form of taxes will bias the responses to interrogation", he felt that "through proper education and proper design of questionnaires or interviews it would seem possible to keep this potential bias small".

Having identified a solution conceptually, Ciriacy-Wantrup never pursued it further. The first significant application was by Davis (1963) which dealt with the economic value of outdoor recreation in the Maine woods; to measure this Davis interviewed a sample of hunters

---

[11] The latter is now also called *non-use value* and *passive use value*.

[12] The same idea was earlier suggested by Bowen (1943) who conceived of surveys as a surrogate for using voting to determine the public's demand for what he called *social goods*. Recently, the term *stated preference* has been used to cover CV and related approaches. They are also known as direct valuation whereas revealed preference approaches are referred to indirect valuation because they do not measure preferences directly but instead infer them from externally observed behavior.

and recreationists and asked how much more they would be willing to pay to visit the area[13]. The next application was by Ridker (1967); to measure the damages from air pollution, Ridker included some questions in a survey about people's WTP to avoid soiling from air pollution. In 1969, a steady stream of CV studies began to appear in the economics literature. Official recognition was given to CV in 1979, when the US Water Resources Council included it along with travel cost as recommended methods of non-market valuation.

The first application of non-market valuation in the USA, the 1957 valuation of recreation benefits from the *California State Water Project*, was a harbinger of things to come: since then many non-market valuation studies have been conducted in the USA in connection with water resources management issues, and the environmental consequences of water projects have come to play a significant role in the design and approval of water projects. These trends emerged slowly in the 1970s and 1980s, driven by developments in the implementation of the 1969 *National Environmental Policy Act* (NEPA)[14]. NEPA required federal agencies to prepare a *detailed statement* of environmental impacts for proposed major actions which significantly affect the quality of the human environment, including the identification of the environmental impacts of the proposed action, alternatives to the proposed action, and any adverse environmental impacts which cannot be avoided should the proposal be implemented.

In consequence, since the mid-1980s it has not been acceptable in the USA to perform an economic assessment of a major water project without including some non-market valuation of the project's environmental impacts. For example, non-market valuations of environmental impacts were included in the Department of Interior's re-assessment of the operation of Glen Canyon Dam on the Colorado River in 1984–1992, and in the Bureau of Reclamation's assessment of the *Central Valley Project Improvement Act* in 1993–1996. In California, they were included in the State Water Resources Control Board's review of the diversions of water from the San Francisco Bay/Delta to the Central Valley and Southern California, conducted in 1987–1994, and in the Board's 1993 Mono Lake Decision requiring Los Angeles to reduce its diversion of water from streams feeding Mono Lake on the eastern side of the Sierra Nevada[15]. In the case of Mono Lake, the Board decided that it was in the public interest to reduce Los Angeles' diversion from Mono Lake by about two thirds, despite the resulting loss of hydropower and water supply (which amounted to over 8% of Los Angeles' total water supply) primarily in order to protect habitat for birds and other wildlife; non-use values associated with habitat protection constituted the main component of environmental benefits (Wegge *et al.*, 1996).

It should be emphasized that the use of non-market valuation applies to positive as well as negative environmental impacts of water projects. The experience in the USA has been that these can generate significant economic benefits associated with water-based recreation, eco-tourism, and the non-use value of ecosystem protection. These environmental benefits sometimes greatly outweigh the benefits from agricultural or even urban water use. In short, in the USA we have now moved from the traditional situation where there was essentially a single objective for large water projects, namely the provision of water for off-stream uses,

---

[13] Probably the first CV study was actually conducted in 1958 for the NPS, which hired a market research company to survey residents of the Delaware River basin about their WTP entrance fees for national parks.
[14] Subsequently, a number of states passed laws imposing similar reporting requirements on agencies of the state government; for example, the *California Environmental Quality Act* was enacted in 1970.
[15] I served as the Board's economic staff for its investigation of both these issues.

to a situation where any new water project must have environmental restoration as an explicit objective along with the provision of any off-stream uses.


## 3   IS WATER DIFFERENT?

Now that the economic concept of value has been explained, the question arises whether it is appropriate to apply this concept to water. Is water an economic commodity, and can it be analyzed using the conceptual framework of economics in the same way as any other commodity?

The answer is contested ground between economists and their critics. One of the four Dublin Principles, adopted at the 1992 *International Conference on Water and the Environment* in Dublin, holds that "water has an economic value in all its competing uses and should be recognized as an economic good". Similarly, Baumann & Boland (1998) write: "water is no different from any other economic good. It is no more a necessity than food, clothing, or housing, all of which obey the normal laws of economics". Per contra, Barlow & Clarke (2002) proclaim it as a "universal and indivisible" truth that "the Earth's freshwater belongs to the Earth and all species, and therefore must not be treated as a private commodity to be bought, sold, and traded for profit ... the global freshwater supply is a shared legacy, a public trust, and a fundamental human right, and therefore, a collective responsibility". Vandana Shiva (2002) writes in a similar vein about a clash between two cultures: "a culture that sees water as sacred and treats its provision as a duty for the preservation of life and another that sees water as a commodity, and its ownership and trade as fundamental corporate rights. The culture of commodification is at war with diverse cultures of sharing, or receiving, and giving water as a free gift".

My own view lies somewhere between these two positions. Baumann & Boland are undoubtedly correct when they point out that food, clothing and shelter, like water, are necessities of life, and they are typically provided through the market without any complaint. Why, they ask, should water be different? I believe there are two reasons why this is so. First, water is clearly viewed by many people as being different. The fact that water, unlike other household commodities, arouses such passion speaks for itself: for better or worse, water is perceived as having a special significance that most other commodities do not possess[16]. This itself has economic consequences. Second, I believe that water has some other *economic* features that make it distinctive. These features make water different from, say, bread or land, *as an economic commodity* yet they are often overlooked by economists. They matter greatly because they affect the demand for water, its value, and the social and institutional arrangements by which it is supplied. To explain them, I need to introduce several more items from the economists' conceptual toolkit.


### 3.1   *Water as a private good, water as a public good*

Since Samuelson (1954), economists have drawn a distinction between conventional market goods – also known as *private goods* – and what are known as *public goods*, "which all enjoy

---

[16] This is true in rich as well as poor countries – in the USA, for example, it is notoriously difficult for publicly owned urban water utilities to obtain political approval for even trivial rate increases while other household utilities such as cable television raise their rates with impunity; Glennon (2004) makes a similar observation.

in common". The two key properties of a public good are *non-rivalry in consumption* and *non-excludability*. With conventional goods, one person's consumption necessarily competes with that of another, in that more consumption by one person renders a smaller quantity of that good available for consumption by anybody else. With public goods, by contrast, more consumption by one person in no way reduces the amount available for others. Conventional consumption goods are excludable in that, if this is so desired, it is physically possible to exclude any person from consuming the commodity. With public goods, by contrast, if the good is available for consumption by anybody, it is available for consumption by all. Examples of a public good suggested by Samuelson were "an outdoor circus or national defense which is provided for each person to enjoy or not, according to his tastes". The abatement of pollution in a lake is another example of a public good, as are other types of environmental improvement: my enjoyment of the clean water in the lake in no way reduces the amount of clean water available for your enjoyment (non-rivalry) and, if the water in the lake is clean for me to enjoy, it is clean for everyone's enjoyment (non-excludability)[17].

In this framework, water is both a private good and a public good. When water is being used in the home, in a factory or on a farm, it is a private good. When water is left in situ, whether for navigation, for people to enjoy for the view or for recreation, or as aquatic habitat, it is functioning as a public good[18]. Moreover, while the water in a reservoir is a private good, the storage capacity of the reservoir *per se* may be a public good. By contrast, most of the other commodities associated with food, clothing or shelter are purely private goods and have no public goods aspect; this is one of the respects in which water is different than these other commodities in economic terms.

Samuelson identified two important consequences of the public good properties. First, while public goods are likely to be supplied collectively, for example through a voting process, rather than through a decentralized market, it is likely that they will be *undersupplied* because people have a selfish incentive to free ride on the collective decision process by understating their true interest in the public good. Second, the valuation of public goods is fundamentally different than that for private goods because a public good can be enjoyed simultaneously by many while a private good can be consumed by only one party at a time. Thus, the value placed on a given unit of a private good is that of a single user – in an efficient market, this will be the user with the highest and best use for the item. By contrast, the value placed on a public good is that of many people, namely all those who care for the item[19]. This is why the non-market

---

[17] In addition to private and public goods, there is an intermediate case where there is rivalry in consumption but not excludability. These are known as common pool resources. Examples include fisheries, forests, grazing grounds, and oil fields. The other intermediate case, sometimes called club goods or quasi-private goods, is where there is non-rivalry combined with the possibility of exclusion. Examples include television frequencies, public libraries, and bridges, for each of which it is possible to exclude access. Furthermore, there may be non-rivalry at low levels of aggregate consumption of a club good, but rivalry at a high level of consumption once the item becomes congested – this can happen, for example, with parks and bridges.

[18] To the extent that water-based outdoor recreation is excludable, this would be a quasi-private good. To the extent that groundwater or water flowing through the distribution system of an irrigation district is non-excludable, these are common pool resources.

[19] This follows from Samuelson's demonstration that the aggregate demand curve for a public good is constructed in a radically different manner than the aggregate demand curve for a private good. With a private good, the aggregate demand curve is the horizontal sum of every individual's demand curve for the good; with a public good, the aggregate demand is the vertical sum; this observation had in

benefits of environmental preservation can sometimes outweigh the use benefits associated with the diversion of water for off-stream agricultural or urban use.

The public good nature of water in situ, historically associated with navigation, has had a decisive influence on the legal status of water. In Roman Law and, subsequently, in English and American common law, and to an extent in Civil Law systems, flowing waters are treated as common to everyone (*res communis omnium*), and are not capable of being owned. These waters can only be the object of rights of use (usufructuary rights), but not of rights of ownership[20]. Thus, even though water and law are often complementary inputs, there is a crucial distinction in that land can be owned, while water cannot.

## 3.2    *The mobility of water*[21]

A distinctive physical feature of water is its mobility. Water tends to move around. It flows, it seeps, it evaporates. When water is applied to plants in the field (or to an urban landscape), a substantial portion either seeps into the ground or runs off the ground as tailwater. In addition, in residential indoor uses and most industrial uses there is usually an outflow of wastewater after the use is completed. The consequence is that there can be several sequential uses of the same molecule of water since water is rarely consumed fully by a given user and what is left is physically available, in principle, for use by others[22].

The mobility of water and the opportunity for sequential use and re-use make water relatively distinctive as a commodity – especially compared to land, for which such multiple, sequential uses are impossible (except in nomadic societies). These properties of water have important economic, legal and social implications. Keeping track of water flows is costly and sometimes difficult. Consequently, it is often hard or impractical to enforce excludability or to establish property rights to return flows. In this respect, water is very different as an asset than land, which is relatively easy to divide and fence. The common solution is to resort to some form of *collective* right of access; in effect, this internalizes the externality associated with the mobility of return flows. A classic example of this is the riparian water right in English and American common law. This permits any landholder whose property is adjacent to a stream or body of water to divert a reasonable amount of water, provided this does not cause harm to other riparian landholders or interfere with their co-equal right to divert a reasonable amount of water. The riparian right to the use of water is not a right to a fixed quantity, and it is a co-relative right shared with all other riparians along the same stream[23].

---

fact already been made by both Bowen (1943) and Ciriacy-Wantrup (1947). In terms of the distinction between use and non-use values, if there is a non-use value for an item this is a public good.

[20] By contrast, groundwater beneath private land and springs or rainwater found on private land is typically treated by the law as being privately owned by the landowner(s) on whose land the spring occurs or under whose land the groundwater lies.

[21] The analysis in this and the following two sections was influenced by reading Young & Haveman (1985).

[22] In the process, however, there can often be some reduction in the quality of the water relative to that in the first use. Because of the solvent properties of water, the return flows are apt to dissolve and absorb chemicals in the media through which they pass.

[23] In American law, there is a further requirement that riparians put the water to a reasonable use. With surface water, the major alternative in American law to the riparian right is the appropriative right, which was developed in the arid West. This permits the diversion of water, regardless of whether the diverter owns the riparian land, in a fixed quantity, subject to the principle of *first in time is first in right*. The

## 3.3  *The variability of water*

In addition to the mobility of water in streams, another crucial feature is the variability of supply in terms of space, time, and often quality. Spatially, water is distributed very unevenly across much of the globe; just six countries – Brazil, Russia, Canada, Indonesia, China and Colombia – account for half of the world's total renewable supply of freshwater (Postel & Vickers, 2004). Even within countries and regions, there is unequal spatial distribution. In California, for example, two thirds of the state's population live in Southern California, but this region receives less than 10% of the state's total precipitation. For any given region, there is substantial variation in precipitation both within the year and between years. In California, for example, approximately 80% of the annual precipitation falls between October and March, while three quarters of the water use occurs between April and September[24]. Beyond this, cycles of wet and dry years occur in California as a function of wider climatic phenomena such as the interannual El Niño-Southern Oscillation and the Pacific Decadal Oscillation. While the annual runoff in California has averaged about 87,500 Mm$^3$ over the past 90 years, it has been as low as 18,500 Mm$^3$ (in 1977) and as high as 166,500 Mm$^3$ (in 1983).

Because of this variability, the major challenge for most large water systems is the spatial and temporal matching of supply with demand. Storage is typically the key to controlling the temporal variability in supply, while inter-basin transfers are used to overcome the spatial mismatch between supply and demand. But, the variability of supply has affected not just the engineering of water resource systems but also the legal, and institutional arrangements for the use of water. The variability of supply is yet one more point of divergence between water and land, and it explains why the property rights regimes are different: it would surely be difficult to apply the ownership rights in land to so variable a resource as water.

Besides the variability in supply, the demand for water may be intermittent, especially in agricultural uses of water where crops need to be irrigated only at periodic intervals rather than every hour of every day. Until the advent of affordable storage, which has mainly been a phenomenon of the 20th century and large-scale diversion of water, the intermittent nature of traditional agricultural demand was an important factor promoting the sharing of access. If there is water in the stream and one member of the group is not currently diverting water

---

theory is that, if the streamflow is inadequate to meet all the diversion requirements, those with a more recent (*junior*) date of initial diversion cede to those with an older (*more senior*) date. On the ground in at least some Western states, the practice seems to be rather different. The precedence of seniority is not self-enforcing without resort to litigation, which is slow and costly. Much of the time, therefore, what actually happens with appropriative rights may be closer to a version of the riparian system. In California, there is not a functioning system to record the actual diversions of water, nor to check these against the quantity associated with the water right. Consequently, much of the surface water use by agricultural occurs outside the formal structure of California appropriative water rights law. This can become an impediment to long-run water transfers as the inadequate documentation casts a shadow of doubt on a seller's specific property right.

[24] The seasonal variability of precipitation in California is exacerbated by the fact that it is an arid region with a Mediterranean climate. But precipitation is distributed unevenly throughout the year in almost all parts of the world, albeit not as severely as in California. In Europe, the major part (46%) of the runoff occurs during April to July, and similarly in South America; in Asia, 54% occurs in June to September; in Africa, 44% occurs in September to December; in Australia and Oceania, 40% occurs in January and April (UNESCO, 2000).

from the stream, other members of the group were allowed to divert the water rather than let it flow to *waste* in the ocean[25]. This is another key difference with land: while the demand for water is intermittent, the demand for land to grow crops or to locate a building is continuous, and there can be no such sharing of the same resource among multiple users. The intermittent nature of the agricultural demand for water is conducive to the collective sharing of a right of access as opposed to individual ownership of a property right.

### 3.4   *The cost of water*

Compared to other commodities, and other utility services, the cost of water has several distinctive features which complicate its supply.

Water is bulky, and expensive to transport relative to its value per unit of weight. Consequently, the transportation infrastructure for water is far less extensive than that for more valuable liquids such as petroleum. Also, compared to electricity, water is relatively expensive to transport, but relatively cheap to store. Therefore, the strategy for averting shortage takes a different form with water than electricity. If there is a sudden shortfall in supply, with electricity this can be made up almost instantaneously by importing power over the grid from a source that could be 1500 km away or more. With water, there is no comparably interconnected transportation grid and, even if there were, it takes longer to move a comparably large quantity of water. Thus, to deal with unexpected outages, one has to either resort to rationing or stockpile sufficient stored water prior to the period of peak use.

Another distinctive economic feature is that water supply is exceptionally capital-intensive compared not only to manufacturing industry generally but also to other public utilities. In the USA, for example, the ratio of capital investment to revenues in the water industry is double that in natural gas, and 70% higher than in electricity or telecommunications. Moreover, the capital assets used in water supply cannot be moved to another location and are generally unusable for any other purpose; they represent an extreme type of fixed, non-malleable capital. Furthermore the physical capital in the water industry is very long-lived. The infrastructure associated with surface water storage and conveyance and the pipe network in the streets can have an economic life of 50–100 or more years, far longer than that of capital employed in most manufacturing industry or in other public utility sectors[26].

In addition, there are significant economies of scale in many components of water supply and sanitation. These are especially pronounced for surface water storage: given a specific

---

[25] Storage changes this, because streamflow can be stored when it is not currently being used. The discussion here focuses on agricultural rather than urban use of water. Urban use is different because it is more continuous in nature, and when there is a piped water supply this is typically pressurized, unlike with agriculture which relies mainly on gravity flow. Gravity distribution fosters sharing of intermittent access, while pressurized distribution fosters simultaneous individual access.

[26] The Roman aqueduct that still stands in Segovia, Spain, is ample testimony to the physical longevity of certain types of conveyance structure. The effective life of a dam is governed by the rate of siltation but can be well over a century. For piping, the American Waterworks Association recommends a replacement cycle of 67 years. A conventional drinking water treatment plant may have a useful life of 40 years, although high technology processes such as reverse osmosis facilities have a shorter life. Compared to surface water, the capital infrastructure associated with groundwater typically has a shorter life; a groundwater pump might typically have a life of about 25 years in the case of an electric pump, or about 15 years for a diesel pump.

dam site, within some range, by increasing the capacity of the dam one can significantly reduce the unit cost of stored water. With a groundwater source, by contrast, the economies of scale in production are much less pronounced. There are also important economies of scale in the treatment and conveyance of drinking water and wastewater[27].

The capital intensity, longevity, and economies of scale mean that water supply and sanitation costs are heavily dominated by fixed costs. In a simple surface-water supply system with minimal treatment of drinking water, minimal treatment of sewage prior to discharge, and a heavy reliance on gravity flow, the short-run marginal cost of water supply and sanitation may be almost zero except for small costs associated with pumping to move water through the system[28]. Even in a modern system with full treatment of drinking water and sewage discharges, the short-run marginal costs are extremely low. There is thus an unusually large difference between short- and long-run marginal cost in water supply[29].

The capital intensity and economies of scale associated with surface water supply have profound economic and social implications. For one thing, because these are classic preconditions for a natural monopoly, they make it more likely that there will be a single provider in any given area. More generally, they foster public provision of a surface water supply rather than individual, self-provision, whether the public provision is by a collective of the users or a monopoly seller[30]. Furthermore, the construction and operation of large-scale surface water storage and distribution systems require a high degree of co-ordination and social control. This was the central thesis of Wittfogel's (1957) study of ancient *hydraulic societies* – civilizations that were dependent on large-scale surface water diversion and distribution systems, such as Mesopotamia, Egypt and China. Wittfogel argued that, in such societies, the effective provision of water required the centralization of power and an *oriental despotism* mode of governance in which a state bureaucracy, headed by an absolute ruler, ruled on the basis of its control of the hydraulic system. Wittfogel's work was subsequently criticized by other

---

[27] With the conveyance of drinking water from the point of production to the point of use, and of treated wastewater to the point of discharge, there are economies of scale with respect to volumetric capacity but not length.

[28] In the USA, the ratio of operating costs to total costs for efficient water firms is about 10%; by contrast, it is 32% for gas utilities and over 57% for electric utilities (Spiller & Savedoff, 1999). In the UK water industry, Armstrong *et al.* (1994) report that operating costs represent less than 20%, and fixed costs more than 80%, of total costs.

[29] When piped water supply was introduced into cities in the 19th century, water agencies chose not to meter individual homes or small non-residential users partly because of an ethos in favor of promoting universal service but also because water was so cheap at the margin that they felt it was not worth the cost of metering it. By contrast, electricity and gas were metered in residential connections. Water service was financed by charges based on the type or value of the property being served. In the USA, metering of residential users did not become common until well into the 20th century, and there were some notable handouts (Denver and New York City did not meter until about 15 years ago). In the UK, nearly all residential water users were unmetered prior to privatization in 1990.

[30] For urban water, the main alternatives to a public supply of surface water are water vendors or household self-provision through pumped wells or rainwater catchments. Where there are water vendors, the unit cost of vended water is always much higher than that of water from piped supply, typically by a factor of 10 or more. The primary reason for the cost differential is economies of scale: it is far more expensive to deliver water in relatively small quantities through multiple trucks rather than in large quantities through a single pipe network. However, the up-front capital investment required for vended water is far lower than for a piped water supply, and this can make vended water a viable alternative.

scholars[31]. Nevertheless, his notion still resonates; it was applied to the American West by Worster (1986), who characterized this as a hydraulic society based on the development and control of water infrastructure by a political elite. This characterization is contested by Kupel (2003), who argues that the history of water projects in Arizona is one of response by civic leaders to requests for service by urban and suburban residents, closely resembling other aspects of the history of modern urban infrastructure. This is not necessarily a contradiction: the commonality in urban infrastructure is capital intensity and economies of scale, with the consequent need for public sector leadership and social co-ordination and, also, the consequent prospect of a handsome increase in land value in the area being served[32].

Another problematic consequence of the capital intensity, longevity of capital, and economies of scale in surface water infrastructure is the propensity to what might be called lumpiness or, less politely, gigantism in these systems. Because of the economics, there is a strong incentive to make a substantial expansion of capacity at a single point in time rather than to plan for a series of incremental changes spread out over time[33]. The drawback is that it may take many years, or decades, before the demand materializes to utilize this capacity (and the willingness to pay – WTP – to finance it). When fully utilized, the project provides water at a low cost; but there is uncertainty whether and when it will be fully utilized, and meanwhile it ties up scarce capital. Large surface water projects are risky, and difficult, inter-temporal balancing acts[34].

### 3.5   *The price of water*

It is important to emphasize that the prices which most users pay for water reflect, at best, its physical supply cost and *not its scarcity value*. Users pay for the capital and operating costs of the water supply infrastructure but, in the USA and many other countries, there is *no* charge for the water *per se*. Water is owned by the state, and the right to use it is given

---

[31] It was pointed out that large-scale irrigation works in Mesopotamia were developed *after* the rise of a centralized state, so that hydraulic society could be the result rather than the cause of state formation. The Maya civilization, where irrigation was of marginal importance, was cited as evidence that centralized states might not always be associated with hydraulic systems. It was also noted that there are several modern communities in Mesopotamia where small-scale cooperative irrigation works without centralized external control.

[32] As noted above, the increase in land value made it possible to finance urban water infrastructure with property taxes rather than through user charges. This may have been economically rational not only because of the very low marginal cost of urban water, but also because of the public good benefits of urban water supply associated with improved fire protection and also what, in the 19th century, were believed to be the public health benefits of washing down streets (Anderson, 1980).

[33] By contrast, systems supplied from groundwater are considerably less lumpy and more scaleable.

[34] This is well illustrated by the experience of the *Central Arizona Project*, the most recent large water project in the USA, which actually went bankrupt (Hanemann, 2002). The two key parameters for the economic viability of a water project are the discount rated use to assess the present value of net benefits, and the rate of growth in the public's ability and, more importantly, WTP for the water. Public agencies can generally borrow at a lower interest cost than private firms, and also are more apt to take a long-term perspective in evaluating investment. The public sector's ability and willingness to apply a low discount rate is a major reason for its predominant role in the provision of water supply infrastructure. For all the rhetoric on privatization of water, the private sector seems more interested in taking over the operation of existing infrastructure rather than financing new infrastructure, except for water treatment facilities which, as noted above, have a shorter life than other water supply infrastructure.

away for free. Water is thus treated differently than oil, coal, or other minerals for which the USA government requires payment of a royalty to extract the resource. While some European countries, including England, France, Germany and Holland, do levy an abstraction charge for water, these charges tend to be in the nature of administrative fees and are not generally based on an assessment of the economic value of the water being withdrawn. Thus, in places where water is cheap, this is almost always because the infrastructure is inexpensive, or the water is being subsidized, rather than because the water *per se* is especially abundant.

In the USA, it has long been noted that the prices charged to farmers are far lower than those charged to urban residents, often by a factor of 20 or more. It is often assumed that this is because the irrigation water has been subsidized by the federal government, but this is not in fact the main reason.

It certainly is true that the federal government has subsidized irrigation in the West by waiving interest charges and other means. Between 1902 and 1994, the federal government spent US$ 21,800 million to construct 133 water supply projects in the West. Although most of the water from these projects is used for irrigation, the cost allocated for repayment by irrigation users was set at US$ 7100 million (33%). Of this, US$ 3700 million was subsequently waived. Of the remaining US$ 3400 million payable by irrigators, only US$ 950 million had actually been repaid as of 1995 (General Accounting Office, 1996). The remaining balance will not be paid off until well into this century, if at all. The combined effect is that recipients of irrigation water from federal projects will have repaid, on average, about US$ 0.10 on each dollar of construction cost. However, these projects account for only about 19% of total irrigation supply in the West. The remainder comes from groundwater or non-federal surface supply projects, none of which is subsidized to a significant degree[35]. While the non-federal irrigation supply is more expensive than the federal supply, it still is much cheaper than urban water supply.

The reason is the sharp difference in the real cost of agricultural versus urban water supply. Unlike urban water, irrigation water is not treated, and it is generally not available on demand via a pressurized distribution system. Moreover, the physical capital used for irrigation supply is often old and long-lived, and it may have been paid off long ago.

There is an additional tendency to under price water in the USA – urban as well as agricultural – because most water agencies set price to cover the historic (past) cost of the system rather than the future replacement cost[36]. There is typically a large gap between these two costs because of the extreme lumpiness and longevity of surface water supply infrastructure. The capital intensity of the infrastructure exacerbates the problem because, after a major surface water project is completed, since supply capacity so far exceeds current demand, there is a strong economic incentive to set price to cover just the short-run marginal cost (essentially, the operating cost), which is typically minuscule[37]. As demand eventually grows and the capacity becomes more fully utilized, it is economically optimal to switch to pricing based on long-run (i.e. replacement) marginal cost, but by then water agencies are often politically locked

---

[35] If there is a subsidy, it is likely to be mainly for the electricity used in pumping water.

[36] This is due partly to the conservatism of the conventional engineering emphasis on cost recovery, and partly to the fact that, since most water supply agencies in the USA are publicly owned rather than investor-owned, there is a strong ethos to avoid making a profit on the sale of water.

[37] Erie & Joassart-Marcelli (2000) argue that this type of water pricing encouraged urban growth, and urban sprawl, in Southern California, but they fail to recognize the economic logic that drives it by virtue of the lumpiness and capital intensity of water supply infrastructure.

into a regime of low water prices focused narrowly on the recovery of the historical cost of construction.

## 3.6    *The essentialness of water*

Water is essential for all life – human, animal, or plant. In economics, there is a concept, also called essentialness, that formalizes this notion. The concept can be applied either to something that is an input to production or to something that is directly enjoyed by people as a consumption commodity. In the case of an input, if an item has the property that *no* production is possible when this input is lacking, the item is said to be an essential input. In the case of a final good, if it has the property that *no* amount of any *other* final good can compensate for having a zero level of consumption of this commodity, then it is said to be an essential commodity. Water obviously fits the definition of an essential final good: human life is not possible without access to 5 or 10 L/d of water per person. Water also fits the definition of an essential input in agriculture and in several manufacturing industries (e.g. food and beverages, petroleum refining, lumber and wood products, paper, chemicals, and electronic equipment) that cannot function without some input of water[38].

However, essentialness conveys no information about the productivity or value of water *beyond the vicinity of the threshold*. It implies nothing about the marginal value associated with, say, applying 76 versus 89 cm of water to irrigate cotton in the Central Valley of California. It says nothing about the marginal value of residential water use at the levels currently experienced in Western Europe or the USA – the latter averages about 455–530 L/d per person, more than two orders of magnitude larger than the minimum quantity that is needed for human survival[39].

The latter statement is not meant to belittle the uses of water by households in Western Europe or the USA. My point is that, in addition to being essential for human life, water contributes in important ways to the enjoyment of the satisfactions of life. Consequently, there are many other residential end uses of water besides its use for drinking. Indeed, if one examines the history of residential water use in the USA from the early 19th century to the present, it is striking how water consumption has grown over time through the steady accretion of end-uses, each representing the discovery of a new way to employ water for people's use and enjoyment. When a piped water supply first became available in the 19th century, the initial household uses were the same ones that had existed when family members had to fetch water from an external source – drinking, cooking, hand washing, and limited bathing[40]. As time passed, many other uses were found – tubs for bathing, water borne sanitary waste disposal, outdoor landscape and garden watering, automatic clothes washers, swimming pools, automatic dish washers, car washing, garbage disposal, indoor evaporative cooling, hot-tubs, lawn sprinklers, etc. The result has been a constantly rising trajectory of per capita household water use.

---

[38] These are the largest water-using industries in the USA in terms of freshwater intake.

[39] Total urban water use in the USA averages about 680–830 L/d per capita, depending on the location.

[40] For example, Blake (1956) notes that out of 15,000 houses with running water in Philadelphia in 1849, only about 3500 were equipped with private baths. In 1871, by contrast, 112,457 Philadelphia houses had running water, and 80,000 of these had bathtubs and fixtures for hot and cold water (Anderson, 1980). The use of water closets to remove human wastes did not become widespread until almost two decades after the introduction of piped water into homes (Tarr, 1979).

Two conclusions can be drawn from this historical experience. First, in developed countries, the fact that water is essential for human life is almost certainly irrelevant when assessing the value of residential water supply because the ways in which water is used are nowhere near the threshold level at which essentialness applies. Second, there is a possibility that some of the developing countries may also move along a rising trajectory of residential water consumption because the things that have made abundant water use an element of a comfortable modern life style in developed countries could also become attractive to people in these countries as their income rises.

For reasons that are entirely understandable, there is a tremendous emphasis in the water literature on the need to secure at least a minimal water supply for the nearly 1100 million people around the globe who currently live without access to an improved water source. In developing this estimate, the United Nations and the WHO used a figure of 20 L/d per person as the minimum human requirement for water for drinking and basic sanitation. Gleick (1999) has argued that this is too low, and has advocated the adoption of a basic human right to 50 L/d per person as the minimum required for bathing and cooking as well as the other basic needs. As noted above, these low levels of consumption are not very different from the initial levels of water use in developed countries when piped water was first introduced in the 19th century. It is important to recognize that, if the efforts to provide improved water supply and sanitation in the developing countries are successful, the future levels of water use that might ultimately emerge in these countries could diverge from these minimum levels, as happened in the developed countries.

In short, while it is obviously appropriate to think in terms of human *needs* for water, it is also appropriate to recognize that people also have *demands* for water as a commodity that generates pleasure by utilizing it in various ways. As they become more affluent, poor people are likely to choose to allocate more of their resources to satisfying not just their needs of the body but also their "wants of the mind, most of them proceeding from imagination", perhaps including some domestic end uses of water that might seem outlandish today.

If and when this broadening of domestic water use beyond the basic minimal level occurs, an important implication is that planners will need to adopt a *behavioral* approach to the analysis and projection of urban demand, as opposed to the engineering/public health approach that dominates the literature on water and poverty today. The behavioral approach focuses not on how much water people need but rather on how much water they are willing to pay for.

The difference between the engineering and behavioral perspectives is well illustrated by the experience of the World Bank over the past fifteen years. Water planners in the Bank originally thought that water and sanitation projects in developing countries were not viable if they required households to pay more than 3–5% of their income for the project services, because this would be more than they would be willing or could afford to pay, rendering the projects infeasible (World Bank, 1975). It became evident from detailed household-level studies sponsored by the Bank in the 1980s and 1990s that, in many developing countries, some households spend considerably more than this on access to traditional, unimproved water and sanitation. Some households purchase water from vendors at prices which can be much higher than the cost of piped water. Where there is piped water supply, many households incur expenses on installing storage capacity in the home to ensure that they have water when the pipes run dry; others undertake a wide variety of practices to treat contaminated water in their home to make it safe to drink. Moreover, some carefully designed contingent valuation surveys by Dale Whittington and his colleagues showed that some households have a WTP for improved water supply that can exceed 3–5% of household income[41].

The issue is not how much a household values access to water versus no access to water at all but, rather, how much it values a piped, public water supply relative to the existing alternatives. In this context, it is interesting to review what is known about how the adoption of water relative to other utility services varies with household income in developing countries. Komives *et al.* (2003) have analyzed data on the percentage of households at different income levels with four utility services: piped water, sewer, electricity, and telephone[42]. As monthly household income increases from very low levels to US$ 300 per month, coverage of all of these infrastructure services increases rapidly; above US$ 300 coverage increases at a slower rate. However, what is most striking is that, for households in this sample, at all income levels, more people have electricity than piped water or sewer. Very few of the poorest households have piped water or sewer, but almost a third of these households have electricity service. In Kathmandu, Nepal, for example, all of the households surveyed had the option to connect to electricity, water, and sewer; the majority of the very poor households chose electricity but not water or sewer. As income increases, the percentage of households choosing water and sewer increases, but the percentage with electricity is always higher. Almost no one, at any income level, has only a piped water service; but many households have electricity and not water. Thus, although most households would certainly like improved water and sanitation services, it is not their most important development priority; given their limited resources, many of them want electricity before an in-house piped water or sewer connection[43].

In short, the fact that water itself is a necessity does not necessarily mean that people prefer piped water over electricity service. Indeed, because water is a necessity, households must already have some access to water supply. The question is thus how much they value an *improved* supply. This will depend on how bad the existing water service is, and how much better the improved service is expected to be[44].

## 3.7    *The heterogeneity of water*

It is common to talk of the value of water as though it were a single, homogenous commodity. This is obviously false: water has many dimensions besides just quantity. These include: (a) location; (b) timing; (c) quality; and (d) variability/uncertainty. To a user, one liter of water is not necessarily the same as another liter of water if it is available at a different location, at a different point in time, with a different quality, or with a different probability of occurrence.

---

[41] The subsequent experience when the Bank went ahead and implemented the projects has borne out the predictions of the contingent valuation surveys quite well (Griffin *et al.*, 1995). It would be wrong to infer too much from the data on vended water purchases because the vast majority of households in developing countries do *not* buy from vendors. This implies that their WTP for vended water is *less* than the cost.

[42] The data come from the World Bank's *Living Standards Measurement Surveys* covering over 55,000 households in 15 developing countries.

[43] This does not mean that water and electricity are not complements – they often are. It simply suggests that people with limited budgets who cannot afford both generally prefer to have electricity first and piped water later.

[44] It may also depend on the quantity of water involved – because of diminishing marginal utility, what a household would be willing to pay per unit for the first 20 L/d per capita is not necessarily the same as it would pay to go from 40 to 60 L/d per capita.

There are two ways to incorporate the multi-faceted nature of water in a formal economic analysis. The first approach is simply to define different types of water as different commodities. For example, the consumption of water in January is represented by $x_1$, that in February is represented by $x_2$, that in March by $x_3$, etc. The consumer is then assumed to have a utility function defined over monthly consumption throughout the year and also over other commodities whose consumption is denoted by $z$ (which can be a vector or a scalar), leading to the formulation:

$$u = u(x_1, x_2, \ldots, x_{12}, z) \tag{1}$$

The significance of this formulation is that it leads to *separate* demand functions for consumption in each month. The demand for water in the ith month will be a function of the price of water in that month, the prices of water in the other months (which may or may not be different), and the price of $z$, as well as the consumer's income, $x_i = h^i (p_1, p_2, \ldots, p_N, p_z, y)$. The differences between one month's demand function and that of another will reflect the different ways in which the two monthly consumptions enter the underlying utility function (1). Given this approach, the annual demand for water is the aggregate of the 12 separate demand functions for the individual months. There is no demand function for annual consumption of water *per se*, except in the special case where the underlying utility function takes the particular form:

$$u = u(x_1, x_2, \ldots, x_{12}, z) = u \left( \sum x_i, z \right) \tag{2}$$

This is the only formulation that generates a well-defined demand function for aggregate annual consumption, $X \equiv \sum x_i$. Note, however, that the formulation in (2) implies that water consumption in any month is a *perfect substitute* for consumption in any other month[45]. More generally, if one discounts the difference between facets of water use and treats water as a single, homogeneous commodity, this is equivalent to assuming that the different types are all perfect substitutes for one another. It is an empirical question whether this is a plausible assumption.

The alternative framework for analyzing differentiated commodities was provided in a somewhat simple form by Lancaster (1966) and then broadened by Maler (1974), and is known as the *characteristics* approach to consumer demand[46]. The Lancaster-Maler model extends the utility model (1) by offering an explicit account of *why* the $x$'s are viewed as separate commodities, based on their specific characteristics. The notion is that there is a set of characteristics or attributes associated with each commodity. Suppose there are $K$ relevant characteristics (attributes), and let $q_{ik}$ denote the amount or level of the kth characteristic associated with one unit of consumption of commodity $i$. The characteristics of each commodity

---

[45] Two commodities are said to be a perfect substitutes in consumption if the consumer is willing to trade-off one for the other at the same, fixed rate of exchange regardless of how much or how little is consumed; in his eyes they can always be used in exactly the same way, with exactly the same outcome. When two commodities are perfect substitutes, they have essentially the same value. The polar opposite of perfect substitute is *perfect complement*. Two commodities are perfect complements if they are valued in fixed proportions to one another; consequently, they will always be purchased together in fixed proportions. In this case, no value is placed on increasing one of the items unless there is a corresponding increase in the other; an old-fashioned example (in England) was tea and milk.

[46] The application of Maler's work to the modeling of differentiated commodities was exposited in Hanemann (1982).

are taken as given by the consumer who is free to vary only the quantity of the commodity, $x_i$. Thus, if the consumer wishes for more of the $k$th characteristic, she accomplishes this not by changing the characteristics of any good, since these are fixed to her, but rather by switching her consumption towards commodities with a high level of this characteristic (i.e. a high value of $q_{ik}$); quality variation is accomplished through quantity variation. If there are $N$ separate differentiated commodities together with undifferentiated consumption, $z$, the utility function takes the form:

$$u = u(x_1, x_2, \ldots, x_N, q_1, q_2, \ldots, q_N, z) \tag{3}$$

where $q_i \equiv (q_{i1}, \ldots, q_{iK})$. The demand functions for commodities now depend on the attributes as well as the prices, and take the form $x_i = h^i(p_1, p_2, \ldots, p_N, q_1, q_2, \ldots, q_N, p_z, y)$, $i = 1, \ldots, N$. Thus, this formulation provides a framework for analyzing the effect of characteristics/attributes on demand – it provides a model of the demand for attributes (i.e. for $q$).

I have focused so far on the multi-faceted nature of water with respect to consumer choice, but this obviously applies to producer choice also. Both of the approaches described above can be incorporated in a production function just as in a utility function. For example, the production analog of (3) is a production function of the form:

$$y = f(x_1, x_2, \ldots, x_N, q_1, q_2, \ldots, q_N, z) \tag{4}$$

where $y$ is output, the $x$'s are forms of water input, and $z$ is a vector of non-water inputs. Suppose, for example, that $y$ is crop production, $N = 2$, $x_1$ is the quantity of groundwater pumped by a farmer, and $x_2$ is the quantity of surface water delivered to the farmer by the irrigation district in which he is located. Even if the price were the same, the experience in California has been that most farmers do not consider groundwater and surface water to be equally attractive. They find groundwater more convenient because they totally control its supply and can obtain it at the flick of a switch while, with surface water, they have to wait until the irrigation district is able to route the water to them through the canal system. On the other hand, in some parts of California there are differences in water quality, with groundwater being more saline than surface water. Thus, immediacy of access and salinity are two of the attributes that enter the crop production function (4) in this case.

In addition to providing a framework for conceptualizing the demand for $q$, whatever this may be, the Lancaster-Maler model also provides a framework for the economic valuation of $q$. It can thus be used to measure water users' WTP for better availability of water, or less saline water, or a more reliable water supply, or more generally water of one type versus another (e.g. the premium on groundwater versus surface water, or on water at one location versus another).

It should be emphasized that the attributes in $q$ that differentiate one type of water from another are by no means limited to the type of physical characteristics mentioned so far, such as location, timing, quality, and reliability. Other aspects, such as how the water is provided, can be the object of people's concern and the focus of their preference. With water, users often care greatly about *fairness* in allocation or payment – what is known as procedural justice – and this may differentiate one source of water from another in their eyes. The Lancaster-Maler formulation permits one to incorporate in $q$ such psychological or sociological attitudes within an economic model of the demand for water, so that one can analyze how these attitudes might generate a different value for water when provided in a particular way or obtained from a particular source[47].

## 3.8  *The fallacy of using average value*

In most policy-related applications of economic valuation involving water, the relevant quantity that needs to be known is the marginal value of water rather than the average or total value. Precisely because water is a necessity of life, most people have *some* access to *some* amount of water, and most policy interventions therefore involve changing the quantity and/or quality of access rather than transforming the situation from no access to some access[48]. *Ceteris paribus*, there is likely to be some degree of diminishing marginal utility for consumers, and diminishing returns for producers; for this reason there can be a substantial difference between the marginal value of an increase in water supply and its average value. This needs to be emphasized because researchers often use an estimate of the average value of water to measure the benefits of a policy intervention; the resulting estimate is likely to be inaccurate.

An example comes from the recent *Spanish National Hydrological Plan*, which proposed a major water transfer from the Ebro River to the Mediterranean coast, from Barcelona in the North to Murcia in the South (MIMAM, 2000). Of the $1050 \, \text{Mm}^3$ to be transferred, $560 \, \text{Mm}^3$ was targeted for delivery to agricultural areas along the coast that have been relying on depleting supplies of groundwater. The correct way to measure the benefit from an increment in water supply for farming in the receiving areas is to estimate the marginal value of water (marginal net profit) in the agricultural uses that would go out of production without the importation of project water. Instead, the economic assessment performed by the Spanish government (MIMAM, 2002) valued the imported water using an estimate of the average value of water in current uses, calculated as the simple ratio of aggregate farming profit in the area divided by aggregate water use. There are two flaws in this approach: 1) it interprets all profit from farming in the area as exclusively a return to water; and 2) it treats the return to water as constant regardless of the amount of water used.

Rather than just being a return to water, the profit from agriculture is likely to be a return to the farmer's investment in land and other fixed assets, and also a return to the farmer's own labor and his family's labor. And, rather than the average value of water being constant in the receiving areas, there are several reasons to believe that the average value declines as more water is supplied, causing the marginal value of water to be less the average value. With varying land quality and the opportunity to grow different crops, farmers in the region are likely to respond to any reduction in water supply by idling their *least productive* land and discontinuing their *least profitable* crops. Furthermore, some users of groundwater also have access to some surface water supply – in varying ways and to varying degrees, the supplies of water for farmers in the receiving area are interconnected so that, within the area, water is a somewhat fungible commodity. Since the imported water is one among several sources of water for irrigation, the relevant demand function is the farmers' demand for project water, *not* their demand for water overall. Because of the availability of substitutes, the agricultural demand for project water is likely to be more elastic than the demand function for water overall.

---

[47] What Frey & Stutzer (2005) refer to as procedural utility can be represented by this model.
[48] By this, I am not implying that interventions are unimportant or trivial in their consequences.

Therefore, the marginal value of the imported water in the region is likely to be substantially less than the existing average value[49].

The key difference between the two concepts is that the marginal value involves the derivative of a relationship, and to estimate this one needs a (formal or informal) model of how water generates value. The average value, by contrast, can be estimated crudely by dividing two quantities without any understanding of how they are related in reality, and without any assurance that this ratio will remain constant. Consequently, the use of an estimate of average value, and the assumption of its constancy, although common, are almost certainly a mistake.

### 3.9    *The benefits of water*

There are numerous ways in which an increment in access to water might produce benefits, whether to those who use the water directly or to others. Examples include: the use of water for agricultural or industrial production, its use for hydropower or for navigation, residential use, flood control, water based recreation, or aquatic habitat. A key tool used by economists in formalizing many of these benefits is the concept of a production function. A production function is conceived as an empirical, causal relationship between the levels of inputs required to produce an output, or an outcome, and the level of output or the outcome that results. One example is the production function for an industrial or agricultural output as a function of water and other inputs to the production process. Another example would be a *health production function* relating inputs (including behavior patterns and levels of resource availability) to the production (attainment) of health status outcomes. A third example would be an *ecological production function* relating inputs and resource endowments to the production (attainment) of ecosystem outcomes.

However, while the notion of a production function is undoubtedly useful as a conceptual tool for organizing one's thought about these matters, it may work less well as a dependable empirical construct. It may work better on a micro-scale (i.e. at a factory level) rather than at the level of an entire regional economy, and it implies a notion of causation that may be oversimplified. In practice, it can often turn out to be surprisingly difficult to measure a production function on a regional scale or, more generally, to measure the specific increment in benefits associated with an increment in water availability; these difficulties are clearly evident in the literature on water and economic development[50].

The notion that water supply contributes to economic growth and development seems intuitively obvious. After all, it is known that many of the world's major cities owe their origin to their location along coasts or rivers where water-borne transportation was facilitated. But, the relevant question is whether an increment in water availability now would generate an increment in economic activity now, and how much. In the USA, federal water projects

---

[49] If one were considering the total elimination of farming across a large portion of the irrigated acreage in the receiving areas, the difference between the average and marginal value could be of little significance, but in this particular case the change involves about 10% of the irrigated farmland and 9% of the water supply in the receiving area. An important piece of evidence suggesting that the NHP was substantially overstating the value of the project water is the fact that the cost of irrigation water in the receiving areas, including the cost of groundwater pumping, is mainly in the range of $0.06–0.27\ \text{€/m}^3$, while the NHP valued the imported water at $0.75\ \text{€/m}^3$.

[50] Similar difficulties are to be found in the literature on water and health.

have long been advocated for their claimed contribution to regional economic development. However, the actual empirical evidence is less obvious and more negative. As Howe (1968) noted, in industrial processes, water costs are a relatively small fraction of total production costs even in water-intensive industries, and there are many examples of firms in such industries choosing to locate plants in water deficient areas because of market or non-water input considerations.

In the USA in the late 1960s, there was a flurry of efforts to conduct formal, ex post statistical analyses of the impact of water availability on economic development; the findings were generally quite negative[51]. The studies were motivated in part by Bower's (1964) hypothesis that the availability of water at the intake end and/or the effluent end is not a major factor in *macro-location* decisions of industry relating to location in major geographical areas or regions, such as river basins, but it can be a major determinant of *micro-location* decisions relating to location within the region or basin. Ben-David (1966) found that employment in the major water-intensive industries was not significantly related to a measure of water availability in a cross-section regression of USA states, but there was a significant positive impact at the county level in a regression of counties within Pennsylvania. Howe (1968) extended this analysis to include all the counties in the USA, and found the evidence for such an effect to be extremely limited. Cox *et al.* (1971) examined counties in the Northeastern USA in which large water projects had been constructed between 1948 and 1958, and found no relationship between project size and economic growth over the period 1950–1960. By contrast, Garrison & Paulson (1972) examined counties in the Tennessee Valley region and found a significant micro-location relation between water-oriented manufacturing employment and a measure of water availability. At the macro-location level, Carson *et al.* (1973) sampled counties in geographic sub-regions from all parts of the country, both rural and urban, and found no significant relationship between federal water resource projects and population growth. Cicchetti *et al.* (1975) extended this study using economic sub-regions as the unit of analysis and found that variables representing federal investment in irrigation facilities had no significant impact on regional income and growth, and only a small and not convincingly significant impact on the value of farm output. There was some relationship between economic growth and federal investments in flood control, hydropower and recreation, but the coefficients were often unstable[52]. A similar study by Fullerton *et al.* (1975) of counties in seven western states found no relationship between water investment and economic growth.

It seems clear that an investment in water supply does not automatically guarantee economic growth. But, what conclusion can be drawn? Is there *never* an economic case for investing in water supply?

I want to suggest that part of the problem arises from the inadequate concept of causation that is being utilized by economists in conceptualizing the notion of a production function.

---

[51] The focus of these studies was long-run employment and economic growth *after* the completion of project construction: it was recognized that there would be a short-term increase in employment during the period of construction.
[52] Some of these results and those of Cox *et al.* (1971) raise the question of the *direction* of causation: do federal investments in flood control, say, cause economic growth, or do growing areas use their political clout to attract federal water investments? Walker & Williams (1982) suggest that the latter is the causal connection.

The philosophy literature makes a distinction between necessity and sufficiency: $X$ could be a necessary but not sufficient condition for $Y$ to occur, or it could be a sufficient but not a necessary condition. However, this distinction is generally ignored in the economic literature, including both the theoretical and empirical analyses of the relationship between water and growth. The production function as conventionally formulated as a relationship along the lines of $Y = f(X, Z, \ldots)$, implies that $X$, $Z$ and the other factors on the right-hand side of the equation are each a *sufficient* condition for producing $Y$: changing any individual element $X$ or $Z$ is sufficient to induce a change in the value of $Y$. Similarly, the conventional forms of regression equation used in the statistical literature imply that the regressors are each sufficient conditions for a change in the dependent variable.

However, the true relationship may be different, and perhaps more complicated. For one thing, it seems plausible that having an adequate supply of water might be a necessary but not a sufficient condition for economic growth. While water does not automatically generate growth, it may be the case that areas which persist in lacking an adequate water supply (regardless of whether or not they started out with adequate water) will not flourish economically. For example, one can expect that people will eventually leave those areas and migrate to other areas that do have an adequate water supply. Thus, lack of water could be a sufficient condition for economic decline or, to put it another way, water may be a necessary but not sufficient for economic growth. But, this is not a relationship that is captured in the existing formulations of production functions and regression equations[53].

In fact, the relationship between water and growth might be even more complicated. It may be that there are *multiple* possible causal pathways, such that while there is some causal linkage between water and growth, the linkage is sufficiently imprecise and variable that water is neither a necessary nor a sufficient condition for growth. In effect, there is *sometimes* a causal linkage, but not always. If this is so, it would require a new formalism to express this type of relationship.

The statistical literature suggests at least two possible methods for estimating a relationship that is heterogeneous in the manner just suggested. Both allow for multiple possible relationships for the determination of $Y$, with a mechanism that is at least partly stochastic determining the specific relationship that applies at any particular instance; within this framework, one of the possible relationships can involve $Z$ as a causal determinant of $Y$ (e.g. as a sufficient condition), while in others $Z$ is not a determinant of $Y$. One statistical approach is the finite mixture model (McLachlan & Peel, 2000), in which it is assumed that different observations

---

[53] One way to operationalize the concept of necessity is through what is known as a switching regression. The general structure of a switching regression, framed around a variable $Z$, is:

$$Y = \begin{cases} f_1(Z, X_1; \beta_1) + \varepsilon_1 & \text{if } Z \leq 0 \\ f_2(Z, X_2; \beta_2) + \varepsilon_2 & \text{if } Z > 0 \end{cases}$$

Here $Y$ is the dependent variable, $f_i(\ )$ represents a possibly non-linear functional relation, $X_1$ and $X_2$ are vectors of explanatory variables (which may or may not be different), and $\beta_1$ and $\beta_2$ are the corresponding coefficient vectors which are to be estimated. In the present context, $Z$ measures the level of water supply and to represent it as essential one would have to impose the condition that $f_1(Z, X_1; \beta_1) + \varepsilon_1 = 0$ whenever $Z \leq 0$.

within the data belong to different groups to each of which a separate statistical relationship applies. Here one estimates both the underlying relationships and also the group membership probabilities for each observation in the data. The other approach is model averaging, of which Bayesian model averaging (Raftery *et al.*, 1997; Hoeting *et al.*, 1999) is receiving considerable attention. This approach assumes that there are many possible models that could apply to the given data set taken as a whole. One proceeds by estimating each model under consideration using the entire data, and then averaging these models. In the Bayesian version, the weights used for averaging are the posterior probabilities that each of the models is the correct one. This approach produces confidence intervals for coefficient estimates that formally account for model uncertainty in addition to estimation uncertainty.

The twin issues of model uncertainty and the need to reflect other causal relations besides sufficiency deserve to receive greater attention in the empirical evaluation of the benefits of water not only for economic development but also for health, fish habitat, and other outcomes[54].

## 4   THE PROBLEM OF WATER FROM AN ECONOMIC PERSPECTIVE

It is commonly said that the problem of water is not one of economics but politics, not one of physical shortage but governance. This is partly correct, but not entirely. The generic problem of water is one of matching demand with supply, of ensuring that there is water of a suitable quality at the right location and the right time, and at a cost that people can afford and are willing to pay.

The difficulty in accomplishing this is partly institutional and certainly includes problems of governance. However, some of the problems of governance themselves have an economic explanation. The omnipresence of fixed costs in surface water supply creates a classic economic problem of cost allocation which has no satisfactory technical solution[55]. The extraordinary capital intensity and longevity of surface water supply infrastructure, and the predominance of economies of scale, create a need for collective action in the provision and financing of water supply that simply does not arise with most other commodities. It has been recognized since Olson (1965) that the provision of goods through collective action may be flawed because of a failure of incentives.

Olson set out to challenge the optimistic notion that individuals with common interests can necessarily be counted on to act voluntarily to further those common interests. The problem arises from harmful coincidences of rivalness/non-rivalness in benefits combined with excludability/non-excludability in costs. Examples are free riding by members of the group who withhold their individual contribution but can still benefit from the results of their colleagues' efforts, and rent seeking by individuals who seek to capture for themselves the benefits

---

[54] A non-water-related health application of Bayesian model averaging is Koop & Tole (2004). A recent book by political scientists has been devoted entirely to the empirical modeling of necessary conditions (Goertz & Starr, 2003).

[55] The most convincing solutions are rooted in bargaining theory and identify a cost allocation based on relative bargaining strength; this is more a political than an economic approach (Young, 1986). An early application of bargaining theory to water is Rogers (1969); for a recent survey see Carraro *et al.* (2005).

of collective action while throwing the cost on others[56]. Consequently, as Olson concluded: "unless the number of individuals is quite small, or unless there is coercion or some other special device to make individuals act in their common interest, rational self-interested individuals will not act to achieve their common or group interest".

The challenge, thus, is to find a suitable non-coercive mechanism that motivates collective action. This has become the subject of vast literature in economics, political science, sociology and game theory. Success can be achieved, in principle, in several different ways. Cultural and social norms shape preferences and may tilt the balance of individual choice in favor of collective action. Homogeneity can help in some circumstances[57]. The nature of the institutional arrangements is crucial[58]: if the rules are simple, transparent and devised locally, if monitoring and enforcement are relatively cheap, with graduated sanctions for non-compliance, if low-cost and fair adjudication is available, then, *ceteris paribus*, successful collective action is more likely. However, the extent to which these conditions are met depends partly on people's outlook and disposition, and partly on the physical reality. For the reasons enumerated above, water supply and sanitation are not always well situated in this regard.

In short, while there clearly are some distinctive emotive and symbolic features of water that make the demand for water different, there are also some distinctive physical and economic features which make the supply of water different and more complex than that of most other goods. This fact has often been overlooked by economists and non-economists alike.

## REFERENCES

Agrawal, A. (2002). Common Resources and Institutional Responsibility. In: E. Ostrom; T. Dietz; N. Dolsak; P.C. Stern; S. Stonich & E.U. Weber (eds.), *The Drama of the Commons*. National Research Council.

Anderson, L.D. (1980). *The diffusion of technology in the 19th century American City: municipal water supply investments*. Ph. D. Dissertation. Northwestern University, Chicago, USA.

Armstrong, M.; Cowan, S. & Vickers, J. (1994). *Economic analysis and British experience*. MIT Press.

Baland, J.M. & Platteau, J.P. (1999). The ambiguous impact of inequality on local resource management. *World Development*, 27(5): 773–788.

---

[56] Spiller & Savedoff (1999) emphasize the possibility of rent seeking by the government. They argue the combination of economies of scale and massive sunk costs of investment encourage governments in developing countries to act opportunistically against the private (and often foreign) companies that operate the water supply system by pressuring them to lower prices, disallowing their costs, requiring them to undertake special investments, controlling their purchasing or employment patterns, or trying to restrict the movement or composition of capital. "All these are attempts by politicians . . . to capture the rents associated with the company's sunk costs by administrative measures". Spiller & Savedoff argue that this leads to a low-level equilibrium of low prices and bad service.

[57] The ambiguous implications of inequality for the success of collective action are noted by Baland & Platteau (1999) and by Bardhan & Dayton-Johnson (2002). A disequalizing shift in the access to common resources has two effects which work in opposite directions. On the one hand, the people who benefit from the change have a larger stake in the resource and, therefore, a greater incentive to act towards the collective good. But, there is a corresponding disincentive to the other individuals whose share of the outcome has been reduced.

[58] This has been emphasized most influentially by Elinor Ostrom; see, for example, Ostrom (1990, 2003). I draw here on Agrawal (2002) who provides a useful summary of the literature.

Bardhan, P. & Dayton-Johnson, J. (2002). Unequal irrigators: heterogeneity and commons management in large-scale multivariate research. In: E. Ostrom; T. Dietz; N. Dolsak; P.C. Stern; S. Stonich & E.U. Weber (eds.), *The Drama of the Commons*. National Research Council.

Barlow, M. & Clarke, T. (2002). *Blue Gold: the fight to stop corporate theft of the World's Water*. The New Press.

Baumann, D.D. & Boland, J.J. (1998). The case for managing urban water. In: D.D. Bauman; J.J. Boland & M. Hanemann (eds.), *Urban water demand management and planning*. McGraw Hill: 1–30.

Ben-David, S. (1966). Impact of water resource development on economic growth. In: *Quarterly Progress Report to the Corps of Engineers*, July. Raleigh, North Carolina, USA.

Blake, N.M. (1956). *Water for the cities*. Syracuse University Press.

Bowen, H.R. (1943). The interpretation of voting in the allocation of economic resources. *Quarterly Journal of Economics*, 58(4), November: 27–48.

Bower, B.T. (1964). The location decision of industry and its relationship to water. Paper presented at the *Conference of the Committee on the Economics of Water Resources Development*. December, San Francisco, California, USA.

Bowley, M. (1973). *Studies in the History of Economic Theory before 1870*. Macmillan.

Carraro, C.; Marchiori, C. & Sgobbi, A. (2005). *Application of negotiation theory to water issues*. Fondazione Eni Enrico Mattei. Working Paper 65, May.

Carson, J.M.; Rivkin, G.W. & Rivkin, D.M. (1973). *Community Growth and Water Resources Policy*. Praeger.

Cicchetti, C.J.; Smith, V.K. & Carson, J. (1975). An Economic Analysis of Water Resource Investments and Regional Economic Growth. *Water Resources Research*, 11(1), February: 1–6.

Ciriacy-Wantrup, S.V. (1947). Capital returns from soil-conservation practices. *Journal of Farm Economics*, 29: 1181–1196.

Clawson, M. (1959). *Methods of measuring the demand for and value of outdoor recreation*. Reprint no. 10. Resources for the Future, Inc., Washington, USA.

Cox, P.T.; Grover, W.W. & Siskin, B. (1971). Effect of water resource investment on economic growth. *Water Resources Research*, February: 32–38.

Davis, R.K. (1963). Recreation planning as an economic problem. *Natural Resources Journal*, 3(2): 239–249.

Dupuit, J. (1844). De la Mesure de l'Utilite des Travaux Publiques. *Annales des Ponts et Chaussees*, 2nd Series, 8. Reprinted in translation (1952) as: "On the Measurement of the Utility of Public Works", in: *International Economic Papers*, 2: 83–110.

Erie, S.P. & Joassart-Marcelli, P. (2000). Unraveling Southern California's Water/Growth Nexus: Metropolitan Water District Policies and Subsidies for Suburban Development, 1928–1996. *California Western Law Review*, 36, Spring: 267–288.

Frey, B.S. & Stutzer, A. (2005). Beyond outcomes: measuring procedural utility. *Oxford Economic Papers*, 57(1), January: 90–111.

Fullerton, H.; Lewis, W.C.; Anderson, J.C.; Keith, J.E. & Willis, R. (1975). *Regional development: an econometric study of the role of water development in effectuating population and income changes*. Report PRRBEO89–1. Utah Water Research Laboratory, Utah State University, Logan, USA.

Garrison, C.B. & Paulson, A.S. (1972). Effect of water availability on manufacturing employment in the Tennessee Valley region. *Water Resources Research*, April: 301–307.

General Accounting Office (1996). *Information on allocation and repayment of costs of constructing water projects*. Bureau of Reclamation (GAO/RCED-96-109).

Gleick, P.H. (1999). The Human Right to Water. *Water Policy*, 1(5): 487–503.

Glennon, R. (2004). The Price of Water. *Journal of Land, Resources & Environmental Law*, 24(2): 337–342.

Goertz, G. & Starr, H. (2003). *Necessary Conditions*. Rowman & Littlefield.

Gordon, B. (1975). *Economic analysis before Adam Smith*. Macmillan Press.

Griffin, C.; Briscoe, J.; Singh, B.; Ramsasubban, R. & Bhatia, R. (1995). Contingent Valuation and actual behavior: predicting connections to new water systems in the State of Kerala, India. *The World Bank Economic Review*, 9: 373–395.

Hamlin, C. (2000). Waters or Water? – Master Narratives in Water History and their implications for Contemporary Water Policy. *Water Policy*, 2: 313–325.

Hammond, R.J. (1960). *Benefit-Cost analysis and water pollution control*. Food Research Institute, Stanford University Press.

Hanemann, W.M. (1982). Quality and Demand Analysis. In: G.C. Rausser (ed.), *New Directions in Econometric Modeling and Forecasting in U.S. Agriculture*. Elsevier/North Holland, Inc., New York, USA: 55–98.

Hanemann, W.M. (1991). Willingness to Pay and Willingness to Accept: how much can they differ? *American Economic Review*, 81(3): 635–647.

Hanemann, W.M. (2002). *The Central Arizona Project*. Department of Agricultural and Resource Economics. Working Paper 937, October. University of California, Berkeley, USA.

Henderson, A. (1941). Consumer's surplus and the compensating variation. *Review of Economic Studies*, 8: 117–121.

Hicks, J.R. (1939). *Value and Capital*. Oxford University Press.

Hicks, J.R. (1941). The rehabilitation of consumer's surplus. *Review of Economic Studies*, 8: 108–116.

Hicks, J.R. (1942). Consumer's surplus and index numbers. *Review of Economic Studies*, 9: 126–137.

Hicks, J.R. (1943). The four consumer's surpluses. *Review of Economic Studies*, 11: 131–141.

Hicks, J.R. (1946). The generalized consumer's surpluses. *Review of Economic Studies*, 15: 27–33.

Hoeting, J.A.; Madigan, D.; Raftery, A.E. & Volinsky, C.T. (1999). Bayesian model averaging: a tutorial. *Statistical Science*, 14(2): 382–417.

Howe, C.W. (1968). Water resources and regional economic growth in the United States, 1950–1960. *Southern Economic Journal*, 34(4), April: 477–489.

Hurwicz, L. & Uzawa, H. (1971). On the integrability of demand functions. In: J.S. Chipman; L. Hurwicz; M.K. Richter & H.F. Sonnenschein (eds.), *Preferences, utility and demand*. Harcourt Brace Jovanovich, New York, USA: 114–148.

Komives, K.; Whittington, D. & Wu, X. (2003). Infrastructure coverage and the poor: a global perspective. In: P.J. Brook & T.C. Irwin (eds.), *Infrastructure for poor people: public policy for private provision*. The World Bank: 77–123.

Koop, G. & Tole, L. (2004). Measuring the health effects of air pollution: to what extent can we really say that people are dying from bad air? *Journal of Environmental Economics and Management*, 47: 30–54.

Krutilla, J.V. (1967). Conservation reconsidered. *American Economic Review*, 57: 787–796.

Kupel, D.E. (2003). *Fuel for growth: water and Arizona's urban environment*. University of Arizona Press, USA.

Lancaster, K. (1966). A new approach to consumer theory. *Journal of Political Economy*, 74: 132–157.

Lockwood, M. (1996). Non-compensatory preference structures in non-market valuation of natural area policy. *Australian Journal of Agricultural Economics*, 40(2): 85–101.

Maler, K.G. (1971). A method of estimating social benefits from pollution control. *Swedish Journal of Economics*, 73: 121–133.

Maler, K.G. (1974). *Environmental Economics: a theoretical enquiry*. Johns Hopkins University Press for Resources for the Future.

Marshall, A. (1879). *The pure theory of (domestic) values*. London School of Economics, UK.

Marshall, A. (1890). *Principles of Economics*. 1st edition. Macmillan, London, UK.

McLachlan, G. & Peel, D. (2000). *Finite mixture models*. John Wiley.

MIMAM (Ministerio de Medio Ambiente [Ministry of Environment of Spain]) (2000). *Plan Hidrológico Nacional: análisis económicos*. September. Madrid, Spain.

MIMAM (Ministerio de Medio Ambiente [Ministry of Environment of Spain]) (2002). *Strategic assessment of the National Hydrological Plan*. Summary Document. January. Madrid, Spain.

Niehans, J. (1990). *A History of Economic Theory: classic contributions, 1720–1980*. Johns Hopkins University Press.

Olson, M. (1965). *The Logic of Collective Action: Public Goods and the Theory of Groups*. Harvard University Press.

Ostrom, E. (1990). *Governing the Commons*. Cambridge University Press.

Ostrom, E. (2003). How types of goods and property rights jointly affect collective action. *Journal of Theoretical Politics*, 15(3): 239–270.

Postel, S. & Vickers, A. (2004). Boosting Water Productivity. In: Worldwatch Institute, *State of the World 2004*. W.W. Norton.

Pribram, K.H. (1983). *A History of Economic Reasoning*. Johns Hopkins University Press.

Raftery, A.E.; Madigan, D. & Hoeting, J.A. (1997). Bayesian model averaging for linear regression models. *Journal of the American Statistical Association*, 92, March: 179–191.

Ridker, R.G. (1967). *Economic costs of air pollution*. Praeger.

Rogers, P. (1969). A Game Theory Approach to the Problems of International River Basins. *Water Resources Research*, 5(4), August.

Samuelson, P.A. (1954). The pure theory of public expenditure. *Review of Economics and Statistics*, 36(4), November: 387–389.

Schumpeter, J.A. (1954). *History of Economic Analysis*. Oxford University Press.

Shiva, V. (2002). *Water wars: privatization, pollution, and profit*. South End Press.

Spiller, P.T. & Savedoff, W.D. (1999). Government opportunism and the provision of water. In: P.T. Spiller & W.D. Savedoff (eds.), *Spilled water: institutional water*. Institutional Commitment in the Provision of Water Services. Inter-American Development Bank. Washington, USA.

Tarr, J.A. (1979). The separate *vs.* combined sewer problem: a case study in urban technology design choice. *Journal of Urban History*, 5, May.

Trice, A.H. & Wood, S.E. (1958). Measurement of recreation benefits. *Land Economics*, XXXIV. August.

UNESCO (2000). *World Day for Water 2000*. See: http://www.unesco.org/science/waterday2000/ Variability.htm

Walker, R.A. & Williams, M.J. (1982). Water from power: water supply and regional growth in the Santa Clara Valley. *Economic Geography*, April: 95–119.

Ward, F.A. & Michelsen, A. (2002). The economic value of water in Agriculture: concepts and policy applications. *Water Policy*, 4: 423–446.

Wegge, T.C.; Hanemann, W.M. & Loomis, J. (1996). Comparing benefits and costs of water resource allocation policies for California's Mono Basin. Economic institutions and increasing water scarcity. In: D.C. Hall (ed.), *Advances in the economics of environmental resources. Volume 1: Marginal cost rate design and wholesale water markets*. JAI Press, Greenwich, Connecticut, USA: 11–30.

Weisbrod, B.A. (1964). Collective consumption services of individual-consumption goods. *Quarterly Journal of Economics*, 78(3): 471–477.

Willig, R.D. (1976). Consumer's surplus without apology. *American Economic Review*, 66: 589–597.

Wittfogel, K.A. (1957). *Oriental despotism: a comparative study of total power*. Random House Inc.

World Bank (1975). *Issues in Village Water Supply*. Washington, D.C., USA.

Worster, D. (1986). *Rivers of Empire: water, aridity, and the growth of the American West*. Random House.

Young, H.P. (ed.) (1986). *Cost allocation: methods, principles, applications*. Elsevier Science Publishers.

Young, R.A. & Haveman, R.H. (1985). Economics of water resources: a survey. In: A.V. Kneese & J.L. Sweeney (eds.), *Handbook of Natural Resource Energy Economics*, Vol. II. Elsevier Science Publishers: 465–529.

CHAPTER 5

# The value of water and theories of economic growth

M.S. Aguirre
*The Catholic University of America, Washington, D.C., USA*

ABSTRACT: Michael Hanemann (this volume) addresses important theoretical and empirical issues that are relevant in ensuring an accurate economic analysis of market and non-market valuations of water, its role in economic growth, and water management and policy issues. This short essay provides some comments on the approach taken by Hanemann. These comments underline, among other issues, the role of water use as well as the need to determine the price and cost of water, the need for intertemporal analysis, the appropriateness of using marginal versus average values, and the importance of taking into account institutional as well as water access conditions. All of these issues suggest that population has a *solution* role in resolving some of the water management problems that we face today.

Keywords: *Water value, economic growth theories, population and resources.*

## 1 INTRODUCTION

At the heart of the water crisis debate lies one question: is there enough water to sustain the present and the future population? How you answer this question depends greatly on whether or not one sees population as a problem. Some would argue that population is a problem in that resources (in this case water) are limited, that they can only sustain a certain number of people (although no one knows what that particular number may be), that the more numerous we become, the less resources each person will have available and it will become more difficult to provide the resources that each consumer desires[1]. Others argue that population is not a problem. They contend that numbers in themselves do not equal scarcity; rather, poorly structured societies as well as bad economic policies and governance bring about scarcity and poverty[2]. How people perceive the issue of population is critical, for it is by these perceptions that international legislative policies are formulated, and local water projects and policies are designed and approved. Thus, it is equally critical that people ensure that

---

[1] Brown *et al.* (1999) capture the typical arguments used in this position. A further corollary of this perspective, some times is identified as Neo-Malthusians. They see people as destroyers of resources and violators of environmental limits. Among the leading representatives of this group are Ehrlich and Hardin. A more recent presentation of their argument can be found in Ehrlich & Ehrlich (1990), and Hardin (1998). For a detailed analysis of their views see Simon (1996a, 1996b), Furedi (1997), and Wolfgram (1999).

[2] See Sen (1981, 1994), Fukuyama (1999), and Kliksberg (2000, 2001). There are also those who attribute the present problems not to population but to the distribution of resources given the present structures. Some of these authors include Dobson *et al.* (1997), Matson *et al.* (1997), Rabkin (1997), Kiester (1999), and Johnson (2000). It is clear that the conclusion of the authors in this volume is that the so call *crisis of water* is a crisis of governance, not of water availability.

their perceptions are grounded not in rhetoric and emotions, but in established scientific and empirical data.

Michael Hanemann (this volume) addresses important theoretical and empirical issues that are relevant in ensuring an accurate economic analysis of market and non-market valuations of water, its role in economy growth, and water management and policy issues. From this perspective, the issues he raises significantly contribute to helping differentiate what is rhetoric and emotions from the real issues in the so-called *water crisis* debate.

This short essay will provide some comments on the approach taken by Hanemann in addressing the issues raised in his chapter, and why they are relevant for economic growth theory.

## 2    SOME COMMENTS

In analyzing the value of water and its relation to the *water and diamond paradox*, Michael Hanemann, recalling Adams Smith and the Classics, notes the difference that exists between the "value in use and the price (value in exchange)", and underlines the important role that marginal analysis plays in the completion of the discussion of the value of a good. This is so because "it highlights the distinction between value and cost, between demand and supply". He thus accurately concludes, that "[t]he distinction between marginal and total [value] is the key to the full resolution of the diamond and water paradox: water may have a smaller value than diamonds at the margin, but it undoubtedly has a larger *total* value". He then goes on to address the valuation of non-market goods and emphasizes that the "use of non-market valuation applies to positive as well as negative environmental impacts of water projects". Yet, while he acknowledges the shift in the argument regarding what determines the source of value from "its inherent usefulness and ability to please man according to rules of reason" to "subjective human preference rather than objective need", Hanemann fails to note the importance that such a shift can have when addressing both the market price and non-market valuation of water. He also fails to acknowledge the importance of the externalities derived from water, such as *environmental restoration* or deterioration. If subjective human preferences are considered without taking objective human needs into account, one can run the risk, and people often do, of using resources inefficiently through waste and/or misallocations. This in turn leads to bad policy and poor water management. The issue becomes even more relevant when this analysis is framed in an inter-temporal analysis, this being within a strictly economic analysis context or that of city planners.

Addressing the *essentialness* of water, Hanemann accurately acknowledges that water "fits the definition" of an essential commodity. It does so as an input of production for "agriculture and also in several manufacturing industries". It also does so as a final good because "human life is not possible without access to 5 to 10 L/d of water per person". He underlines the fact that water, "in addition to being essential for human life, [it] contributes in important ways to the enjoyment of the satisfaction of life". From here, he draws two conclusions: (1) "the fact that water is essential for human life is almost certainly irrelevant when assessing the value of residential water supply [because the] … ways in which water is used are nowhere near … level[s] at which essentialness applies"; and (2) "developing countries might also move along a rising trajectory of residential water consumption because … life styles in developed countries could also become attractive to people in these countries as their income rises".

Furthermore, he argues that "while it is obviously appropriate to think in terms of human *needs* for water, it is also appropriate to recognize that people have *demands* for water as a commodity that generates pleasure by utilizing it in various ways". Therefore, he concludes, "planners will need to adopt a *behavioral* approach to the analysis and projection of urban demand, as opposed to the engineering/public health approach that dominates the literature on water and poverty today". Along these lines, Hanemann points to the fact that the question is "how much [the household] values *improved* [water] supply" and not whether the "household values access to water *versus* no access to water at all". Since water is an essential good, everyone has some way to access it. Otherwise, they would not survive[3].

While we agree to the benefits of a *behavioral* approach, we find the analysis of the role of water essentialness in the assessment of the residential water supply's value, only acceptable if no intertemporal analysis is taken into account. Such an assumption, however, is certainly mistaken in the context of water policy design and management. In an intertemporal analysis of water supply, *need* becomes a very relevant factor, not only because the basic consumption of present and future generations must be met, but also because it affects the optimum point of present and future consumption as well as pricing. To ensure the access to clean water of future generations, costs are involved and thus, necessity and use, cost and value, are not and should not be disconnected (Asano, this volume, clearly speaks to this point). From this point of view, one can also appreciate why *access to* not just *improved* water supply is relevant, especially when considering this issue within the context of developing countries. Furthermore, it speaks to the fact that, in some policy-related applications of economic valuation of water, the marginal and average values need to be taken into account. This is true especially, but not exclusively, in developing countries. Many households in developing countries might have access to *some* water, but that *access* might not be humanly acceptable from an economic and health point of view (for further comments on this point, see: Bergkamp, this volume; Sullivan, this volume). This is a case in point where the "heterogeneity of water", which Hanemann accurately addresses in his chapter, clearly applies. Not all water is the same.

Finally, Michael Hanemann raises an issue that is very relevant in the context of growth when he states that "water could be a necessary but not a sufficient condition for sustained economic growth" and recommends the use of a "discrete mixture approach" for handling the "causal, but not always" direct role that water has in real economic growth. As has been previously mentioned, institutions and political stability as well as human, social, and moral capital, play an important role in economic development and growth.

## 3   ECONOMIC THEORIES OF GROWTH

Based on what constitutes the main determinant of real economic growth, one can roughly divide the economic theories of growth into four groups. Two of them have a negative

---

[3] When addressing this issue, Hanemann recalls the case studied by Komives *et al.* (2003) and he concludes that there is a preference of electricity over improved water sources in developing countries. He fails to acknowledge, however, that these are not necessarily exclusive of each other but can be complementary. For example, access to electricity could allow consumers to access groundwater through electric water pumps rather than piped water. The first one often is significantly cheaper than the second one.

perception of population while the other two have a positive view. All schools often resort to economics and environmental sciences to validate their theory.

The argument to support a positive view of population in the water-population-environment relationship is mainly centered in the concepts of human, social, and moral capital, and they find validation in empirical data (for a review of the literature, see: Simon, 1996a, 1996b; Wolfgram, 1999; Eberstadt, 2000). On the other hand, the steps in the argument used to support a negative view of population in the water-population-environment relationship can be reduced to mainly three. (1) Rapid growth on population means the spread of scarcity (less water available for each person or to satisfied the subjective human preferences for water), and it is a main obstacle to economic growth in poor countries (as water is essential for the process of growth), because it reduces or cancels potential improvements in living standards and aggravates conditions such as poor health, sanitation, and malnutrition. (2) The political implications of such trends threaten government stability in developing countries, and encourage the confrontation between developed and developing countries. And (3) it pushes future generations to scarcity, and an unsustainable environment (water) carrying capacity. However, literature from both disciplines suggests opposite conclusions (Simon, 1996a, 1996b; Wolfgram, 1999; Eberstadt, 2000).

### 3.1    *Classical Economic Growth: Malthusian Theory* (this section relays heavily on Aguirre, 2002)

Thomas Robert Malthus proposed in his 1798 *Essay in the Principle of Population* a relationship between population growth and what he termed subsistence. The first grew geometrically while the second increased only at an arithmetic ratio. Consequently, he claimed the existence of an inverse relationship between population growth and economic development derived from the law of diminishing returns[4]. Malthus' problem was that he failed to explore his theory against historical experience; no theory can be said to be scientifically proven if that theory cannot be verified by empirical evidence.

Malthus' inverse relationship between population and growth leads *Classical Economic Theory* to the following arguments: (a) *The consumption effect* (for a given amount of water, population growth affects consumption directly); (b) *The production effect on private and public goods* (population growth affects consumption indirectly through the effect on production per worker. With fixed capital, average production per worker will suffer diminishing returns. Along the same lines, with a fixed level of revenue, a larger population will increase the demand for public services – utility services, education and health care – thus reducing the quality of these services and indirectly hindering development through the reduction of funds allocated to infrastructure); (c) *Age-Distribution effect* (a faster-growing population implies a larger proportion of children and, given the amount of resources, a smaller output *per capita*); and (d) *Dilution of Capital* (with a fixed income, population growth reduces savings and human capital – education per person – and therefore reduces physical and human investment.

---

[4] This law implies that more people mean fewer goods (including water) for each person; thus, as population grows, scarcity of water inevitably increases. He believed that man's ability to increase his food (including water) supply was constrained in three particular ways: through land (water) scarcity, limit productive capacity of cultivated land, and the law of diminishing returns.

In summary, and applying it to water, the classical theory of population growth, assuming a fixed level of water, predicts a decrease in *per capita* water availability (and therefore less economic growth or income) in two ways: more consumers divide any given amount of water, and each worker produces less because there is less capital, private and public, per worker. In addition, the growing number of young children poses an additional burden in the reduction of consumption because they consume but they do not produce. Finally, population growth hinders economic growth because, by reducing savings and water services, it reduces investment in water technology. The key made in this theory is the *ceteris paribus* condition (other things being equal) where resources (including water) are given and therefore constant.

However, when challenged, this theory fails both theoretically and empirically. When applied to the case of water, this failure is also present. Analyses at both levels suggest that there is no statistically proven simple relationship between population growth and economic growth, population size and economic growth, population size and resources, or population growth and environment[5]. The absence of a correlation contradicts the conventional Malthusian deductive conclusion. The only persuasive argument in the face of this absence of correlation, as Simon (1996b) states, is a plausible scenario in which one or more specified variables that have been omitted from the analysis would, in fact, lead to a negative relationship between population growth and economic growth. Some of these factors can include poor water governance, new technologies developed or those raised by J.A. Allan in the political arena. In this context, Hanemann's assessment on the type of modeling that should be used for water analysis seems to be especially relevant.

In fact, since the seminal work of Coale & Hoover (1958), several studies have followed supporting or contradicting the negative population view. The 2003 World Water Assessment Programme concluded that "this crisis is one of water governance ... caused by the ways in which we mismanage water". On the other hand, we do encounter problems of scarcity around the world, problems of poor consumer selection (water waste), or utility services' design such as the one mentioned by Hanemann regarding water and electricity (Llamas & Martínez-Santos, this volume, address this issue).

### 3.2   *Neo-Classical theory*

The Neo-Classical model of growth, as presented by Solow (1956) as well as some models of technological diffusion focuses on economic growth through investment, ignoring any link between population and the economy. That is, adjustments in growth take place due to the behavior of investment in physical capital (water technology and infrastructure in our case). In these models, growth is a worldwide process and country characteristics determine the relative level of income. Thus, low persistence is consistent with shocks of any size. Shocks may only play a minor role in determining the long-run path of output, despite being an important determinant of variance in decade-long growth rates.

Although these models have been able to explain the experience of developed countries in certain cases, they have failed to explain the worldwide experience. Dorwick & Nguyen (1989) present results for OECD countries that support the Neo-Classical theory. One may be tempted to think also on Singapore, South Korea, and other Asian countries as well. Yet, in the

---

[5] These works include Denison (1985), Rosemberg & Birdzell (1986), Scully (1988), Barro (1989), Simon (1992, 1996a), Birdsall (1995), Eberstadt (1995), Agenor & Montiel (1999), Agenor (2000).

case of the first two countries mentioned, human as well as physical capital investment took place as Blackburn & Ravn (1993) suggest.

From the point of view of this theory, achieving efficiency in water use and recycling through new technologies and infrastructure is the key to economic growth and water sustainability.

### 3.3   *Human capital theory*

Gary Becker advanced a model that relates the concept of human capital to economic growth. In doing so, he proposes an alternative to Malthusian models. In this theory, human capital is introduced as an important source of economic development that depends on advances in technological and scientific knowledge. A key assumption of this model is that the rate of return on investments in human capital rises rather than declines as the stock of human capital increases; man is creative and therefore the education of today implies more production in the future. For this reason, resources are not necessarily fixed and may increase as the population increases. Furthermore, Becker *et al.* (1993) found that population growth, when studied in the light of human capital theory, leads to multiple equilibrium points: an underdeveloped steady state with high birth rates and low levels of human capital, and a developed steady state with low fertility and high stocks of human and physical capital. They concluded that this means that history and luck are also critical determinants of a country's growth experience[6].

This implies that population growth is not the only determining factor in economic development, as the Malthusian theory claims. Training and educational programs together with physical capital investment are the important factors though. But what about diminishing returns? Becker found the answer to this issue in the increase of labor productivity due to education, consequently rejecting the Malthusian assumption of fixed resources. Following the concept of human capital, recent works have proposed models that relate population to growth[7]. They set forth an alternative to the Malthusian and Neo-Classical models of economic growth by introducing human capital as an important source of economic development, a source that depends on both technological and scientific knowledge[8].

Economic development has not been solely explained by the expansion of physical capital per worker, as the Neo-Classical school has proposed, or by the decrease in population as Malthus suggested. The human capital approach criticizes both perspectives because they underestimate the dignity and creativity of the economic agent, thereby failing to acknowledge this economic agent as the *ultimate resource*. Other issues, such as terms of trade, service of the debt, the cost of intermediate goods, and institutional features of each country, including political stability, are important for economic growth as well. Yet, it has been the introduction of human capital that has shed new light on the understanding of the development process. It is worth noting that the evolution of the discussion around economic development and water supply also reflects this evolution. During the 1970s, the focus was on infrastructure, and in

---

[6] Concerning the issue of luck, history and growth see Barro & Lee (1993), De Long & Summers (1993), and Easterly *et al.* (1993).

[7] Some of these works are analyzed within an overlapping generation model. See: Barro (1974), Becker (1974, 1991), Razin & Ben-Zion (1975), Willis (1985), Becker & Barro (1988), and Becker *et al.* (1993). King (1993) includes the proceedings of a conference on population and economic growth sponsored by the World Bank.

[8] Such findings have been long sustained by Julian Simon, Norman Macrae, Aaron Wildavsky, Ben Wattenberg, Karl Zinsmeister, and others.

the 1980s it was on economic liberalization. It was not until the 1990s when the focus shifted towards institutions and water governance.

Within this view, one can appreciate the relevance of the distinction made by Hanemann of "water as a necessary *versus* a sufficient condition for economic growth". It also speaks of the importance of incorporating intertemporal analysis in the determination of both the demand and the supply of water, as has been previously mentioned.

### 3.4   *Neo-Malthusian theory*

Since the Agenda 21 of Rio (UNEP, 1992), the Malthusian theory has once more gained an audience in the population debate. Supporters of this theory claim that population growth is unsustainable not only with regard to food, but also with regard to resources such as oil, minerals, land, and water. In 1968, two influential *Neo-Malthusian* works, Ehrlich's *Population Bomb* (1968) and Garrett Hardin's *Tragedy of the Commons* (1968), issued warnings about the limits of sustenance and of resources, which they claimed were doomed. Within this theoretical framework, there are two main sub-categories: The *Limited Resource Perspective* and the *Socio-Biological Perspective*. The former takes the classic Malthusian argument and applies it to all natural resources, while the latter, almost acting as a sub-set of the former, treats the environment as a limited resource and regards people as a threat to the biodiversity and ecological balance of that resource.

Both perspectives have failed to produce sound projections because they lack sound data and logic. For example, Paul Ehrlich claimed in 1968 that "hundreds of millions" of people would die of starvation by the 1970s, that 65 million Americans would starve, that the population of the USA would decline by 22.6 million persons, and that England would cease to exist by 2000 (Ehrlich, 1968). More recently, Ehrlich & Ehrlich (1990) renewed these predictions, although with more *caveats*, since the original predictions failed to materialize. Statistical data regarding water provided by international bodies such as the United Nations and the World Bank as well as by the experts' evidence presented in this volume, however, fail to support their position. The world is nowhere near the mass starvation or water crisis predicted by Ehrlich or Brown, yet much policy has been designed based on their assumptions not only in the area of water management but on other resources as well.

Supporters of market forces (Fullerton & Stavins, 1998, among others), argue that the market will correct for inefficiencies, and that carefully constructed initiatives can help to guide the market, particularly in the area of environmental-water protection. There are also those who disagree with this perspective because they attribute the present problems not to population but to the distribution of resources given the present structures[9]. What we know is that among poor countries there are some that have a high rate of population growth and others have too little population. The overall decrease in population growth (the number of children per women worldwide has decreased from 3.6 in 1980–1985 to 2.7 in the present) has not helped some countries to overcome poverty. The population has not become poorer in spite of having increased, but it has produced beyond the subsistence level. The productivity of the earth's land and water availability has grown more quickly than the world's population. In addition, the quality of life has increased also in less developed countries according to the *2003 Human Development Report*.

---

[9] See footnote 2 for references.

It is within this theory where, I believe, the importance of addressing the impact of the shift from objective need to subjective human preference in determining the value of water is made more obvious, whether it is analyzed at the margin (where it should be analyzed in most cases) or at the level (total value which should be use when the quality of the access to water is being considered). When subjective human preferences (level of consumption, comfort, or expectations) are stressed over the objective human need, economic and supply-planning water analysis can mislead policy decisions. A clear case is the wasted water seen in developed countries: an average of 200 to 600 L/d of water consumption versus 47 L/d in developing countries (UN Statistics Division, 2003, http://unstats.un.org/unsd/environment/), or the use of leaks in water pipes for the provision of water at a higher cost, leading the poorest to high price differentials. Such situations speak to the importance of addressing the need to focus not only on the management of the supply, but also of the demand. A good way to begin to address this problem, as Llamas & Martínez-Santos (this volume) point out, is the elimination of *perverse subsidies*, which disturb the market price.

## 4    CONCLUSION

The economic value of any good, including water, has experienced an evolution accompanied by both theoretical and empirical tools of economic policy analysis. As Michael Hanemann (this volume) has indicated, some of these developments have been helpful while others have not. Whether one does or does not support the market as a sole means to achieve efficiency in water governance, it is clear that population plays a key role in achieving it as it is the foundation for the growth of human, social and moral capital.

From an economic development perspective, it can be said that Hanemann addresses important issue regarding the values of water which are relevant for the economic growth theory and the empirical analysis used for water management and policy. Yet, precisely in view of the reality faced by those in developing countries as well as in the theoretical underpinnings of the different models of economic growth, some comments have been offered regarding the author's assessment of the value of water and its consequences. These comments underline the role of water use as well as the need to determine the price and cost of water, the need for intertemporal analysis, the appropriateness of using marginal *versus* average values, and the importance of taking into account institutional as well as water access conditions. All of these issues suggest that population has a *solution* role in resolving some of the water management problems that we face today.

## REFERENCES

Agenor, P.R (2000). *The Economics of Adjustment and Growth*. Academic Press, San Diego, California, USA.
Agenor, P.R. & Montiel, P. (1999). *Economic Development*. Princeton University Press, New Jersey, USA.
Aguirre, M.S. (2002). Sustainable development: why the focus on population? *International Journal of Social Economics*, 29: 12.
Barro, R. (1974). Are government bonds net wealth? *Journal of Political Economy*, 82(4): 1095–1117.

Barro, R. (1989). *Economic Growth in Cross Section Countries.* Working Paper n. 3120. NBET, Cambridge, UK.

Barro, R. & Lee, J.W. (1993). International Comparisons of Educational Attainment. *Journal of Monetary Economics*, 32(3): 363–394.

Becker, G. (1974). A Theory of Social Interactions. *Journal of Political Economy*, 87(4): 1063–1093.

Becker, G. (1991). *A Treatise on the Family.* Harvard University Press, Cambridge, UK.

Becker, G. & Barro, R. (1988). A Reformulation of the Economic Theory of Fertility, *Quarterly Journal of Economics*, 103(1): 1–25.

Becker, G.; Murphy, K. & Tamura, R. (1993). Human Capital, Fertility, and Economic growth. In: G. Becker, *Human Capital: a theoretical and empirical analysis, with a special reference to Education.* Chapter XII. Third Edition. Chicago University Press, Chicago, USA.

Birdsall, A. (1995). Economic Approaches to Growth and Development. In: H. Chenery & T. Srinivasan (eds.), *Handbook of Development Economics.* North-Holland, Amsterdam, the Netherlands.

Blackburn, K. & Ravn, M.O. (1993). Growth, Human Capital Spillovers and International Policy Coordination. *Scandinavian Journal of Economics*, 95(4): 495–515.

Brown, L.; Gardner, G. & Halweil, B. (1999). *Beyond Malthus: Nineteen Dimensions to the Population Problem.* Worldwatch Institute. Washington D.C., USA.

Coale, A. & Hoover, E. (1958). *Population growth and economic development in low-income countries.* Princeton University Press, New Jersey, USA.

De Long, J.B. & Summers, L.H. (1993). How strongly do developing economies benefit from equipment investment? *Journal of Monetary Economics*, 32(2): 395–415.

Denison, E. (1985). *Trends in America Economic Growth, 1929–1982.* Brookings Institute, Washington, USA.

Dobson, A.P.; Bradshaw, A.D. & Baker, A.J.M. (1997). Hopes for the Future: Restoration Ecology and Conservation Biology. *Science*, 277 (25 July): 515–522.

Dorwick, S. & Nguyen, D.T. (1989). OECD Comparative Economic Growth, 1950–1985: catch-up and convergence. *American Economic Review*, 79(5): 1010–1030.

Easterly, W.; Kremer, M.; Pritchett, L. & Summers, L.H. (1993). Good policy or good luck? Country Growth Performance and Temporary Shocks. *Journal of Monetary Economics*, 32(3): 459–483.

Eberstadt, N. (1995). *Tyranny of Numbers: Mismeasurement and Misrule.* American Enterprise Institute, Washington, D.C., USA.

Eberstadt, N. (2000). *Prosperous Paupers and Other Population Problems.* Free Press, Washington, D.C., USA.

Ehrlich, P. (1968). *The Population Bomb.* Ballantine Books, New York, USA.

Ehrlich, P. & Ehrlich, A. (1990). *The Population Explosion.* Simon & Schuster, New York, USA.

Fukuyama, F. (1999). *The Great Disruption.* The Free Press, New York Fullerton, USA.

Fullerton, D. & Stavins, R. (1998). How Economists see the Environment. *Nature*, 395 (October): 433–434.

Furedi, F. (1997). *Population and Development.* St. Martin's Press, New York, USA.

Hardin, G. (1968). The Tragedy of the Commons. *Science*, 162: 1243.

Hardin, G. (1998). *The Ostrich Factor.* Oxford University Press, Oxford, UK.

Johnson, G. (2000). Population, Food, and Knowledge. *American Economic Review*, 90(1): 1–14.

Kiester, E. (Jr.) (1999). A Town Buries the Axe. *Smithsonian Magazine*, 30(4) (July): 70–79.

King, R.G. (ed.). (1993). National Policies and Economic Growth: a World Bank Conference. *Journal of Monetary Economics.* Special issue, 32(3): 359–575.

Kliksberg, B. (2000). *The role of Social Cultural Capital in the Development Press.* Latin American Studies Center. College Park, University of Maryland, USA.

Kliksberg, B. (2001). *The Social Situation of Latin America and its impact on Family and Education.* Organization of the American States. Washington, D.C., USA.

Komives, K.; Whittington, D. & Wu, X. (2003). Infrastructure coverage and the poor: a global perspective. In: P.J. Brook & T.C. Irwin (eds.), *Infrastructure for poor people: public policy for private provision.* The World Bank: 77–123.

Matson, P.A.; Parton, W.J.; Power, A.G. & Swift, M. (1997). Agricultural Intensification and Ecosystem Properties. *Science*, 277 (25 July): 504–509.

Rabkin, J. (1997). *Greenhouse Politics.* American Enterprise Institute, Washington, D.C., USA.

Razin, A. & Ben-Zion, U. (1975). An intergenerational model of population growth. *American Economic Review*, 65(2): 923–933.

Rosemberg, N. & Birdzell, L.E. (1986). *How the West Grew Rich: The economic Transformation of the Industrial World*. Basic Books. New York, USA.

Scully, G. (1988). The Institutional Framework and Economic Development. *Journal of Political Economy*, 96(3): 652–662.

Sen, A. (1981). *Poverty and Famines: an Essay on Entitlement and Deprivation*. Clarendon Press, Oxford, UK.

Sen, A. (1994). Population and Reasoned Agency: Food, Fertility, and Economic Development. In: K. Lindhal-Kiessling & H. Landberg (eds.), *Population, Economic Development, and the Environment*. Sage Publications, New Delhi, India.

Simon, J. (1992). *Population and Development in poor Countries: Selected Essay*. Princeton University Press, New Jersey, USA.

Simon, J. (1996a). *The Ultimate Resource 2*. Princeton University Press, New Jersey, USA.

Simon, J. (1996b). *The State of Humanity*. Blackwell, New York, USA.

Solow, R. (1956). A contribution to the theory of economic growth. *Quarterly Journal of Economics*, 70(1): 65–94.

UNEP (1992). *Report of the United Nations Conference on Environment and Development: Agenda 21*. E.93.I.8 (Agenda 21). 3–14 June, Rio de Janeiro, Brazil.

Willis, R. (1985). *A Theory of the Equilibrium Interest Rate in an Overlapping Generations Model: Life Cycle, Institutions and Population Growth*. Discussion Paper 85–8. Population Research Center, NORC and University of Chicago, USA.

Wolfgram, A. (1999). *Population, Resources, & Environment: a Survey of the Debate*. Web Page: (http://faculty.cua.edu/aguirre).

III

Irrigation

# CHAPTER 6

# Irrigation efficiency, a key issue: *more crops per drop*

K.D. Frederick

*Resources for the Future, Washington, D.C., USA (currently retired)*

ABSTRACT: Since the start of the green revolution in the 1960s, irrigation has accounted for about 80% of the increase in agricultural production. Irrigators now use about 70% of all water withdrawals and 80% of the water consumed worldwide to grow 40% of the world's food and fiber. Current water uses, however, are depleting and degrading water resources and creating doubts as to the adequacy of supplies to meet future demands. About 10% of the world's agricultural food output depends on non-renewable groundwater supplies.

Existing water uses and irrigation developments emerged largely in an era when water was not viewed or treated as a scarce resource and the environmental impacts of water projects and diversions were ignored. Salinity and water logging have severely affected 20–30 million irrigated hectares and adversely impacted another 60–80 million. Sedimentation is reducing storage capacity of existing reservoirs. And the combination of high financial and environmental costs of developing new supplies and the growing demands of non-agricultural water uses limit development of new irrigation water supplies. A growing appreciation for the services provided by aquatic resources and concerns about irrigation's adverse impacts on these resources are additional obstacles to expanding agricultural as well as other non-environmental water uses.

Past trends as well as current irrigation water uses are unsustainable. Yet, even under optimistic projections for increasing dryland farming, continued growth of irrigated production is required to meet future food and fiber demands. Global population is expected to increase by about 2000 million people in the next quarter century with almost all of the increase coming in developing countries where malnutrition is common.

The opportunities for and obstacles to reducing irrigation water use while increasing production to avert future water and agricultural crises are examined. Potential opportunities include institutional reforms to provide incentives for more efficient water use and development of higher yielding and more nutritious plant varieties, crops that can be grown in more hostile environments or with lower quality water, and varieties that yield more harvestable biomass per unit of water. Potential obstacles include vested interests that benefit from current water laws and inefficiencies, concerns about the potential environmental impacts of genetically modified food, and the high costs of the research and infrastructure investments needed to make these potential opportunities possible. The implications of increasing levels of atmospheric $CO_2$ on plant photosynthesis and future climate conditions add uncertainty to the challenge of meeting the agricultural demands of a larger and, hopefully, more affluent and better-fed population in an environmentally benign and sustainable way.

Keywords: *Agricultural water use, irrigation efficiency, water scarcity and agricultural production, climate change, irrigation water*

## 1 THE ROLE OF IRRIGATION

Irrigation is the dominant use of water and essential to the production of the world's food and fiber. Irrigators use about 70% of all water withdrawn from streams, lakes, and groundwater aquifers and about 80% of the water consumed worldwide. The 250 million irrigated hectares produce 40% of the world's food and fiber on only 17% of all cropland (Shortle & Griffin, 2001: ix).

Irrigation in combination with high yielding crop varieties and the application of fertilizer and pesticides have accounted for about 80% of the increase in agricultural production since the start of the green revolution in the 1960s. Synergies between irrigation and the green revolution technologies contributed to this growth. Reliable supplies of water are essential for achieving most of the benefits of these technologies. And the higher yields that are possible with these technologies encourage the spread of irrigation.

The 5-fold increase in irrigated land in the last century has stressed water supplies in many areas around the world, creating doubts about the adequacy of supplies to meet future agricultural as well as other demands for water. The combination of its current water use, its importance to meeting current and future demands for food and fiber, and concerns about the sustainability of agricultural water uses give irrigation a special role in determining whether a water crisis is or will become a myth or reality.


## 2   SIGNS OF STRESS

Current water uses are depleting and degrading some water resources. Water diversions for irrigation and other uses have severely degraded many rivers and lakes around the world. Salinity and water logging from poor drainage of irrigated lands pose additional problems for meeting projected water and agricultural demands. The Food and Agriculture Organization (FAO) estimates that salinity and water logging have severely affected 20–30 million irrigated hectares and adversely impacted another 60–80 million hectares (Rosegrant *et al.*, 2002a: 2).

About 10% of the world's agricultural food output now depends on non-renewable ground-water supplies. Water tables are falling a meter or more annually in parts of Mexico, India, China, the USA, and several other countries (World Commission on Water for the 21st Century, 2000: 287). In the USA, groundwater use for irrigation exceeds recharge on about 4 million hectares, 20% of its irrigated land. Rising water costs due to declining well yields and increasing pumping depths have reduced irrigation from the Ogallala aquifer in the USA High Plains by more than a million hectares since the mid-1970s. Projections indicate that rising water costs will force more than 2 million irrigated hectares in this region to be abandoned or returned to dryland farming by the year 2020 (National Research Council, 1996: 130–133).

The ecological, economic, and human health damages attributable to the expansion of irrigation in the Aral Sea basin are extreme but not isolated examples of the costs of excessive water diversions, poor drainage, and runoff contaminated by agricultural chemicals. Since 1960 the sea has lost about two-thirds of its volume and become too saline to support the former thriving fishing industry. The productivity of the delta ecosystems of the basin's two main rivers has suffered. High salt levels forced about 1 million hectares out of production and reduced yields an estimated 60% on the remaining lands. And contamination of drinking water supplies by pesticides and toxic salts blown by dust storms from the exposed seabed have increased mortality and morbidity rates in the lower basin (Frederick, 1991; Postel, 1999).

The costs of developing additional water supplies are rising for several reasons. Dams and reservoirs have been the primary means of converting naturally varying water resources into more reliable and controlled supplies. However, there are limits to and diminishing returns to the safe yield produced by successive increases in a river's reservoir capacity. Because the best reservoir sites are developed first, subsequent increases in storage require larger investments. For example, a study of decadal changes in reservoir capacity produced per unit volume of dam for the 100 largest dams in the USA suggests that average capacity per m$^3$ of dam declined 35-fold from the 1920s to the 1960s (United States Geological Survey, 1984: 33). A stream's maximum possible yield is limited by its average annual flow. But reservoir evaporation losses offset the gains from surface storage well before this maximum is reached. The social costs of storing and diverting water also increase as the number of free-flowing streams declines and society attaches more value to water left in a stream.

Sedimentation is estimated to be reducing reservoir storage capacity by about 1% per year. Replacing this capacity with new reservoirs might cost US$ 10,000–13,000 million a year if enough reservoir sites could be found. But dredging sediment to preserve the capacity of existing reservoirs could be 10-times more expensive according to Postel (1999: 86). In the USA, reservoir losses from sedimentation have probably exceeded additions to storage from new construction since 1990 (Frederick & Gleick, 1999: 28).

Evidence of diminishing returns and rising water costs in the developing countries is abundant. Irrigation costs more than doubled from 1970 to 1990 in India, Indonesia, and Pakistan. Costs increased 3-fold in Sri Lanka, more than 50% in the Philippines, and 40% in Thailand in recent decades. High costs combined with declining cereal prices result in low economic returns for new irrigation projects in Asia. Developing more water for irrigation is likely to be even less attractive economically in Latin America and Africa where construction costs tend to be higher than those in Asia (Rosegrant *et al.*, 2002a: 3–4).

Wetlands are valuable components of hydrologic systems. They purify water by filtering and settling pollutants and sediments, and they reduce flooding by dispersing high water flows over time and area. They are among the earth's most productive ecosystems, providing habitat and food for fish and wildlife and a variety of harvestable resources such as timber, berries, fish, fur, and peat. Irrigation projects, however, have contributed to the loss of hundreds of millions of hectares of wetlands that have been drained and filled for cropland, flooded behind dams, and dried up as a result of streamflow depletion.

Existing water uses and irrigation developments emerged largely in an era when water was not viewed or treated as a scarce resource and the environmental impacts of water projects and diversions were ignored. But a growing appreciation for the services and amenities provided by wetlands, streams, lakes, and other aquatic resources and concerns about irrigation's adverse impacts on these resources are now additional obstacles to expanding agricultural as well as other offstream water uses. The combination of high financial and environmental costs of developing new supplies and the growing demands of non-agricultural water uses constrain development of new irrigation water in many areas of the world.

Pressures to reallocate supplies from irrigation, which is a relatively low-value water use, to other uses are increasing as the demand for non-agricultural water rises with population and income growth and as water becomes scarcer and more costly. Globally, withdrawals for domestic and industrial uses rose 4-fold from 1950 to 1995, nearly twice as fast as agricultural uses (Rosegrant *et al.*, 2002a: 2). To date, transfers of water supplies once used for agriculture to domestic, industrial, or environmental uses have posed little problem to meeting growing food and fiber demands. Adoption of more efficient irrigation practices and green revolution technologies have produced more crops per drop. To meet future agricultural and water demands, continued increases in the productivity per unit of water for both dryland and irrigated agriculture will be essential.

## 3   THE CHALLENGE

Global population, about 6000 million at the start of this century, is expected to reach nearly 8000 million by the year 2025. Almost all of this growth is expected to occur in the developing countries where 1300 million people currently lack access to safe water, 2600 million lack sanitation facilities, malnutrition is common, and average per capita caloric consumption is less than 80% of the average for the OECD countries. Increasing agricultural production to meet the demands of a larger and, hopefully, more affluent and better-fed population in an environmentally benign and sustainable way is an imposing task in view of the water and agricultural stresses detailed above.

The World Commission on Water for the 21st Century (2000) concluded that we are on an unsustainable path leading toward a water crisis. The linkages between water and agriculture suggest this could also be a path toward a food and health crisis for much of the developing world. With business as usual being unsustainable and likely to result in unacceptable production and environmental outcomes, the challenge is to identify an alternative path that is sustainable and will meet growing water and agricultural demands and then to adopt the policies required to move to this path. The *Impact-Water* model developed by Rosegrant *et al.* (2002a) described below attempts to identify a strategy that will achieve environmental sustainability and meet the water and food demands of the larger and more affluent population.

## 4   THE IMPACT–WATER MODEL

Rosegrant, Cai, and Cline developed a global model of water and food supply and demand to examine the implications of alternative water and investment policies on water supplies and use, agricultural production, and the environment between 1995 and 2025. Global demand for cereals and meat is projected to grow by 46% and 56% respectively over these 30 years. In the developing countries, which are expected to account for more than 80% of the world's population by 2025, demand is projected to increase by 65% for cereals and more than double for meat (Rosegrant *et al.*, 2002a: 2).

The *Impact-Water* model (IM) combines an international model of agricultural production and trade with a water simulation model. The agricultural *Impact* model examines the effects of various food policies and rates of agricultural research on crop productivity, and the impacts of income and population growth on food supply and demand. The supply, demand, and prices for 16 agricultural commodities are determined in 36 regional or country submodels that are linked through trade. The impacts of water availability on agricultural production are introduced through the water simulation model that accounts for renewable water supplies, non-agricultural water demand, water infrastructure, and economic and environmental policies related to water development and management. Water availability is introduced as a stochastic variable with observable probability distributions based on the 30-year climate record from 1961–1990. The IM model is used to consider the future of water and food in the year 2025 with a projected world population of 7900 million, an increase of nearly 2000 million from the 1995 base year, under several alternative scenarios (Rosegrant *et al.*, 2002a: 18–31).

## 4.1   *Business as usual*

The business as usual scenario assumes a continuation of current trends in water and food policy and management. This scenario and its implications are described in Rosegrant *et al.* (2002a: 35–38, 61–108). Institutional and water management reforms are limited and piecemeal, and government investments in agriculture and irrigation are reduced. Under this scenario, water withdrawals are projected to increase 22% globally and 27% in the developing countries from 1995 to 2025. Total consumptive use increases 16% globally and 18% in the developing world. Irrigation consumptive use rises only 4% and accounts for only 20% of the global increase.

Water scarcity intensifies in the developing world, which accounts for 93% of the total increase in water consumed for irrigation. Economic incentives encourage some farmers to adopt more efficient irrigation practices. But political opposition from those concerned about the impacts of higher water prices and entrenched interests that benefit from existing water allocation deters adaptation in some areas. Overall, irrigation efficiency and river basin management improve slowly. Groundwater mining continues and there is no change in the share of water for environmental uses.

Water scarcity leads to slower growth of food production and shifts in where it is grown. Irrigated and rainfed agriculture each account for about half of the production increase. Greater food production depends largely on higher yields because of the slow growth in the area devoted to food crops. But growth in yields declines from the 1.5% average from 1982 to 1995 to 1% from 1995 to 2025 because many of the actions that produced past yield grains are not easily repeated and public investment is reduced. The growth of food production in developing countries slows as a smaller fraction of their irrigation water demand is met. Relative crop yield (the ratio of actual projected yield to the economically attainable yield without water stress) for cereals in the developing countries declines from 0.86 in 1995 to 0.75 in 2025. The decline attributable to increased water stress is equivalent to an annual production loss of 130 million metric tons, China's annual rice production in the late 1990s. Consequently, developing countries become much more dependent on food imports under the business as usual scenario.

## 4.2   *A crisis scenario*

A moderate worsening of current trends in water and food policy and investment results in a genuine water crisis according to the authors of the IM model. The water crisis scenario and its implications are described in Rosegrant *et al.* (2002a: 38–40, 109–136). Deteriorating infrastructure and poor management lead to a decline in water-use efficiency. Productivity growth in rainfed areas declines, particularly in marginal areas. Erosion and sediment loads in rivers rise as upper watersheds are deforested by people forced to turn to slash-and-burn agriculture. More reservoir storage is lost to sedimentation, wetland losses increase, and the health of aquatic ecosystems is further compromised as the water reserved for environmental purposes declines. Groundwater mining accelerates until about 2010 when declining well yields and rising pumping depths start to make irrigation from some key aquifers in China, India, and North Africa too expensive. In comparison to the business as usual scenario, global water consumption is 13% higher in 2025, total cereal production is 10% lower, food prices are sharply higher, and food security in the developing countries deteriorates. Irrigation accounts for virtually all of the increase in water use between the crisis and business as usual scenarios as farmers attempt to compensate for lower efficiency and water losses.

### 4.3   *Toward sustainability*

A sustainable water outcome under the IM model requires increased investment in crop research, technological change, and water management reform. The sustainable water use scenario and its implications are described in Rosegrant *et al*. (2002a: 40–44, 109–136). Water prices to the agricultural sector are gradually increased, providing incentives to conserve and funds to maintain and build new infrastructure. Water markets provide irrigators opportunities to profit from conserved water and reallocate supplies to alternative uses. Communities and water user associations are given responsibility for operating and managing irrigation systems. On-farm investments in irrigation and water management technology increase along with the efficiency of both irrigation systems and basin wide water use. Higher water prices for municipal and industrial uses also encourage investments in conservation and recycling. Groundwater overdrafts are gradually phased out through regulations, stricter enforcement, and market-based approaches that assign groundwater rights.

In comparison with business as usual, global water consumption is 20% lower by the year 2025 under the sustainable water scenario. Irrigation use also declines by 20% and accounts for nearly three-fourths of the overall reduction. The increased priority given to environmental water uses initially reduces the reliability of water for irrigation. Over time, however, more efficient water use helps offset the loss of water and the reliability of irrigation water supplies improves. But irrigated cereal production under the sustainable scenario is 5% lower than with business as usual.

Total cereal production in 2025 under the sustainable scenario is 48% above the 1995 baseline and 1% higher than business as usual. Irrigated cereal production rises 49% over the 30 years in spite of a 17% decline in the amount of water consumed for irrigation. The developing countries increase irrigated cereal production 50% consuming 19% less water. The developed countries produce 46% more cereals using 5% less water.

Underlying the 1% difference in cereal production between the sustainable and business as usual scenarios are significant changes in how and where crops are grown. Global cereal yields are 7% higher on rainfed lands and 2% lower with irrigation due to the lower reliability of its water supply in comparison to the business as usual scenario. Cereal production is 29 million tons higher in the developing countries and 10 million tons lower in the developed countries, implying a reduction in net food trade under the sustainable scenario.

Rosegrant *et al*. (2002b: 16) conclude that the IM model sustainability scenario "shows that with improved water policies, investments, and rainfed cereal crop management and technology, growth in food production can be maintained while universal access to piped water is achieved and environmental flows are increased dramatically". Improved productivity of rainfed agriculture is critical to meeting agricultural demands and reducing the pressures on water supplies. Under the sustainable scenario, rainfed agriculture accounts for 57% of the projected increase in global cereal production from 1995 to 2025. This is in contrast to developments over the last third of the 20th century when the spread of irrigation and the green revolution technologies accounted for about 80% of the increase in agricultural production. Infrastructure investments that link remote farmers to markets and policy changes that reduce the risks of rainfed agriculture and encourage on-farm investments contribute to dramatic yield increases even in drought-prone and high-temperature areas.

Research expands the opportunities for increasing crop and water yields on both irrigated and dryland farms and institutional reforms provide the incentives to conserve water and adopt better agricultural and water management practices. Advances in water harvesting systems and adoption of advanced farming techniques such as contour plowing, land leveling, and minimum-till and no-till technologies enable more effective use of rainfall in crop production. These improved farming techniques result in less erosion and reservoir sedimentation, reducing the need for new reservoir capacity.

The major role of dryland production projected in the sustainable water use scenario reduces the changes in irrigation required to ensure food security and water sustainability. But even with the increased reliance on dryland farming, global irrigated cereal production is projected to grow by nearly 50% over the 30-year period while consuming 17% less water (Rosegrant *et al.*, 2002a: 110–123).

## 5    SUSTAINABILITY: OPPORTUNITIES AND OBSTACLES

With irrigation consuming about 80% of all water withdrawn worldwide and current uses depleting and degrading supplies, adoption of more water-conserving and environmentally-benign irrigation practices are critical for achieving sustainable water use. On the other hand, meeting future food and fiber demands requires continued growth of production from irrigated lands even under the IM model's optimistic projections for increasing dryland agricultural output. Postel (1999: 10) concludes that future production increases will depend more on irrigated than on dryland farming. She is skeptical about the prospects for achieving large increases from dryland farming because most of the agricultural lands endowed with abundant and reliable rainfall, such as the USA cornbelt and Western Europe's wheat areas, are already producing close to their maximum potential.

The opportunities for and obstacles to reducing irrigation water use while increasing production sufficiently to avert future water and agricultural crises are examined below.

### 5.1    *Water use efficiency*

Irrigation involves withdrawing water from a surface or groundwater source and delivering it to a plant. The portion consumed in the irrigation process consists of water that is: (1) transpired by the plant; or (2) evaporated from soil and plant surfaces and lost as a result of infiltration and runoff to locations where it cannot be economically recovered. The portion of the water withdrawn that is not consumed adds to groundwater recharge or downstream supplies and is available for other uses.

#### 5.1.1    *Plant transpiration*
Transpiration is affected by many factors, including plant type and cover, stomatal behavior, and atmospheric carbon dioxide concentrations. The irrigation method employed to deliver water to the crops usually does not affect transpiration unless it fails to provide the plant with sufficient supply on a timely basis. Reducing plant transpiration usually results in water stress and reductions in the quantity and perhaps the quality of the yield (National Research Council, 1996: 106). Researchers have consistently found a linear relationship between a crop's transpiration and its yield of plant matter up to the point at which water is no longer a limiting factor. This suggests that for any given crop in a given location,

more plant production results in more water consumed through transpiration (Postel, 1999: 172–173).

There are differences in the efficiency with which plants use water. Consequently, changes in the crops that are grown and development of new varieties might produce more crop per drop. Moreover, exposing plants to higher concentrations of carbon dioxide increases their rate of photosynthesis, which normally leads to higher yields, and can reduce the amount of water transpired by the plant (Mendelsohn & Rosenberg, 1994: 24).

Plants such as corn, sorghum, millet, and sugar cane ($C_4$ plants) are naturally more efficient photosynthesizers than $C_3$ plants, which include most of the world's small grains, legumes, root crops, cool season grasses, and trees. Laboratory and greenhouse studies indicate that doubling the level of carbon dioxide increases the yields of $C_3$ plants by 15% to 20% and of $C_4$ plants by 5% (Adams *et al.*, 1999: 12). The enriched carbon environment also results in partial closure of a plant's stomates (pores), reducing transpiration from a given leaf area. If these results hold up under open field agriculture, significant gains in output per unit of water might result from the increases in atmospheric carbon dioxide concentrations that are anticipated over this century.

A number of complicating factors could offset these potential gains in water efficiency. The interaction between temperature levels and evapotranspiration is particularly relevant since carbon dioxide is the principal greenhouse gas contributing to global warming. Shugart *et al.* (2003: 18) note that "even if increased atmospheric $CO_2$ could potentially allow a tree to keep the stomata of its leaves closed for longer periods, it might still need to continue to leave the stomata open for the purpose of evaporative cooling to maintain heat balance. As a consequence, the improvements in water use efficiency could be offset by the need to maintain heat balance".

### 5.1.2  *Irrigation efficiency*

Irrigation water losses to evaporation, infiltration, and runoff are common and often large at the farm level. In developing countries on-farm irrigation efficiency is generally between 20–40%. However, water-use efficiency for the basin as a whole can be much higher than on individual farms because some of the water that is not consumed by evapotranspiration at the farm can be an important source of supply for downstream users. In Egypt's Nile basin, for example, overall basin efficiency is estimated at 80% even though individual water efficiency averages only 30% (Rosegrant *et al.*, 2002a: 6). Because of these and other externalities associated with irrigation practices, the benefits and costs of alternative irrigation systems and cultivation practices are likely to differ at the farm and basin levels.

On-farm water losses and perhaps higher crop yields can be achieved through investments in more efficient irrigation systems and adoption of improved management practices. Lining canals or installing gated pipe can reduce losses in delivering water from the farm gate to the head of the field. Surface irrigation systems such as flood and furrow are the most common methods of delivering water to crops. With these systems large quantities of water are usually applied to ensure the entire field is irrigated. More than half of the water may run off the field or infiltrate into the soil beyond the root zone of the plant. Land leveling, shorter furrows, and construction of borders enable a more uniform distribution of water to the crops, reducing on-farm water losses and the amount of water required to ensure all the plants are irrigated. Sprinkler systems can reduce or eliminate surface runoff, but they may increase evaporation losses in a hot windy environment. Drip irrigation applies water directly to the root zone of the crop, which can virtually eliminate field water losses.

Irrigation systems that apply water more uniformly to the field can increase crop yields as well as reduce on-farm water use. For example, studies in India, Israel, Jordan, Spain, and the USA show that drip irrigation has increased crop yields by 20–90% with 30–70% less water (Postel, 1999: 174). The state of the art for increasing the amount of crop per drop would combine microirrigation with sensors to monitor soil moisture and computer programs that analyze when and how much to irrigate. Since drip and other highly efficient irrigation systems are used on only 1% of the world's irrigated area, the possibilities for increasing yields to water are large. However, the economic, management, and institutional obstacles to widespread adoption of these methods are also large.

Increasing on-farm irrigation efficiency usually implies higher costs as well as benefits to the farmer. The net benefits or costs to the farmer of investing in a given irrigation system or tillage practice vary depending on factors such as soil conditions, capital and labor costs, management skills, and water costs. The net social and private benefits of the investment will differ depending in part on its impacts on the quantity and quality of water available to downstream users. Investments that reduce consumptive use on the farm while maintaining or increasing agricultural output increase effective water supplies. Moreover, any practice that increases irrigation efficiency generally decreases the sediment, salts, and chemicals that can pollute downstream supplies. Reducing erosion helps protect a farm's long-term productivity as long as salts do not accumulate in the root zone.

In the short run, drip irrigation might be able to reduce on-farm consumptive water use to little more than the amount transpired by the plant. However, sustainable irrigation requires applying enough additional water to leach salts from the root zone and drainage to remove water from beneath the field. All surface and groundwater supplies contain salts that are left behind when water is transpired by plants and evaporated from fields. Allowing salts to accumulate in the soil is detrimental to plant growth. Inadequate drainage results in rising groundwater tables and eventually to waterlogging, which destroys the field's productivity by cutting off a source of oxygen to the plants and by concentrating salts near the surface.

## 5.2    *Biotechnology*

The hybrid seed varieties that led to the green revolution and the rapid expansion of irrigated agriculture starting in the 1960s increased the output per unit of land and water largely by increasing the harvestable proportion of a plant's biomass. According to Postel (1999: 192–193), many plant breeders believe that the most fruitful possibilities for increasing the harvestable proportion of the major irrigated grains have already been exploited.

Advances in genetic engineering in recent decades have generated hopes that genetic modification of plants will be a panacea for many of the world's agricultural problems and provide the tools for another green revolution. Technological optimists anticipated more than a decade ago that biotechnology would lead to the rapid development and introduction of new higher-yielding and more nutritious plant varieties, crops that can be grown in more hostile environments or with lower quality water, and varieties that yield more harvestable biomass per unit of water. Economic considerations and concerns about the potential environmental impacts of genetically modified (GM) crops have dampened some of the initial high expectations for these crops.

The area planted with GM crops has grown from virtually nothing in the mid-1990s to about 60 million hectares, roughly 4% of the world's arable land. To date, however, GM crops have had little, if any, impact on how much crop is produced per drop or the plant's ability to utilize lower quality water supplies. To justify the costs of developing and testing new GM plants, companies have focused their research on developing seeds farmers will purchase new every year. Four crops – maize, soya, canola, and cotton – account for virtually all the GM planted area. And the major contribution of the GM seeds from a farmer's perspective is that they provide greater protection against insects and herbicides, reducing the need to purchase agricultural chemicals. When farmers in rich countries are being paid to take land out of cultivation and irrigation water around the world is subsidized, there is little incentive to develop genetically altered crops that provide resistance to drought or that can be grown using brackish and saline water (The Economist, 2003).

Economic factors are not the only obstacle to the development and introduction of GM crops that produce more crop per drop. Since 1998 the European Union has banned the planting or importation of any new genetically engineered crop. Environmental activists have trashed fields planted with GM crops. Despite widespread hunger, Zambia has refused to accept USA food that contained genetically modified corn. And Mexico has prevented CIMMYT, an international organization dedicated to creating better crops for farmers in the developing world, from testing GM corn outside a greenhouse (Charles, 2003).

Although current conditions may not encourage investments designed to develop seeds that increase the crop per drop, this should change as the need to produce more food and fiber with less irrigation water mounts. In spite of the lack of strong economic incentives, researchers have identified genetic traits involved in protecting plants from salt, cold, and drought. As water becomes scarcer and more expensive and GM crops more widely accepted, these characteristics are likely to be incorporated into plants suitable for field application. The holy grail for meeting future agricultural demands with less water would be to bioengineer plants to photosynthesize water more efficiently without sacrificing other desirable traits of the plant. Research on one of the proteins involved in photosynthesis has improved its productivity in lab tests. Some scientists, however, are pessimistic about our ability to eventually alter photosynthesis, a process that has remained virtually unchanged for 2000 million years of plant evolution and natural selection (Postel, 1999: 192–193; The Economist, 2003).

## 5.3  *Institutional reforms*

The institutions that establish the opportunities and incentives to use and abuse water resources are largely relics of an era when water was viewed as a free resource. They currently inhibit the development and adoption of water conserving technologies.

Irrigators are often the principal beneficiaries of the laws and subsidies influencing water use. They are the largest users and have high-priority rights for much of the readily accessible supplies. They pay nothing for the water itself, and the costs of the projects that store and deliver water to their farms are often subsidized. Farmers receiving water from government projects rarely pay more than 20% of water's real cost and generally much less (Postel, 1999: 230). Low prices and lack of opportunities to profit from conserved water provide little incentive to invest in efficient irrigation systems.

There is a growing consensus that greater reliance on economic principles in managing and allocating water is critical for more efficient and sustainable use. The report of the World

Commission on Water for the 21st Century (2000) advocates adopting full cost pricing, the polluter and user pays principle, and water marketing to encourage water conservation research and investments. Higher water prices are a critical element of the sustainable water scenario developed by Rosegrant, Cai, and Cline described above. Under this scenario, agricultural water prices would be twice as high in developed countries and three times as high in developing countries by 2025 compared to business as usual.

Raising water prices and reducing subsidies can be politically risky. Water is commonly viewed as too important to be subjected to market forces, and attempts to eliminate or reduce agricultural subsidies have generally failed in the face of strong resistance. A politically more acceptable approach for introducing economic incentives to conserve water is to provide irrigators with opportunities to sell unused supplies. They would then value water in terms of its opportunity cost – the value they could get by selling water – rather than at the subsidized price they pay for it. Selling saved water would provide funds for water-conserving and yield-increasing investments.

Charging the full costs of water and developing active water markets also requires overcoming practical barriers such as lack of equipment to measure how much is delivered or to terminate deliveries to non-paying farmers. Since it is expensive to move water out of existing conveyance facilities, the current infrastructure limits the opportunities to economically transfer water. The laws and regulations that define water rights and limit where and how it is used may also need to be changed to establish viable water markets.

While water markets can be an important step for introducing economic principles into irrigation decisions, they are not a panacea for achieving efficient use. Efficient markets require well-defined, transferable property rights, and the full costs and benefits of a transfer must be borne by the buyer and seller. Both the nature of the resource and the institutions that allocate and manage water can make it difficult to meet these conditions. Transferring water from one use or location to another often affects third parties as well as the buyer and seller. The variety of public as well as private services it provides and the interdependencies among users limit the potential for establishing unfettered and efficient water markets. Nevertheless, in view of the current inefficiencies and lack of incentives to conserve, most any changes that facilitate transfers in response to changing supply and demand conditions and provide irrigators with incentives to conserve are likely to result in a more socially efficient water use (Frederick, 2001).

## 5.4    *Global warming*

The prospect of a greenhouse warming is another reason for introducing institutional changes that facilitate voluntary water transfers. Global warming would alter precipitation, evapotranspiration, and runoff patterns, as well as the magnitude and frequency of extreme events. However, the nature, timing, and even the direction of the impacts on the regional and local scales of primary interest to water planners and managers are uncertain. Some regions are likely to experience more intense precipitation days, as well as increased flooding. Others may have more frequent and severe droughts. And some may experience both more intense floods and droughts.

The impacts of a greenhouse warming on water supplies are likely to create winners as well as losers. On balance, however, the impacts are likely to be negative in part because water-supply systems are designed and operated on the assumption that future climate will look like the past climate. The prospect of an unusually rapid change in the climate adds uncertainty and risk to the design and operation of new water projects. Consequently, developing institutions that facilitate and encourage voluntary water exchanges through markets would provide a system that is both more efficient and better able to adapt to whatever the future might bring.

## 6    CONCLUSIONS

A recent report of the Food and Agriculture Organization (FAO) concludes that chronic hunger increased in the last half of the 1990s to a total of 842 million people. About 95% of these people live in poor, developing countries (Lynch, 2003). Poverty, disease, wars, and agricultural trade policies, not global food production, are the major causes for the recent increases and current levels of chronic hunger in the world. Indeed, overproduction and lack of demand have characterized agricultural markets in the USA and Europe in recent decades. Nevertheless, meeting the food and fiber demands of nearly 8000 million, hopefully more affluent, people in the year 2025 in an environmentally benign and sustainable way is an imposing task.

Current water use is unsustainable and likely to lead to future water and agricultural crises. However, meeting future demands while reducing irrigation water use to sustainable levels appears to be achievable with major changes in business as usual. Large increases in investments in research, infrastructure, and efficient irrigation systems are needed. Institutional changes that encourage treating water as an economic good are also essential. But even with changes that produce more crop per irrigated drop and utilize lower quality water for which there is little demand from non-agricultural users, irrigation cannot be expected to continue producing the vast majority of the increases in agricultural output required to meet the demands of a growing and more affluent global population. Rainfed agriculture may need to account for half or more of the increases in production required to balance future agricultural supplies and demands.

Meeting future food and fiber demands depends on the quantity and quality of production that actually reaches the consumers and producers that use the farm output. Inefficiencies in agricultural marketing can result in large losses. Reducing transportation and storage losses is another means of increasing the effective marketable output per unit of water transpired.

Water and food crises tend to be regional and local in nature, rather than global. Changes in precipitation, and temperatures associated with a greenhouse warming could exasperate future regional problems. The ability to respond to these regional and local problems depends on both the adequacy of global agricultural production and international trade. In Africa, which is already burdened by political instability, extreme poverty, and poor infrastructure, agricultural production is particularly vulnerable to climate changes. Meeting their future food needs is likely to depend on the largess of the international community, which in the past has often fallen short of Africa's needs. China's food production is also highly dependent on benign precipitation conditions as well as non-sustainable groundwater irrigation. But in contrast to the situation in Africa, China's dynamic economy enables them to import any food shortfalls as long as there is adequate global production.

Longer-term sustainable food and fiber production requires dealing with population growth well beyond the year 2025 used for the *Impact-Water* model projections and adapting to the impacts of a greenhouse warming which are ignored in the modeling results.

# REFERENCES

Adams, R.M.; Hurd, B.H. & Reilly, J. (1999). *Agriculture and Global Climate Change: A Review of Impacts to U.S. Agricultural Resources.* Pew Center on Global Climate Change. Arlington, Virginia. USA.

Charles, D. (2003). Corn that Clones Itself. *Technology Review*, 106(2): 32–41.

Frederick, K.D. (1991). The Disappearing Aral Sea. *Resources*, 102: 11–14.

Frederick, K.D. (2001). Water Marketing: Obstacles and Opportunities. *Forum for Applied Research and Public Policy*, 16(1): 54–62.

Frederick, K.D. & Gleick, P.H. (1999). *Water & Global Climate Change: Potential Impacts on U.S. Water Resources.* Pew Center on Global Climate Change. Arlington, Virginia, USA.

Lynch, C. (2003). Chronic Hunger is Increasing. *Washington Post*, November 27, A20.

Mendelsohn, R. & Rosenberg, N.J. (1994). Framework for Integrated Assessments of Global Warming Impacts. *Climatic Change*, 28: 15–44.

National Research Council (1996). *A New Era for Irrigation.* National Academy Press. Washington D.C., USA.

Postel, S. (1999). *Pillar of Sand: Can the Irrigation Miracle Last?* WW Norton & Company. New York, USA.

Rosegrant, M.W.; Cai, X. & Cline, S.A. (2002a). *World Water and Food to 2025: Dealing with Scarcity.* International Food Policy Research Institute. Washington D.C., USA.

Rosegrant, M.W.; Cai, X. & Cline, S.A. (2002b). *Global Water Outlook to 2025: Averting an Impending Crisis.* International Food Policy Research Institute. Washington D.C., USA.

Shortle, J.S. & Griffin, R. (eds.) (2001). *Irrigated Agriculture and the Environment.* Edward Elgar Publishing. Northhampton, Massachussets, USA & Cheltenham, UK.

Shugart, H.; Sedjo, R. & Sohngen, B. (2003). *Forests and Global Climate Change: Potential Impacts on U.S. Forest Resources.* Pew Center on Global Climate Change. Arlington, Virginia, USA.

The Economist (2003). Climbing the helical staircase: A survey of biotechnology. March 29: 1–24.

United States Geological Survey (1984). *National Water Summary 1983 – Hydrologic Events and Issues.* Water Supply Paper 2250. U.S. Government Printing Office. Washington D.C., USA.

World Commission on Water for the 21st Century (2000). A Report of the World Commission on Water for the 21st Century. *Water International*, 25(2): 284–302.

CHAPTER 7

# Irrigation efficiency, a key issue: *more crops per drop.* *Observations and comments*

P. Arrojo

*Department of Economic Analysis, University of Zaragoza, Spain*

ABSTRACT:   Some samples of the current unsustainability are: the breakdown in the biodiversity and specially in fisheries; the general degradation of rivers, lakes, wetlands and aquifers; the grave social impacts in poor communities by large dams and by the general degradation of aquatic ecosystems; impacts on deltas, coastal areas and coastal platforms; degradation of irrigated lands diminishing productivity; or the breakdown of the water cycle and its influence on climate changes. The growth of irrigation is the key of the crisis of sustainability in the aquatic ecosystems, and at the same time the key in the fight against hunger. Sustainability of ecosystems cannot be anymore considered as a restriction but a need for fighting successfully against hunger.

There are different functions of water: water for life, water as public service and general interest, water for business. These different functions of water ask for different management approaches. We must recognise priorities among these functions.

This chapter also focuses on in tendencies, paradoxes and socio-economic contradictions that currently exist in water management, related to hydraulic works, markets, subsidies, prices, or WTO (World Trade Organization) policies.

Beyond recognising the value of some traditional production models, we need to design and promote a new water culture (the culture of sustainability): better life with less resource (not only by technical and market efficiency) and less and different way of consumption. How to produce can be even more important than how much. It is necessary a new cultural approach of consumption: *quality* instead of *quantity*. Assessment of eco-social efficiency given by integrated traditional models of living and producing in sustainable ways is needed.

Keywords:   *Sustainability, eco-social efficiency, values of water, new water culture*

## 1   THE CRISIS OF WATER ECO-SYSTEM SUSTAINABILITY

The spectacular growth in irrigated agriculture – its surface area increasing fivefold over the course of the 20th century – has undoubtedly been one of the key factors in the fight against hunger around the world. However, at the same time it has also been – and still is – the key factor in the crisis of unsustainability which is putting the health of most continental water eco-systems at risk, degrading the biological productivity of rivers, lakes, wetlands and coastal platforms, and aggravating the food crisis it aims to resolve, thus completing the vicious circle of unsustainability.

As Professor Frederick (this volume) indicates, 10% of food production is based on ground waters, the aquifers of which are at their recharge limits. This is only the tip of the iceberg of the unsustainable logic upon which we have based our present development model.

In his chapter, Professor Frederick deals with some of the clearest and best known cases of unsustainability, such as the case of the Aral Sea, or the large aquifers of Mexico, India, China or the USA. However, as an indicator of the dynamics of this unsustainable exploitation, we should also note the breakdown in the bio-diversity of continental aquatic ecosystems and the serious effects on marine fisheries, owing to the breakdown of these continental water systems.

The very concept of sustainability for the continental aquatic ecosystems is currently under debate and still being agreed upon, from both a biological (sciences of nature) and also a social perspective (corresponding commitments). However, it seems evident that from this dynamic process, the trend points towards diagnosing the current degradation of our rivers, lakes, wetlands and aquifers as being unacceptable, chiefly owing to the dumping of pollutants and excessive water extraction. Beyond ecological degradation, if we consider the serious effects on human health and on the basic resources of many communities, as well as the undermining of the human rights of entire settlements affected by large dams (between 40 and 80 million people have been forcefully evicted from their homes, according to estimates in the *World Commission on Dams* report (WCD, 2000), we can see a clearer picture of the eco-social complexity of *unsustainability*.

The most dramatic ecological, social, and environmental problems for continental aquatic ecosystems occur in arid or semi-arid steppe areas, as a result of excessive diversions. However, it is also true that in more humid climates, impacts on the continental hydrological cycle are produced by a kind of inverse phenomena. Particularly in densely inhabited areas, and often as a result of agricultural development, drainage and processes to achieve water-tightness tend to dry out aquifers and decrease the level of humidity in the ground, thus accelerating the dynamics of drainage to the sea. The hypothesis as to the possible effect of these breakdowns in the continental hydrological cycle upon regional climate changes (drop in average rain-fall, increased irregularity, rise in risks of flooding, etc.) are currently leading to extremely interesting studies.

The comments on implied costs due to reservoirs silting up are extremely pertinent but, should nevertheless be completed with the mention of other costs which are even more relevant; those of the environmental and economic impacts that this sediment retention produces on the deltas and beaches along the coastlines. We are now beginning to recognise and value the importance of the natural flow of these sediments, in order both to compensate for the natural subsidence suffered by river deltas, as well as the dynamic recharging of beaches and coastal sediment flows, which are chiefly dependent on river flows. The combination of these impacts, with the rise in sea level due to the melting polar ice caps, as a consequence of the current process of global warming, makes the effects of this silting up produced by large dams ever more extreme and the costs higher.

The progressive drop in the productivity of a large proportion of recently developed irrigation-based agriculture deserves specific analysis, since this is a worrying tendency and a direct result of the current development model in which short-term interests override medium and long-term ones, and kill off any chance of sustainability. The final report by the World Commission on Dams draws attention to the seriousness of increasing salinisation in many of these new irrigated lands which at present affects 20% of irrigated surface areas. Furthermore, the resulting soil degradation usually leads to abandonment of land which would otherwise have been productive for dry farming. This abandonment is often linked to erosion processes which, in turn, lead us to irreversible cycles of desertification.

Special mention must also be made of the direct and indirect impact of drying out large areas of wetlands worldwide. The argument behind draining and drying out these wetlands is often based on the need to make them more productive in the fight against hunger and poverty. Nevertheless, we are gradually becoming more aware of the importance of these ecosystems, both as generators of biodiversity, and hence of protein (especially fisheries), which is key to the diet and subsistence of many communities. With the destruction of these wetlands, their function of regulating and regenerating the quality of water has been upset, seriously affecting the availability of water resources. Moreover, serious damage has been done to the rich biodiversity of these ecosystems.

Very often many of these sources of food (fishing, hunting, forestry and small-scale agriculture etc.), closely linked to the sustainability and good health of ecosystems, are not accounted for in official statistics, since they do not form part of the usual market structure. Nevertheless, they have enormous importance in the real economy of many communities, especially in impoverished or developing countries.

## 2   TRENDS, PARADOXES AND SOCIO-ECONOMIC CONTRADICTIONS

Prof. Frederick's arguments regarding the economic problems which large, new hydraulic projects all over the world have come up against are both interesting and accurate. His treatment of the relentless effects of the law of *rising marginal costs*, particularly in relation to large dams, is similar to that of the final report of the World Commission on Dams (WCD, 2000). Nevertheless, we must also take into account the complementary effect of the law of *falling marginal profits*, which can be seen clearly in new irrigation projects (poorer quality of irrigable lands, higher altitudes and/or poorer geo-climatic conditions, etc.). The combined effect of the two laws leads to frequent negative cost-benefit balances, especially in hydraulic projects, the chief aim of which is to develop new irrigation. It must also be pointed out that the systematic loss of agriculture's relative profitability has contributed to this situation on a worldwide level, as most countries have seen that rises in crop prices are well below inflation rates.

Furthermore, rationalisation of water management seems necessary, especially as regards its use in irrigation, creating incentives which would promote saving and the modernisation of infrastructure. I must state here most categorically that such rationalisation does not necessarily have to be managed by market dynamics, however, as that approach is inappropriate for water management, except in specific cases of strongly regulated markets, such as the Water Banks in California. Without Professor Frederick's work specifically defending the *free market* as a strategy for economic rationalisation, he does make comments which I feel it is fitting to question or, at least qualify. Specifically, it seems appropriate to underline that this rationalisation does not have to be directed by free market patterns which, in general, are inappropriate for handling the ethical, social and environmental values of water, beyond that of purely economic wealth. I do, however, believe in the advisability of promoting (strictly regulated) public markets for water rights, such as the California Water Banks, in order to improve the management of *business-water* in periods of drought (Arrojo & Naredo, 1997).

Subsidies for irrigation water in both developed and developing countries have clearly negative implications for efficient water use. However, the option of abandoning such subsidies raises fears about the economic viability of many family farms and the impact this would have on food prices.

The worldwide contrast between experiences of irrigation with subsidised surface water and – generally – unsubsidised groundwater highlights the pressing need to review this kind of subsidy. In fact, most of this groundwater irrigation, following the criteria of full cost recovery, is not only viable but much more efficient and profitable than most irrigation using surface and subsidised water. This review must be undertaken on a case-by-case basis, guided by caution at the decision-making level, and consistent with the paradigm of sustainable development. In the most extreme cases of impoverished farmers, direct subsidies to the farmer would be preferable to general water subsidies. That type of direct, equivalent subsidy would avoid the financial impact on farmers, while at the same time encouraging them to save water by improving efficiency.

When applying economic rationalisation measures and processes to modernise irrigation, no generalisations should be made. Each case should be studied individually, from both the technical and ecological point of view, and particularly from the social one. Many traditional irrigation methods provide overall efficiency (including river ecosystems), albeit with scarce technical effectiveness in terms of running. In addition, many of these traditional irrigation systems constitute basic sources of subsistence for many communities, communally managed and highly socially effective, despite the fact that they clearly do not form part of the market's global dynamics. In these cases, irrigation flows cannot be considered as business-water but rather as life-water for these communities. To attempt to impose modernisation processes (both technical and economic) in these cases may be not only unwise but also counterproductive.

With respect to the increase in food prices, this could have positive rather than negative consequences. This price increase would be an incentive for sustainability, as it would reflect the real value of water scarcity. In so far as these possible subsidies would reflect the real value of water, not only as regards recovering investments, energy and running costs, but also the scarcity value of water itself and the resulting environmental costs, a rise in prices could provide incentive for a more reasonable and sustainable use of available resources. The key to the question is who would absorb the repercussions of these costs. Obviously from an international context such as the one imposed by the World Trade Organization (WTO), in which the poorest countries and weakest sectors of society are particularly vulnerable to a market in which the most powerful impose their rules, the impacts on these weakest sectors could be extremely negative. In this context, measures should be considered which would compensate for and avoid these negative impacts. Similarly, the marginalized social classes (impoverished consumers) in urban areas of most countries should be cushioned against these effects.

In the medium term, the solution to poverty must depend upon countries' *own development* and even their own *food sovereignty*, if *political sovereignty* is not guaranteed by international institutions through a worldwide democratic order.

We must not, however, forget the root of the problem if we wish to resolve it:

- Firstly, a large proportion of the problems of poverty are due to the destruction and break-down of the rural social fabric, further aggravated by the fall in international agricultural prices as a consequence of massive subsidies to industrial-scale agriculture in the most developed countries, particularly the USA and the European Union (EU).
- Secondly, the solution to poverty must be based on the development of these impoverished countries themselves which generally involves consolidation of their primary sector.
- Lastly, in many cases so-called food sovereignty becomes an indisputable objective, particularly in so far as political sovereignty is not clearly guaranteed by international institutions.

For all these reasons a true assessment is needed of the rural environment in general and of agriculture as a livelihood in particular, not only as regards accepting the real costs of irrigation water but also as regards agricultural production as a whole, including its underlying social and environmental values.

However, when the discussion centers on food scarcity, international *free market* scenarios and low crop prices, a major paradox becomes apparent. Scarcity, huge demand and low prices without doubt form a strangely paradoxical triangle. If we take a closer look, we shall surely find profound contradictions, as we shall later explain.

*To start with, we have to consider whether the solution to hunger, poverty and inequality can be successfully dealt with at all by using a free market strategy without environmental, social, or ethical rules which regulate it.*

## 3    THE ROLE OF THE MARKET IN THE STRUGGLE AGAINST HUNGER AND FOR SUSTAINABILITY: POSSIBILITIES AND TRAVESTIES

The present irreversible process of globalisation without doubt offers new approaches to dealing with world problems of hunger and poverty. The *trading of virtual water* is one such approach. However, the logic of the *free market* is riddled with falsehoods and must be controlled and regulated from ethical and sustainable perspectives.

Free trade could solve many problems of water scarcity by importing products whose production requires the intensive use of water (mainly agricultural products). In this way, countries with water scarcity would save their own water resources and put them to use for products with higher added value (thus increasing economic efficiency in water use). Very often, however, cutting down on – or even cutting out – staple food crops basic to national needs might imply dangerous strategic weaknesses, leading to dependence on foreign countries and thus endangering their security and independence.

Current unilateral international trends, as led by the present USA government, are weakening international institutions such as the United Nations (UN), as well as trust in these strategies. This unilateral approach and the lack of multilateral guarantees for countries, which could later face possible trade embargoes, is dissuading them from developing *virtual water trade strategies*.

If we look closely at the agricultural market's liberalisation policies imposed by the WTO, the aforementioned paradoxes and contradictions reach a scale difficult to understand and accept, particularly when we consider the protectionism of the most powerful countries (especially the USA and the EU), based on production and export subsidies.

The resulting combination of this liberalisation (ignoring ethical, social and environmental values) and the protectionist policies of the great powers is merely worsening the present problem. It not only fails to deal with the questions of hunger and poverty but is also destroying the traditional structure of agricultural production, accelerating the abandonment of rural areas, and encouraging the rapid increase in urban growth, as well as increasing these countries' dependence on international trade for their food supplies.

The key question is often not *how much is produced* but rather *how it is produced*. Traditional production methods are not usually the most efficient ones from an international market viewpoint but are generally extremely efficient in managing values such as *equity* and *sustainability*. As these methods are based on wise, deeply rooted community-based principles, they are often *highly eco-socially efficient.* It is clearly inconsistent, though, to expect international free markets to be able to manage these values without suitable legislation and regulating institutions.

To present the problem of hunger in the world as a simple physical equation of *production versus needs* not only falls short of its complexity but can also even lead to errors. These errors will become serious if the WTO free market model is taken on board as the way of managing this equation. This model, far from offering solutions, is in fact aggravating the problems of hunger and environmental unsustainability in the world from the structural standpoint.

With these comments, I do not, however, intend to deny the value of the technical arguments presented by Prof. Frederick, who focuses on improving the efficiency of irrigation and revitalising rain-fed crops on dry lands. Nevertheless, I believe that these changes should be made within a new international socio-economic context which differs greatly from the present one, and this is not reflected in the *Towards Sustainability* scenario proposed by Professor Frederick.

Furthermore, beyond resolving the serious contradictions implied in the USA and EU's protectionist agricultural policies, we must question whether or not merely allowing developing countries fairer opportunities to sell their agricultural products on the world market would automatically lead to better conditions for the majority of rural poor. The general answer must be *NO*, unless several important questions, such as the following, are clarified first:

(1)  Who currently has or will have access to international markets in developing countries in the future? In most cases, only the richest farmers and large companies reach these complex markets.
(2)  What are the political conditions of these countries? Democracy, human rights, public information and protection for the social and environmental rights of communities.
(3)  The impact on rural areas when industrialised agriculture is imposed, taking into account the destruction of traditional agriculture and the social fabric, leading to fragile and unbalanced development with a dangerous dependence on single crops.

Following *free market principles*, the WTO has left the social and environmental aspects of production to one side. This gives an advantage to large-scale farmers and companies and threatens traditional livelihoods and the rights of communities.

Hence, it is imperative that the WTO should recognise that social and environmental values are safeguarded by traditional production methods, or other modern techniques based on ecologically friendly agricultural methods. It is essential to set down regulations to protect these livelihoods and forms of socio-economic organisation, which are so highly efficient in reaching the goals of eco-social sustainability, and are so often formally proposed and accepted in international forums. International institutions such as the UN must be strengthened, so as to guarantee basic human rights such as the right to food, health or drinking water, over and above the rules of the *free market* or, better still, to guarantee these rights as regulations and limits for the *free market*.

International agricultural prices should be increased, as is shown in the sustainable scenario presented by Prof. Frederick. This would mean following the laws of the market, whereby what is scarce but necessary becomes more expensive. This market must, however, be suitably guided by ethical, social and environmental objectives (eco-social sustainability), as has been the case in the developed countries' approach to attaining a high level of social welfare. Food production must become more profitable, in the general context of a worldwide economy, thus bringing income improvements essentially for small scale producers and family farms. This

would be an important impulse for the rural economy, increasing food production and facili-
tating efficient modernisation processes. Defending the opposite – i.e. the present situation
of pitiful prices under pressure from subsidies and export aid for USA and EU farmers as a
strategy to fight against world hunger – is simply sarcasm.

## 4   ENVIRONMENTAL SUSTAINABILITY AS A KEY FACTOR IN FOOD PRODUCTION

The problem of the sustainability of ecosystems is often presented as restrictions which hinder
economic development and impede solutions to serious issues such as world hunger. This
analysis is as perverse as it is mistaken. In Professor Frederick's own documents, he presents
significant data which questions this so-called contradiction between *environmental sustainability and economic development*, with special reference to the ecological disaster of the
Aral Sea.

Although this case is probably the most widely-known, it must be pointed out that it is
merely one more example of a wide range of problems concerning the impoverishment of
agricultural and biological productivity, as a consequence of unsustainable intensive practices. These practices are deteriorating the quality of soil, water and other natural resources in
the medium and long-term. Salinisation of lands due to new irrigation is particularly worrying. This problem already affects about 20% of the irrigated land surface (WCD, 2000) and,
coupled with processes of deforestation and the abandonment of traditional agricultural areas,
is opening up the floodgates to erosive phenomenon which irreversibly complete the cycle of
desertification in many regions.

We must also consider the mid and long-term consequences of extensive contamination caused by modern agricultural techniques based on the widespread use of pesticides
and fertilisers. These impacts, together with drying wetlands and serious effects on deltas and
mangrove swamps, are causing a breakdown in the natural production of protein (fishing and
hunting) at land and sea. These impacts are often the result of complex synergies, which make
precise cause-effect analysis difficult. This leads to evasion of responsibilities. Nevertheless,
the combined result of all the impacts resulting from the multiple breakdown of aquatic and
soil ecosystems is devastating. What is even worse is the way in which this is breaking down
the traditionally stable relationships between many communities and their natural environment, provoking the destruction of the rural fabric. Perhaps the most serious indirect impact
of the destruction of rural livelihood is the impact of the breakdown in traditional models
of food production (agriculture, forestry, fishing and hunting, etc.), firmly integrated within
these societies and, as previously stated, extremely efficient in the resolution of social and
environmental problems (*high eco-social efficiency*).

In this sense, the linkage between water and agriculture of which Prof. Frederick speaks,
quoting arguments and conclusions from the *World Commission on Water for the 21st Century*
(2000), must be characterised as a *relationship between agriculture and environmental sustainability* involving both the sustainability of aquatic continental and littoral ecosystems and
the sustainability of land. This is, without doubt, the consistent view from which Rosegrant,
Cai, and Cline (quoted by Professor Frederick) attempt to find "an alternative path that is
sustainable and one which will meet growing water and agricultural demands": the path that
must be conceptualised as *eco-social sustainability*.

I would like to express my strong agreement with the re-evaluation and development of rain-fed agriculture in Professor Frederick's *Toward Sustainability* scenario, as being a key factor. It is essential to do away with the myths surrounding irrigation and recognise the value of rain-fed agriculture, due to its greater agro-environmental suitability for sustainability scenarios.

With reference to the sustainability of aquifers, although it would often be necessary to review and lower the scale of water drawn, I have serious doubts that the potential of the integrated use of ground and surface waters has been correctly quantified. The lack of data on aquifers and the priority which most governments place on the use of surface waters via large hydraulic works, make *integrated use of surface and groundwater* one of the options with greatest potential for development in many areas.

## 5   THE NEED FOR A *NEW CULTURE* AND THE SIGNIFICANCE OF EDUCATION

Certainly, the limited space of a scientific paper does not permit Professor Frederick a detailed explanation of the methodology used to design the scenarios presented. All the same, I would still express a certain degree of scepticism regarding the extensive world figures and data often used in the design of this kind of long-term scenario. These statistics tend particularly to leave aside both scattered traditional economies, which have not been integrated within national and international markets, and also underground economies. This problem has much greater relevance in poor or developing countries, especially if the questions we wish to study are linked to underdevelopment, poverty or food sufficiency. In these cases, key elements and data overlooked may play a more decisive role in both the diagnosis and possible solutions.

One of the key issues for the medium and long-term future is undoubtedly the cultural and educational factor. The world is becoming increasingly more complex and globalised, and large-scale solutions will rarely be successful if they are not designed and applied through participation-based processes involving the communities affected in each region.

The very evolution of populations, which is usually presented as an exogenous and inexorable element, determining the forecasts of any imaginable future scenario, will depend on cultural and educational factors which must be tackled. On this point, the educational and cultural level of women is considered to be a particularly essential factor in this respect.

The change in eating habits, which is currently pushing meat consumption (with high costs for *virtual water*) above healthy limits in many countries, may indeed alter agricultural strategies and their respective water needs.

Amongst the lines of action needed to develop a collective intelligence based on a *New Culture of Water and Sustainability*, the need to develop suitable policies for regional and urban planning stands out as a salient feature. I wonder if the models and scenarios of Rosegrant, Cai and Cline, quoted by Prof. Frederick, include this factor.

*How to produce* can and should be as important, or more, as *how much to produce*, since *eco-social efficiency* can, and does, in fact, differ greatly, depending on present and possible different models and lifestyles.

## 6   THE ROLE OF TECHNOLOGICAL DEVELOPMENT

Obviously one of the key elements implied in the *more crops per drop* slogan that stands out in Prof. Frederick's work, is to be found in the application of new technologies. Nevertheless,

I believe it would be useful to emphasise the significance of technological development in the *qualitative dimension*, beyond its undoubted importance for the increase in *quantitative* efficiency emphasised in this slogan.

The new technologies of semi-permeable membranes, combined with energy recovery techniques in desalinisation plants, are leading to increasingly lower energy costs. These costs are presently between 0.40 and 0.50 €/m$^3$, while energy requirements are around 3.5 Kwh/m$^3$ (Arrojo, 2003a). This is revolutionising the question of viable alternatives, especially as regards supplying urban coastal areas. The long term implications of these technological advances are even greater if we bear in mind the high quality and reliability of the drinking water thus obtained. It should be pointed out that already now, and more and more in future forecasts, the most conflictive situations will be linked to growing urban population density in coastal areas.

The decreasing cost of these new technologies means that desalinisation can compete advantageously with the regulation and long-distance transportation of surface water via large-scale infrastructures, the cost of which shoots up above 1 €/m$^3$ for distances over 500–600 km (especially if environmental and even security costs are accounted for). This offers alternatives which reduce the pressure and impacts on continental aquatic ecosystems, as are now being embarked upon in Spain, after large-scale water diversion from the river Ebro was rejected and desalinisation of briny and sea water has been made a priority.

This technological development is in tune with the growing priority of the *quality factor* rather than *quantitative problems*, which appear to prevail in Prof. Frederick's work. Already today, and in the future, the most pressing problems of water in the world concern quality more than quantity, although the two aspects are certainly linked.

*Avoiding contamination*, by both ecological production models and suitable technology for the treatment of returns will become the priority, surpassing *the polluter pays* principle. From this new perspective (which gives pride of place to the preservation of quality), the conservation of ecosystems and of their natural capacity for regeneration becomes a priority. This is the logic which tends to be imposed in advanced countries. The possibility of increasing water reuse must enable much greater availability of resources without increasing the pressure on ecosystems.

The technologies involving re-use of urban water for agriculture or even for secondary uses (lavatories, gardens, car-washing, washing machines, etc.), through a double network (easy to implement in new urban areas) provides raw water of sufficient quality at costs below 0.12 €/m$^3$ (Arrojo, 2003b).

Perhaps the greatest advantage of these new technologies resides in its modularity and flexibility as opposed to the rigidity of the major traditional hydraulic systems. It is fundamental that we should fit quality water production to the real needs of each moment using modular strategies, and once and for all abandon grandiose, oversized public works which are rarely sufficiently justified and always backed by declarations of somewhat suspect general interest. Through such modularity and flexibility, greater transparency in cost allocation would be achieved, bringing with it an easier introduction of *demand management* strategies.

## REFERENCES

Arrojo, P. (2003a). Spanish National Hydrological Plan: reasons for its failure and arguments for the future. *Water International*, 28(3): 295–303.

Arrojo, P. (2003b). *El Plan Hidrológico Nacional: una cita frustrada con la historia*. RBA Editores-Integral, Barcelona, Spain.

Arrojo, P. & Naredo, J.M. (1997). *La gestión del agua en España y California*. Colection Nueva Cultura del Agua, 3. Editorial Bakeaz, Bilbao, Spain: 192 pp.

WCD (World Commission on Dams) (2000). *Dams and development. A new framework for decision-making*. Earthscan Publications Ltd., London and Sterling, UK and USA.

World Commission on Water for the 21st Century (2000). A Report of The World Commission on Water for the 21st Century. *Water International*, 25(2), June: 284–302.

IV

Virtual Water

CHAPTER 8

# Virtual water – Part of an invisible synergy that ameliorates water scarcity

J.A. Allan

*School of Oriental and African Studies (SOAS), London University, UK*

ABSTRACT: The purpose of the chapter will be to demonstrate that there are a number of economic processes that have the capacity to ameliorate local water scarcity. A number of arid and semi-arid regions and economies encountered water scarcity in the past thirty years. Many more will encounter water scarcity in the next three decades. The analysis will review briefly a threefold synergy of ameliorating processes – first, the contribution of unaccounted water in soil profiles [known as effective rainfall to hydrologists and engineers] to the production of food staples; secondly, the global role of virtual water in ameliorating water scarcity in dry regions; and thirdly, the impact of socio-economic development on water management options. All three processes have the characteristics of being economically invisible and politically silent. But they ameliorate conflict in the easily politicised domain of water allocation and management. Their impacts are determining with respect to solving local water deficits. The role of a fourth ameliorating technology, desalination, will also be examined.

The status of the world's water resources will also be reviewed and estimates of the proportion of global water resources, which become involved in virtual water transactions will be provided. The main focus of the chapter will be on the extent to which virtual water in trade has successfully met the past and current needs of water deficit regions as part of the threefold synergy – soil water, virtual water and socio-economic development. There will be a very preliminary evaluation of the role of virtual water in meeting the future needs of water scarce regions during the demographic transition of the 21st century.

It will be shown that the water, food and trade nexus is not easy to model because of the dynamics of the political economies in the North and the South. Water sector policy-making is subject to evolving discourses, which can easily de-emphasise the underlying environmental and economic fundamentals. The chapter will conclude that invisible and silent virtual water in the invisible threefold synergy will provide the *big water*, that is the food-water for water scarce regions. Its invisibility and silence will, however, have the effect of attenuating the pace of water policy reform. Both reforms addressing water use efficiency and those giving consideration to the environmental services provided by water will be attenuated. A range of demographic, economic, social and political theory will be used to frame the discussion.

Keywords: *Virtual water, soil water, socio-economic development, the threefold synergy, global water-food-trade nexus, incidental benefits, water policy, political economy*

*Not all waters are equal: some are more evident than others.*

## 1  INTRODUCTION: SOLUTIONS TO CLOSURE AND WATER SCARCITY

The title of the programme/book to which this chapter is a contribution asks the question is there a current or a future global water crisis? Is there enough freshwater – surface and groundwater, and soil water to meet the demands of current and future populations? Within the question there is an implied additional question – are some regions and some river basins facing current water crisis and will they face irreversible water crises?

The purpose of the chapter will be to show that there is sufficient freshwater and soil water in the world to meet current and future water needs if we accept the demographic predictions in currency since the *UN World Population Conference* in Cairo in 1994. The low estimates of the world's population in the late 21st century are about 8000 millions. High estimates are about 11,000 millions. The present world population is about 6500 millions requiring about 6500 km³/yr of freshwater and soil water. The future population estimates imply annual requirements of between 8000 and 11,000 km³ of freshwater and soil water.

The conceptual and analytical contribution of the chapter, in a review of the future adequacy of global water – the subject of the book – is to show that trade in commodities has already achieved a pivotal position in enabling the comparative advantages and disadvantages of regional water endowments to be balanced. A commodity such as grain, one of the major traded commodities, requires for its production amounts of water 1000 times the weight of the commodity. An economy receiving a tonne of grain does not have to face the economic, and more important, the political stress of mobilising 1000 tonnes (1000 m³) of water. The future role of virtual water in remedying the increased regional water deficits will be proportionately more important as populations increase in water deficit regions.

Pessimists such as Lester Brown (2003) and Sandra Postel (1999), and Postel & Richter (2003) argue that the world does not have the water resources to meet the needs of future populations. They point to the strains facing the water scarce Middle East and especially the situation in heavily populated South Asia and China where water quantity and quality are worsening problems and groundwater resources are under severe pressure.

Optimists such as the author of this chapter (Allan, 2001), and most of the international agency staff and many scientists intimate with demographic and water resource problems, (World Bank, 2003; IFPRI, 1995, 1997; Dyson, 1996; Rosegrant *et al.*, 1995; Rosegrant & Cai, 2002; Brichieri-Colombi, 2004) judge the challenge of providing enough food to meet global needs as addressable, albeit with problems in sub-Saharan Africa. These problems are attributable to poverty rather than to water poverty.

It will be shown that one of the important consequences of the existence of the trade associated with the concept of virtual water is to make it possible for water users, water policy-makers and the general public to embrace simultaneously optimism and the pessimism. It has been noted that "pessimists are wrong but useful; optimists are right but dangerous" (Allan, 2001). Optimists and pessimists interpret the underlying fundamentals differently. In practice the underlying environmental and economic fundamentals are rarely effective starting points from which to campaign for changes in water policy. Governments and peoples in severely water scarce regions are in denial about their water resources and their policy options.

An understanding of the politics and local history of water policy-making processes is a much better starting point for a water policy reform campaign. This is not a new idea. Marx pointed out the determining role of politics. The historically evolved *abstract* domain of politics overwhelms the information coming from science and sub-optimal markets in what Marx termed the *concrete* economic domain (Fine & Saad-Filho, 2003). The concrete provides the foundations for the (political economy) superstructure, where those with power make allocative decisions. Evidence on the concrete is only very selectively included in policy-making discourses. Scientists rarely draw up or impact policy-making agendas. Nor are they normally much involved in developing politically controversial allocative policy. They can help to raise awareness of how things are. But they normally have to struggle mightily to bring about significant shifts in perception.

One of the reasons that water science and water professionals have limited purchase on policy-making is because there are economic processes outside the water sector and its water-sheds, which solve water scarcity problems. The analytical tools of hydrologists, environmental scientists, and hydraulic engineers are not effective in these *problemsheds* which lie beyond the watersheds of the water sector. The agenda of the meeting to which this chapter was a contribution is evidence of the norm amongst water professionals and scientists that water problems are mainly viewed to be the concern of professionals and scientists associated with water and the productivity and quality of water.

The main reason that water professionals have limited influence on water policy discourses is their limited experience of advocating and implementing water policy reform in the late 20th century. Before the 1970s water scarcity was little in evidence. Technical, supply management measures were economically appropriate and had not yet been identified as environmentally hazardous. By the 1980s the negative impact of supply management water policies on the environment required that water professionals tell the truth to power. Technical remedies were no longer always appropriate. By the 1990s the message that water should be valued required that politically costly re-allocative measures be part of the unwelcome truth to be digested by the powerful. In practice water scientists and professionals did not tell truth to power. That controversial task was undertaken by environmental civil movement groups. Many water scientists fell in behind the campaign and have reinforced the environmental project.

After two or more decades of reorienting their own approaches to managing water scientists and professionals are still undecided about how to cope with the politics of water policy-making. For centuries, even millennia, water professionals solved the water problems of the societies managed by political elites. They did this without much consideration for the value of the environmental services of water or of the economic value of the water inputs to society and agriculture. Recognizing these two issues as new fundamentals was challenging. Conveying their unavoidable importance to water users existing in rural poverty and to the politicians running their economies has proved to be very difficult. Water users with few or no alternative livelihood options to irrigated farming, and water policy makers wanting to avoid paying high political prices associated with re-allocative reforms cannot manage water demand. The outcome was to take water policy discourse into realms of painful uncertainty.

Water professionals had little experience of a world where problem solving was social and political rather than technical. The skills needed to operate in the political domain are unfamiliar. They require the capacity to deal with ambiguity and uncertainty. In another policy domain, Teller, the nuclear physicist, had no such scruples and knew how to influence politics. "He knew that it was more like magic than logic, and in spinning his spells, he was both dishonest, wasteful, and at times dangerous. He never saw this as evil or immoral. He once corrected Oppenheimer's famous comments that 'scientists have known sin'. As far as Teller was concerned 'scientists have known power' and what it demands." (Goodchild, 2004).

Water scientists and water professionals are still unsure whether they should know power, and remain uncomfortable telling new truths to power. Activists from the green movement have been much more successful in shifting the debate and water policies than the science community and water professionals. The activists are regarded by many water professionals and scientists as *dishonest, spinners of spells* and *at times dangerous*. But they have influenced the discourse and associated policy reform. They have not achieved their goals but they have had a very significant discursive impact on the ideas in currency and especially on the valuation of the water services in the environment.

## 2   SOIL WATER AND VIRTUAL WATER: BRIEF DEFINITIONS OF TWO NON-EVIDENT WATER SECTOR PROCESSES

"Things do not always appear as they are." (Fine & Saad-Filho, 2003: 4, quoting Feuerbach & Marx).

The reason that food production and water have been emphasised thus far is because water for food is the dominant use of water. About 90% of the water needed by an individual is devoted to food production. The remaining 10% is needed for drinking, domestic use and the support of non-agricultural livelihoods.

The three departures from conventional approaches to water allocation and management integral to the message of this chapter are the following. The first generally ignored feature, though hydrologically and economically fundamental, is soil water. The second is virtual water. Virtual water is a new concept to the water sector. But the term is merely an example of Ricardo's powerful, two century old, notion of comparative advantage. Wichelns (2004) has added a purist economics definition insisting that virtual water is an example of *absolute advantage* because it does not have embedded in it *optimal* production and trading strategies. He recognizes that "the virtual water metaphor addresses resource endowments, but it does not address production technologies or opportunity costs. Hence, the metaphor is not analogous to the concept of comparative advantage. The metaphor can be helpful in motivating public officials to consider policies that will encourage improvements in the use of scarce resources, but comparative advantages must be evaluated to determine optimal production and trading strategies".

Prior to human consumptive water-use natural water resources supported only biological and hydrological systems. All the consumptive use of water by human populations has to be taken from the diverse environmental water resources available to their political economies. Available water includes surface waters, groundwater, soil water and atmospheric water. Surface waters, groundwater and soil water are the waters that the human population use for their economic, social and amenity purposes. Surface waters and groundwaters are taken into account by water scientists and water policy-makers insofar as they can be quantified. Soil water is generally ignored. *Things do not always appear as they are.*

### 2.1   *Soil water: the first invisible*

Soil water is not accounted in national water budgets. The Shiklomanov (2000) research group is an honourable exception in that soil water appears in all its datasets. Nor does soil water figure much in the statistics of the international agencies. These datasets focus on freshwater at the surface and in the aquifers (FAO, 2003; FAO, 2004 – Aquastat; World Bank-WRI, 2003). The Aquastat database does include a reference to the rainfall that contributes to the generation of national water resources. But it is not possible to derive an estimate of the soil water element of a national budget.

This discussion will not emphasise the water services provided by the water in the natural environment. The absence of such discussion does not mean that such water services are regarded as unimportant. There is not space here to address the topic adequately.

## 2.2 *Virtual water: the second invisible*

Virtual water has been conceptualised and defined elsewhere (Allan, 2003). The concept has also been critiqued and debated (Merrett, 2003). Virtual water is the water associated with the production of commodities. Research has focused on the water required to produce grain staples such as wheat. It requires $1000 \, m^3$ of water (1000 tonnes) of water to produce a tonne of wheat. Importing a tonne of wheat has a very favourable impact on the economies that endure water deficits. It follows that virtual water should have a place in reviews of global and regional water resources and in strategizing water policies. The water sector and the freshwater in a nation's rivers and groundwaters are not a sufficiently comprehensive basis for quantifying, analyzing and optimizing the allocation and management of water resources.

*To see things as they are* it is necessary to review water resources and water transferring processes comprehensively. Water allocation and management is a global issue and a global challenge. The same is true of many natural resources, for example oil. The global hydro-economic system can solve the problems of water scarce regions. Non-hydrological systems, such as trade, are as problem solving as the water and water flows within watersheds captured by hydrology. The proportion of the water re-distribution challenge successfully addressed by virtual water processes is already impressive. It has the potential to be of even greater significance. Hoekstra & Hung (2002: 25) estimate that $695 \, km^3$ of freshwater and soil water entered international trade in virtual form out of the $5400 \, km^3$ of water used to produce crops in the 1995–1999 period.

Livestock products are particularly water intensive. A study of the virtual water content of the 1995–1999 trade in livestock products (Chapagain & Hoekstra, 2003a, 2003b, 2003c) calculated the figure to $336 \, km^3/yr$. Most livestock products are raised on soil water. But a proportion of such products are produced from irrigated fodder. By adding the annual global virtual water figures for crop production – $695 \, km^3$ – to the $336 \, km^3$ for livestock products, we get a total global figure of $1031 \, km^3/yr$. There is some double accounting in this number but the calculation nevertheless does confirm the very important part played by virtual water, mainly derived from soil water, that enables sustainable water resources management across the globe.

All the volumes of water discussed above are situated in rising long term trends, although there is a current temporary leveling off in the trade in major grains. The leveling off is the result of improved efficiency as measured by returns to water in both the Northern and the Southern economies (Dyson, 1996, 2001).

Meanwhile estimates of current global freshwater use in irrigated farming suggest that the total mobilized is about $1430 \, km^3/yr$ – that is 26% of the total water used in crop production (Rosegrant & Cai, 2002). The same authors suggest that by 2025 the use of irrigation water could rise to $1480 \, km^3/yr$. Clearly the virtual water solution, associated with 13% of global water used in crop production is significant as a problem solving process. The trade in livestock products adds to this global percentage. Virtual water is especially significant in addressing those problems, which local technological solutions cannot address in water scarce regions. Just as important as its proportion of total water use in crop production and its capacity to solve otherwise un-addressable problems in water scarce regions, is its flexibility. This quality of flexibility will be discussed below.

### 2.3    *The concept virtual water*

Virtual water is a powerful concept as well as a powerful if invisible operational solution, in that, like comparative advantage, it is both an intensive and an extensive concept (Weber, 1904, 1917). Such concepts help us communicate, even if an intensive/extensive notion is at first unwelcome to water policy-makers and especially to politicians riding the politicized water sector tiger. First, in the case of water, the concept links the *intensive* process of combining water in for example the agronomic and technical processes of food crop production. Secondly, it is an *extensive* concept that it links water with the transactions that move and trade water intensive commodities such as grain. Without such a concept the activities of production and trade have to be conceptualised separately as recommended by Merrett (2003). Happily the concept of comparative advantage is a useful precedent in commending the integration of the intensive and the extensive processes to professional economists.

This intellectual inspiration of comparative advantage led to the Heckscher-Ohlin (H-O) Theorem of trading factors of production (Hakimian, 2003). The notion is something of a mega-intensive/extensive trade theory. The theory has intuitive relevance to water, as water can readily be defined as a factor of production. Although its role in economic systems can be very difficult to track. Unfortunately attempts to deploy the H-O Theorem have at best met with mixed success. The absence of success is partly the result of inadequate datasets. Even more important are other factors than resource endowments which impact directions and levels of trade. For example economies of scale, trade policies and demand factors can be important (Bowen *et al.*, 1987; Krueger, 1977; Harkness, 1978, review extensions and empirical tests; also Lawler & Seddighi, 2001; Krugman & Obstfeld, 1991).

One of the outcomes for the author of looking at the literature on trade in factors of production is to deter too much commitment to quantifying virtual water processes. The work of both Earle (2001) and Hakimian (2003) shows that while the metaphor remains convincing it is extremely difficult to emerge with persuasive numbers when attempting empirical confirmation.

The same is true, but to a lesser extent, of the preliminary work on water by the scientists at the Institute of Hydraulic Engineering (IHE) in Delft (Hoekstra & Hung, 2002). Their research on virtual water came up with global numbers that fit quite well with what specialists expect to see. 5400 km$^3$/yr of water are estimated to be used in crop production according to the heroic use by Hoekstra & Hung (2002) of the best available global and national datasets. The figure makes sense in relation to the estimates of global population of say 6500 millions at 2000. A rough estimate of annual domestic and non-agricultural livelihood water for the global population would account for a further 650 km$^3$. The food/agricultural water plus the domestic/non-agricultural livelihoods water would then total 6050 km$^3$/yr. This total is remarkably close to the number that a quick and dirty estimate of global water consumption based on a consumption of 1000 m$^3$/yr per person.

The Hoekstra and Hung work is of particular value because it has provided numbers that confirm the big water/small water idea. They show that big water, the water used to produce food and other crops, accounts for about 90% of the water used by the human population. The number remains counter intuitive for rural communities where populations are rising. It is especially challenging to rural societies living in poverty where non-agricultural livelihood options are absent. The idea is seriously unacceptable to politicians managing economies where creating more jobs per drop is difficult.

The national numbers can only be as good as the datasets on which they are based and these are not yet reliable. The IHE work has been extremely useful, however, in that it has calculated a very much needed *first approximation* of global and national virtual water transactions. Just as important as the quantification is the interest it has attracted in a diverse range of water scientists and professionals in all parts of the world. That interest has turned into a significant epistemic event, reflecting a new level of engagement by scientists and professionals involved in water science and water policy-making in both the North and gradually in the South (see Hoekstra, 2003; Chapagain & Hoekstra, 2003a, 2003b, 2003c; Haddadin, 2003; Zimmer & Renault, 2003; El-Fadel & Maroun, 2003).

A very important aspect of the both the productive water related transactions and the trading transactions discussed in this section is that it is soil moisture rather than freshwater that is used to produce the bulk of the food that is traded (Hoekstra & Hung, 2002). It is soil water, mainly in humid temperate zones, that ameliorates the water deficits of regions enduring freshwater and/or soil water deficits, in the arid and semi-arid tropics. Farmers and traders can mobilize a process with a global reach to achieve a remedy, of which hydraulic engineers cannot dream.

### 2.4   *The value of virtual water*

The issue of the value of virtual water has been usefully discussed by Renault (2003). He pointed out that water can be differently perceived and differently valued. The perception of the value of water in economies that export water intensive commodities can be very different from that perceived in economies that import water intensive commodities. The opportunity value of the water might be close to zero in a soil water rich environment such as the temperate regions of North America and Europe. The equivalent embedded water in the water intensive commodities imported into the water scarce economies of the Middle East would be much more valuable. In some circumstances the marginal water could be desalinated water with value of between US$ 0.50 and US$ 1.00. The valuation of virtual water is an important issue but there is no space to discuss it here beyond drawing attention to the importance of the economic context in which such valuation is made.

## 3   SOCIO-ECONOMIC DEVELOPMENT: A THIRD INVISIBLE SOLUTION FOR THE WATER SCARCE

Distant soil water and virtual water trading processes have, until recently, not been recognized as being significant in addressing the problems of the water scarce. They have not been a part of explanations of *how things are* for the water scarce in for example the Middle East region. The concepts remain very much part of a minority outsider discourse even in the Middle Eastern epistemic science and policy communities associated with water allocation and management.

The third non-evident process that contributes to the amelioration of water deficits is socio-economic development. The achievement of more *jobs per drop* rather than more *crops per drop* has been key to the successful amelioration of water scarcity except in sub-Saharan

Africa. This ameliorative process is not even on the analytical agenda of most water scientists and water policy-makers in the South.

Socio-economic development associated with the strengthening and diversification of economies has achieved impressive levels of allocative efficiency in association with virtual water related processes. Overpopulated economies and overpopulated regions of some national economies in dry regions have become diverse and prosperous despite being extremely water short. The water poverty of a region has not determined that a region would endure poverty. The very desirable allocative efficiencies achieved in the water sector in the past half century have, however, been incidental benefits resulting from the socio-economic development.

These allocative water efficiencies were not part of water policies. They were *hidden-hand* (after Adam Smith) processes. The hidden hand enabled the perversely hydrocentrically (Brichieri-Colombi, 2003) inclined water policy-makers and their water professionals to achieve sustainable economic outcomes. Job creation in industry, services and the public sector silently and invisibly reallocated water at a scale beleaguered politicians yearn for. Job creation associated with economic diversification has proved to be a mighty demand management tool. In addition job creation outside agriculture has generated incomes and revenues with which political economies have been able to access virtual water in the global system. Economic transformation achieved in the East Asian economies, and in for example Israel (Allan, 2001: 249), demonstrate how economies can achieve water security through economic diversification.

Without economic diversification there would be no trade in the commodities associated with virtual water. Without trade in these commodities there would be no demand for the soil water in water surplus regions. Economic diversification enables a water scarce economy to articulate its demand for water in a system that can provide it, namely the trade in water intensive commodities.

Virtual water transactions and diversified job creation have enabled five decades of relatively conflict free global co-evolutionary transition from water abundance to water scarcity. For the past 25 years potential regional water scarcity has been managed within sustainable regional and global water management regimes. The very effective invisible *threefold synergy* of soil water access, virtual water processes and socio-economic development has provided a very effective and a politically stress free regime.

## 4    THE TIMELY AVAILABILITY OF THE FLEXIBLE THREEFOLD SYNERGY: SOIL WATER, VIRTUAL WATER AND DEVELOPMENT

The worsening water scarcity experienced in some regions of the world in the second half of the 20th century was encountered at a fortuitous moment in world economic history. The first good fortune of the water scarce was the availability of a comprehensive global trading system in food. This is not to say that the food trade is new. The system has been operating effectively for millennia. Rome in antiquity was fed by grain and other commodities raised in North Africa. Comparative advantage is not static. As Wichelns (2004) argues in each case it is related to the production systems and the opportunity costs that obtain. By the late 20th century economies such as Egypt, the main North African food exporter in antiquity, were importing nearly half their food. Two thousand years ago Egyptian agricultural exports to Italy were substantial. Semi-arid economies facing water scarcity encountered their scarcities at a moment in history when, fortuitously, the world's temperate regions had become highly industrialized. They produced substantial crop surpluses throughout the second half of the 20th century.

A second fortuitous, and invisible, non-water sector, process assisting the water scarce since the middle of the 20th century has been the economically perverse agricultural policies of the European Union (EU) and the USA. Their production and export subsidies on wheat for example, ensured that half-cost grain was available on the world market. The pivotal global grain importers, the Middle East and North Africa economies and Japan, were both well able to purchase imports. They have been significantly advantaged by the EU/USA subsidies. Together they accounted for about 150,000 Mm$^3$ of virtual water in crops traded in the global system in the last five years of the 20th century. These major importers accounted for 23% of the total. The author is at the same time aware of the negative economic impacts of low world prices for grain on economies with very low levels of Gross Domestic Product (GDP) per head, especially those in sub-Saharan Africa. These economies enjoy no potential comparative advantage except in crop production. Their farmers are seriously impacted when they cannot obtain a local market price for their crops because of imports of subsidised staples.

### 4.1   *Variable water demands: the threefold synergy is also flexible and responsive*

Virtual water is part of an important *threefold synergy* – soil water, virtual water and socio-economic development – that enables communities, nations and river basins to access sufficient water to meet their variable water needs. A further very important feature of the threefold synergy is its unmatched flexibility. Since the 1970s it has enabled the extension of a form of supply management miracle across the globe. The synergy has been capable of meeting varying regional and emergency food/water demands. Variable demand is normally associated with drought and its consequences. The threefold synergy can successfully augment policies that in themselves cannot achieve local sustainable water allocation or address water scarcity emergencies.

For at least three decades the threefold synergy has proved its capacity to respond to systemic trends associated with increased demands associated with rising populations – for example in the Middle East. The system has coped with the *closure* of the Middle East and North African water systems and the resulting increased demands for food imports. It has also coped with the variable demands coming from the former Soviet Union and the Russian Federation. The biggest variable demands have come from China. These demands have occurred despite China's own extraordinary increases in production since 1961. The fluctuating demands for grain imports from China have also been accommodated. That there is a mix of local and international solutions is revealed in the decline in volumes of grain entering international trade in the last five years of the 20th century. Despite the progressive increases in the world's population the volumes of grain traded have been declining since the late 1990s.

### 5   THE THREEFOLD SYNERGY AND THE ENVIRONMENT

A further benefit of the threefold synergy is its amelioration of arid and semi-arid regions' competition for water between agricultural use and environmental services. By reducing the pressure on water resources in water scarce regions the threefold synergy has alleviated, or at least it has had the potential to alleviate, progressively higher demands on the freshwater and the soil water in the environments of the water scarce. The availability of the threefold synergy has enabled demands on local surface and ground waters to level off and in some cases to be reduced. This is especially the case in Northern semi-arid economies – for example in Israel. Water has been returned to the environment in Northern semi-arid economies such as those of the USA western states to reinstate, to some extent, the environmental services provided by natural hydrological systems (Allan, 2001: 146–148). In the South, the availability of virtual water has not yet had the same impact.

The shift towards precautionary and green water policies in the North has not been a response to the economic processes that made it possible. The advocacy of the green movement has been the important factor in achieving environmentally sensitive water policy-reform in the North. In this Northern discourse the new resource managing circumstances afforded by the threefold synergy has enabled pressure on the water in the environment to be alleviated in these Northern political economies.

The green social movement is poorly developed, however, in most Southern political economies. In Southern semi-arid economies the apparent amelioration of scarcity by the threefold synergy can play a negative, rather than a positive, role in strategising the allocation of water *vis-à-vis* the environment. Old practices and policies that are economically nor environmentally appropriate can be left in place because to reform them would only be achieved at unacceptably high political prices. Paying such prices can be avoided in the *apparent* water security provided by the invisible threefold synergy. Water abstraction practices associated with the livelihoods of poor farmers with no alternative job options tend to remain in place, including those, which damage the water environment (Allan & Olmsted, 2003).

## 6   CONCEPTUALISING THE SOIL WATER, TRADE AND TECHNOLOGY ELEMENTS OF THE THREEFOLD SYNERGY

Figure 1 is helpful in conceptualizing a number of essential ideas that could usefully be adopted by the water science community as well as by the diverse epistemic group associated with making and advising on water policy.

The first idea is that of *big water* and *small water*. The big water is the 90% of *freshwater and soil water* needed to produce the food needs of human populations. Small water is the 10% of freshwater – *it can only be freshwater* – needed to provide drinking, domestic and municipal water as well as the water needed for non-agricultural livelihoods.

Secondly, Figure 1 is useful in conceptualising the role of soil water. Two political economies with similar populations – Egypt and the United Kingdom – located in very different environments, can be compared. The diagram indicates the significance of their respective water endowments. Egypt has to use almost all its average annual availability of between 55,000 and 60,000 Mm$^3$ of freshwater in the Nile River. The figure includes reuse. It has negligible soil water. Egypt accesses virtual water to meet its rising water deficit. The UK uses about 15,000 Mm$^3$ of surface and ground waters. Soil water is not negligible in the UK water budget. About 25,000 Mm$^3$ of local soil water and access to virtual water enables the UK to meet its food-water needs. In both cases the virtual water calculus is a net figure in that both economies export food commodities as well as importing them.

The calculated water footprints of both economies are quantitatively very similar and negative (Hoekstra & Hung, 2002). The concept of the water-footprint is a development of the much more complex concept of the ecological footprint and captures the extent to which a political economy needs to access resources beyond its boundaries or alternatively the circumstance where an economy can export its natural resources (Wakemagel & Rees, 1996).

In addition Figure 1 is useful in conceptualizing and highlighting the role of the two important solutions for any economy entering water deficit circumstances. First, desalinated water, and secondly, virtual water. The ability to access both of these solutions depends on the performance of the economy. Both the production of manufactured water and access to virtual water depend on the capacity to mobilize financial resources. In the case of water manufacture it is necessary to service the capital investment and the operating costs of

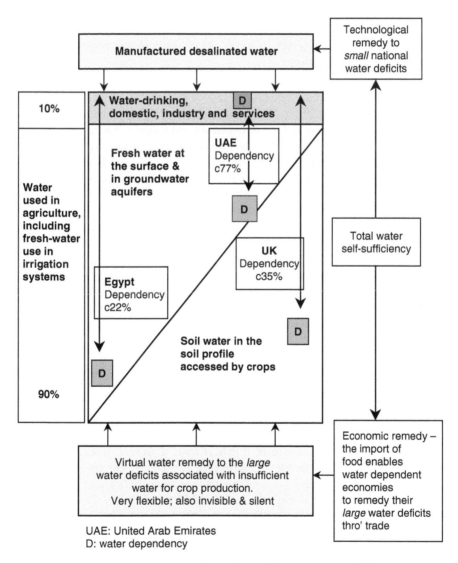

UAE: United Arab Emirates
D: water dependency

Figure 1. (1) The extent of the water dependency (D) of different water deficit economies (Hoekstra & Hung, 2002: 55–59). (2) How local water can come from freshwater sources and soil water sources. (3) How manufactured water and virtual water can remedy the water deficits.

desalination technologies. In the case of the virtual water solution it is necessary to have the capacity to purchase commodities such as food on the world market.

Desalination is a topic, which deserves a separate discussion. During the past decade one of the desalination technologies – reverse osmosis – has experienced a dramatic reduction in operating costs as a result of the fall in the cost of the filtering membranes. Unfortunately the capital, and especially the operating, costs of both multi-stage flash and reverse osmosis systems are notoriously difficult to estimate (Stauffer, 2004). The problem arises because the cost effectiveness of combined power generation and desalination plants depends on the seasonal interaction of the load factors in the power and desalination domains.

## 7   VIRTUAL WATER IN RELATION TO GLOBAL FRESHWATER RESOURCES AND THEIR MANAGEMENT

It has been pointed out above that virtual water moved by trade accounts for about 20% of the water used in global crop production. Meanwhile most water policy is made on the assumption that freshwater – being the only water that engineers and farmers can pump and economists can account for – is the only water that should enter into the planning and policy-making processes. The Russian hydrologists (Shiklomanov, 2000) and the Chinese water resources professionals [personal observation] do recognize the role of water in both the hydrological and the economic cycles. Shiklomanov (2000) suggest that about 5500 Mm$^3$ of water are involved in agricultural production. He includes production from soil water.

The 1031 km$^3$ of the water used to produce the crop and livestock commodities that enter world trade account for 19% of the agricultural water use. Research at the *Value of Water Research Unit* at the IHE in The Netherlands has provided first approximations of the virtual water associated with crops (Hoekstra & Hung, 2002) and with livestock products (Chapagain & Hoekstra, 2003a, 2003b, 2003c) that enter world trade. The IHE research estimates that 695 km$^3$ of water are associated with the crop commodities traded (Hoekstra & Hung, 2002) and 336 km$^3$ are associated with international trade in livestock and livestock products (Chapagain & Hoekstra, 2003a, 2003b, 2003c).

The volumes of water associated with international trade in agricultural commodities including livestock products are impressive. They are even more impressive when they are compared with the volumes of freshwater managed in engineered systems for economic use. Some of the biggest storage structures are in Africa, the Kariba Dam with 180,000 Mm$^3$ storage, the Aswan High Dam with a nominal storage capacity when full of 169,000 Mm$^3$ and a live storage of about 60% of this volume, and the Akosomba Dam in Ghana with a storage capacity of 150,000 Mm$^3$. All these reservoirs have more than four times the capacity of the Hoover Dam in the USA. Other major structures such as the Three Gorges Dam in China have a total storage of 80,000 Mm$^3$ and a capacity to manage about 40,000 Mm$^3$ of water partially to absorb the initial surge of the annual monsoon flood. The three major dams on the Euphrates in Turkey have a total storage capacity of about 110,000 Mm$^3$ and a live storage of about half this volume. Most of the 45,000 storage structures (World Commission on Dams, 2000) are very small compared with these major reservoirs.

If we accept the Rosegrant & Cai (2002) figure of 1430 km$^3$/yr of water used in irrigation systems, including the 600 km$^3$/yr of water pumped from aquifers, mainly for crop production, then the engineering effort to store water in surface systems affects a relatively limited, though potentially very productive, proportion of the 5.5 km³/yr used in agriculture. Soil water is the major source of water for agriculture and is also the major source of the water associated with virtual water.

## 8 THE GLOBAL FRESHWATER RESOURCE, ITS CURRENT AND ESTIMATED FUTURE AVAILABILITY AND USE

The discussion so far has focused on the economic and trading systems in which water using activities are embedded to provide solutions to the problems of the water scarce. The history of water use and management to date across the economies of the world has shown both the water resources and the systems which remedy the problems of the water scarce have been sufficient. The other question to answer is whether there will be enough water to meet future aggregate global water needs using these water managing and trading systems kept in place by diversifying local economies?

Viewed globally very little of the world's freshwater is mobilised by human systems. At the global level a recent study by Rosegrant & Cai (2002: 174), suggests that both the developed and the developing worlds' uses of freshwater will only reach 10% of total renewable water by 2025. It was estimated to be 8% in 1995. Such gross numbers are not safe nor are they of much relevance for water policy and decision-making.

Table 1. Underlying factors influencing water supply in 1995 and in 2025. Selected countries and regions.

| Countries/ Regions | Storage increase km³ | Annual rate (%) | SMAWW increase km³ | Annual rate (%) | GMAWW increase km³ | Annual rate (%) | Basin efficiency 1995 | 2025 |
|---|---|---|---|---|---|---|---|---|
| **Countries** | | | | | | | | |
| China | 157 | 0.62 | 145 | 0.87 | 27 | 0.71 | 0.54 | 0.68 |
| India | 135 | 1.55 | 134 | 0.80 | 14 | 0.23 | 0.57 | 0.70 |
| USA | 0 | 0.00 | 48 | 0.41 | 3 | 0.09 | 0.72 | 0.78 |
| **Regions** | | | | | | | | |
| South Asia | 176 | 0.90 | 62 | 0.66 | 1 | 0.05 | 0.54 | 0.65 |
| SE Asia | 37 | 0.84 | 78 | 1.27 | 9 | 1.15 | 0.47 | 0.55 |
| Sub-Saharan Africa | 74 | 0.61 | 50 | 2.14 | 20 | 1.01 | 0.44 | 0.50 |
| Latin America | 62 | 0.47 | 87 | 1.15 | 12 | 0.61 | 0.44 | 0.51 |
| MENA | 81 | 0.72 | 51 | 0.66 | 2 | 0.12 | 0.68 | 0.75 |
| Developing world | 577 | 0.66 | 635 | 0.88 | 86 | 0.46 | 0.54 | 0.64 |
| Developed world | 44 | 0.16 | 128 | 0.47 | 20 | 0.28 | 0.64 | 0.71 |
| World | 621 | 0.45 | 763 | 0.77 | 105 | 0.41 | 0.56 | 0.65 |

*Source*: Rosegrant & Cai (2002: 172).
SMAWW: surface maximum allowed water withdrawal; GMAWW: groundwater maximum allowed water withdrawal; MENA: Middle East and North Africa.

Table 2.   Underlying factors/driving forces influencing water demand: projected change 1995 to 2025 for selected countries and regions.

| Countries/Regions | Population increase | | GDP increase | | Irrigated area increase | | Livestock production increase % |
|---|---|---|---|---|---|---|---|
| | Million | % | US$/capita | % | Million ha | % | |
| **Countries** | | | | | | | |
| China | 261 | 21.3 | 2390 | 355 | 4.9 | 6.9 | 122 |
| India | 395 | 42.5 | 1123 | 281 | 9.7 | 25.1 | 143 |
| USA | 58 | 21.9 | 24,405 | 93 | 0.7 | 6.4 | 31 |
| **Regions** | | | | | | | |
| South Asia | 630 | 50.7 | 909 | 237 | 11.2 | 19.1 | 143 |
| SE Asia | 201 | 42 | 2332 | 198 | 1.7 | 8.6 | 136 |
| Sub-Saharan Africa | 525 | 98.4 | 125 | 45 | 3.5 | 8.6 | 157 |
| Latin America | 212 | 45.0 | 3942 | 110 | 2.3 | 25.7 | 87 |
| MENA | 215 | 66.1 | 1495 | 88 | 1.9 | 18.4 | 87 |
| Developing world | 2801 | 47.3 | 1808 | 167 | 26.0 | 13.5 | 116 |
| Developed world | 57 | 4.6 | 17,787 | 98 | 2.4 | 5.1 | 16 |
| World | 2858 | 37.9 | 3562 | 74 | 28.4 | 11.9 | 56 |

*Source*: Rosegrant & Cai (2002: 173).

Nevertheless, Rosegrant & Cai (2002) have provided a useful overview first, of driving forces; and secondly, of surface and groundwater resources in relation to current and projected sectoral demand (see Table 1 and Table 2). They have made such estimates particularly useful by incorporating estimates of how increases in water use efficiency will impact future levels of water use. They include in their models estimated improvements in productive/technical efficiency (Allan, 2002: 129–130) in irrigation, and especially in economic/allocative efficiency (Allan, 2002: 130).

Table 1 shows that developing economies face different problems from those that affect developed economies. Developed economies have installed almost all of their surface water storage structures and will build few such structures in future. Developing economies will substantially increase the volume of surface water storage by 2025. Surface freshwater sources are already and will continue to be the main sources of freshwater. Rosegrant & Cai (2002) estimate that surface freshwater will be between six and seven times as important as groundwater in both developing and developed economies in the coming two decades. Both developing and developed economies will improve their water use efficiency. Major advances will be achieved in developing economies because of the much larger volume of freshwater mobilised in developing economies and the lower base level efficiencies in these economies.

The different demographic driving forces on developing and developed economies are shown in Table 2. In the North, water demand will not increase significantly in the coming century because populations have reached a late phase in their demographic transition. The populations of Southern economies will level off but not until the middle and later decades of the century (IFPRI, 1995). As a consequence professionals at International Water Management Institute (IWMI) and the International Food Policy Research Institute (IFPRI) estimate that there will be significant expansion of irrigated areas in the South and also in the production of livestock products. The latter are significant as they are very water intensive and are many times more water intensive in terms of calorie provision than staple crops such as cereals (see Table 2).

Table 3.   Total water withdrawal and total withdrawal as a percentage of total renewable
water (%). Estimated 1995 and projected 2025 for selected countries and regions.

| Countries/Regions | Total withdrawal contributing to supply | | | Total withdrawal as a percentage of renewable water (%) | | |
|---|---|---|---|---|---|---|
| | 1995 | 2010 | 2025 | 1995 | 2010 | 2025 |
| *Countries* | | | | | | |
| China | 679 | 771 | 858 | 26 | 30 | 33 |
| India | 674 | 750 | 813 | 30 | 33 | 36 |
| USA | 497 | 524 | 533 | 24 | 25 | 26 |
| *Regions* | | | | | | |
| South Asia | 1027 | 1142 | 1235 | 24 | 27 | 29 |
| SE Asia | 203 | 242 | 289 | 4 | 4 | 5 |
| Sub-Saharan Africa | 128 | 166 | 215 | 2 | 3 | 4 |
| Latin America | 298 | 355 | 411 | 2 | 2 | 3 |
| MENA | 236 | 263 | 297 | 69 | 77 | 87 |
| Developing world | 2762 | 3145 | 3528 | 8 | 9 | 10 |
| Developed world | 1144 | 1232 | 1274 | 8 | 9 | 10 |
| World | 3906 | 4378 | 4794 | 8 | 9 | 10 |

*Source*: Rosegrant & Cai (2002: 174).

Table 3 provides estimates of existing and future levels of withdrawal of freshwater in
developing and developed economies (Rosegrant & Cai, 2002: 174). Those economies that
are known to be most short of freshwater, the economies of the Middle East and North Africa
are already utilising almost 70% of their renewable water. By 2025 they could be using almost
90%. The major economies of Asia, China and India, are using about 30% currently and
by 2025 will be using between 33 and 36%. Sub-Saharan Africa, Southeast Asia and Latin
America have substantial unused water resources. The position for many regions is much worse
than suggested by these data. Most freshwater in natural hydrological systems is inaccessible
to water users. Some significant rivers such as the Nile in Northeast Africa and the Yellow
River in China are subject to such heavy use that they are dry, or almost dry, by the time they
reach the sea. The Nile delivers to the Mediterranean less than 10,000 Mm$^3$ of water out of
the average flow of 84,000 Mm$^3$. The Yellow River does not reach the sea in most years.

   While moderately sized rivers endure very intense use the big rivers deliver to the sea
between 50 and 90% of their natural flows. The Yangtse in China, the Ganges/Brahmaputra
in South Asia and the Congo in Africa have average flows of over 1000 km$^3$/yr. The massive
Amazon River is in a league of its own draining over 7000 km$^3$/yr of water to the Atlantic
which is over 95% of its natural flow.

   The irrigation sector is the major user of freshwater in any economy where it is decided
to deploy irrigation technologies. It is such a big user that it is possible to generalize, that
wherever it is decided to install irrigation systems the economy will run out of freshwater to
meet the needs of the irrigation sector. As a consequence of the increase in the use of freshwater
in irrigation competition for access occurs at some point unless it is possible to meet the food
needs of a population by food and virtual water imports. The competition is commonly most
evident between water for agriculture and water for the environment if a social movement has
given a voice to environmental water.

Table 4.   Potential and actual consumptive use of water for reliable irrigated agriculture. Estimated 1995 and projected to 2025 for selected countries and regions.

| Countries/Regions | Potential irrigation consumption | | | Actual irrigation consumption | | | Irrigation water supply reliability index (IWSR) | | |
|---|---|---|---|---|---|---|---|---|---|
| | 1995 | 2010 | 2025 | 1995 | 2010 | 2025 | 1995 | 2010 | 2025 |
| *Countries* | | | | | | | | | |
| China | 280 | 283 | 291 | 244 | 227 | 233 | 0.87 | 0.77 | 0.76 |
| India | 400 | 442 | 466 | 321 | 320 | 329 | 0.80 | 0.72 | 0.69 |
| USA | 133 | 134 | 131 | 124 | 118 | 120 | 0.93 | 0.88 | 0.90 |
| *Regions* | | | | | | | | | |
| South Asia | 605 | 657 | 691 | 484 | 489 | 498 | 0.80 | 0.73 | 0.71 |
| SE Asia | 98 | 103 | 106 | 85 | 88 | 91 | 0.86 | 0.82 | 0.83 |
| Sub-Saharan Africa | 69 | 78 | 87 | 50 | 55 | 62 | 0.73 | 0.65 | 0.67 |
| Latin America | 107 | 122 | 129 | 88 | 89 | 96 | 0.82 | 0.73 | 0.75 |
| MENA | 156 | 170 | 184 | 122 | 126 | 137 | 0.78 | 0.72 | 0.71 |
| Developing world | 1445 | 1549 | 1617 | 1162 | 1165 | 1214 | 0.80 | 0.73 | 0.71 |
| Developed world | 313 | 314 | 308 | 268 | 263 | 275 | 0.86 | 0.84 | 0.89 |
| World | 1758 | 1983 | 1924 | 1430 | 1429 | 1485 | 0.81 | 0.75 | 0.74 |

*Source*: Rosegrant & Cai (2002: 177).

As discussed above water scarcity is the result of the allocative decisions made by communities on the use of land. Water scarcity for the small volumes of water needed for municipal, industrial and domestic water use is unusual. A few economies in the Arabian Peninsula, Kuwait, Abu Dhabi, Oman, Qatar, cannot meet such needs but this is exceptional. And they can easily afford the desalination solution. Cities located at high elevations in the same region – Damascus (Syria), Amman (Jordan), Sana'a and Ta'iz (Yemen) – also endure current shortages and face much more serious challenges as sea water desalination is not an economic option.

Water scarcity to meet the big volumes of water needed to service irrigation schemes is, however, very common. High water demands for irrigation especially impact the water needed to sustain the environmental services provided by water, which underpin the security of all economies. Table 4 shows the potential and actual consumptive use of water for reliable irrigated agriculture. Again the difference between developing and developed economies is evident.

The South and the North have different levels of consumption of domestic water and the economies in the South and the North will increase their levels of use by different proportions. The use of water and the efficiency of that use in the different livelihoods developed in the South and the North are very significant with respect to both the volumes of water used and especially with respect to the capacity of a political economy to achieve economic self-sufficiency. Once the economic self-sufficiency is achieved, it is possible for communities and nations to address water scarcity.

The discussion of water availability and water use and the different trajectories of use and development evident in the South and the North have a number of recurring themes. These recurring issues are the consequence of the different levels of population change and of socio-economic development in the South and the North. The South has five sixths of the world's population. The North has only about one sixth. The South has limited adaptive capacity reflected in average GDPs per head (see World Bank: www.worldbank.org) and Human Development Indices (HDI, see UNDP: www.undp.org) which are only 10% of those enjoyed in the North. Adapting to water scarcity has proven to be possible in the diverse and strong economies of the North whether they enjoy secure or insecure water. The economies of the South are poor and it is this poverty which impairs their capacity to adapt to water scarcity.

The demographic and resource numbers and the ability to estimate the pace and the effectiveness of socio-economic development mean that analysts of water futures face uncertainty and therefore ambiguity. Pessimists emphasise the uncertainty that the global water resource will be sufficient to meet the water demands of the global population. Optimists observe the declining trends in population increase and the relatively modest challenge of increasing food production by 30% or at worst 50%. Yield increases achieved in the second half of the 20th century if extended to the poorly performing farms of sub-Saharan Africa would easily meet future food needs.

## 9   VIRTUAL WATER AND THE THREEFOLD SYNERGY ATTENUATE THE MYTH OF *THE WATER CRISIS*, BUT PROVIDE SIMULTANEOUSLY ONE OF THE MAJOR SOLUTIONS

The purpose of the chapter has been to highlight some important underlying fundamentals in the domain of water use and water policy-making. Three ignored and de-emphasised elements of the hydrological and the hydro-economic system have been identified. This threefold synergy – soil water, virtual water and the socio-economic development – enables economies to trade their way to water security. All three elements are ignored by water policy-makers and especially by those who promote the idea that there is severe and possibly inaddressable global water crisis. First, soil water is ignored as part of water budgets. Secondly, virtual water is de-emphasised because if it were to become prominent in the political discourse of water scarce economies it would bring negative and destabilising political consequences. Those affected would be the major water users in the irrigation sector and the water policy-making elite in the poor economies that cannot enjoy the policy-making benefits deriving from economic diversification. Thirdly, economic diversification which is not intuitively seen as directly relevant to the water sector as it plays a *hidden-hand* role in enabling virtual water transactions. The technical professionals who generally manage water do not see solutions outside the water sector. As a result they are very exercised about the inadequacies of their local watersheds.

The politically rational de-emphasis and lack awareness of these underlying fundamentals is sub-optimising for the water using and water policy-making processes. They aggravate the already existing, seriously *second best*, water managing operations, especially in developing economies. They also enable water policy makers to avoid necessary, but politically costly, re-allocative reforms. Such reforms would both improve the economic returns to water as well as enhance the levels of environmental service provided by water in the natural water environment.

The threefold synergy in the global political economy of water has not been highlighted until the writing of this chapter. The author senses that the identification of three economically invisible and politically silent processes in the water sector has already been identified as normal processes in highly diversified and developed political economies. Certainly it is not difficult to share these ideas with water users from developed economies. In economies where poverty prevails, the ideas have no purchase because the power of economic diversification has not been experienced.

Myths establish themselves in circumstances of uncertainty. They flourish and endure as they have been constructed to accord with cultural preferences. Contradictory myths commonly exist at the same time. The myth of future unmanageable global scarcity has been generated in the risk aware neo-liberal North in late-modernity – that is since the late 1970s. The global water crisis myth is proclaimed at the same time that the farmers and politicians of Middle Eastern economies claim that all they need is a little more water, then they will manage it more carefully, and everything will be all right. It has been argued here that both these myths are wrong. They have been constructed in the first case by scientists with a partial understanding of the complex hydro-economic global system. They emphasise demographic indicators and water resource availability and ignore the remedies outside the hydrological systems. In the second case where farmers and their political leaders face palpable water shortages, in for example the Middle East, there is a convergence on the myth that water resources are adequate if differently managed.

Myths usually endure because people and politicians want to ignore bad news and avoid change. This chapter has highlighted that such myths are kept in place in water scarce countries because there are invisible economic processes that make it appear that water is not a problem. The myth of global water scarcity is a different type of myth. It is the result of hyper-risk aware Northern environmentalists and water scientists relating local hydrological and demographic data without taking into account remedies available in technology, economic development and trade. The purpose of the chapter has been to show that non-hydrological global systems have been ameliorating local water scarcity for half a century. They have the capacity to meet the water demands of foreseeable future populations.

## REFERENCES

Allan, J.A. (2001). *The Middle East water question: hydro-politics and the global economy.* London: I B Tauris.

Allan, J.A. (2002). Hydro-peace in the Middle East: Why no water wars? A case study of the Jordan River Basin. *SAIS Review*, Vol. XXII: 2 (Summer–Fall, 2002): 255–272.

Allan, J.A. (2003). Virtual water: the water, food, and trade nexus. Useful concept or misleading metaphor? *Water International*, 28(1): 4–11.

Allan, J.A. & Olmsted, J. (2003). Trading virtual water, the MENA water deficit and the consequences for water policy reform. In: Lofgren, H. (ed.), *The Middle East water economy*. IFPRI, Washington.

Bowen, H.P.; Leamer, E.E. & Sveikauskas, L. (1987). Multicountry, multifactor tests of the factor abundance theorem. *The American Economic Review*, 77(5), December: 791–809.

Brichieri-Colombi, J.S.A. (2003). *Food security, irrigation and water stress: logical chain or environmental myth?* Occasional Paper 57, King's College London/SOAS Water Research Group. London, UK. (See http://www.kcl.ac.uk/kis/schools/hums/geog/water/occasionalpapers/home.html#trade).

Brichieri-Colombi, S. (2004). *Who speaks for the river?* Unpublished PhD Thesis of the University of London.

Brown, L. (2003). *Plan B: rescuing a planet under stress and a civilization in trouble.* WW Norton, New York, USA.

Chapagain, A.K. & Hoekstra, A.Y. (2003a). Virtual water between nations in relation to trade in livestock and livestock products. *Value of Water Research Report Series*, 13. IHE, Delft, The Netherlands.

Chapagain, A.K. & Hoekstra, A.Y. (2003b). The water needed to have the Dutch drink coffee. *Value of Water Research Report Series*, 14. IHE, Delft, The Netherlands.

Chapagain, A.K. & Hoekstra, A.Y. (2003c). The water needed to have the Dutch drink tea. *Value of Water Research Report Series*, 15. IHE, Delft, The Netherlands.

Dyson, T. (1996). *Population and food: global trends and future prospects*. Routledge, London, UK.

Dyson, T. (2001). World food trends: a neo-Malthusian prospect? *Proceedings of the American Philosophical Society*, 145(4), December: 438–455.

Earle, A. (2001). *The role of virtual water in food security in Southern Africa*. Occasional Paper 33. King's College London/SOAS Water Research Group, London, UK. (See http://www.kcl.ac.uk/kis/schools/hums/geog/water/occasionalpapers/home.html#trade).

El-Fadel, M. & Maroun, R. (2003). The concept of virtual water and its applicability in Lebanon. In: A.Y. Hoekstra, Virtual water trade: a quantification of virtual water flows between nations in relation to international crop trade. Value of Water Research Report Series, 12. IHE, Delft, The Netherlands: 171–182.

FAO (2003). *Review of world water resources by country*. Water Report, 23. FAO, Rome, Italy.

FAO (2004). Aquastat database. FAO, Rome, Italy (On-going on: http://www.fao.org).

Fine, B. & Saad-Filho, A. (2003). *Marx's Capital*. Fourth edition. Pluto Press, London, UK.

Goodchild, P. (2004). Meet the real Dr. Strangelove. *The Guardian*, G2 Supplement, 2 April 2004: 8–9.

Haddadin, M.J. (2003). Exogenous water: a conduit to globalization of water resources. In: A.Y. Hoekstra, Virtual water trade: a quantification of virtual water flows between nations in relation to international crop trade. *Value of Water Research Report Series*, 12. IHE, Delft, The Netherlands: 159–169.

Hakimian, H. (2003). *Water scarcity and food imports: an empirical investigation of the* virtual water *hypothesis in the MENA Region*. Occasional Paper 46. King's College London/SOAS Water Research Group. London, UK. See (http://www.kcl.ac.uk/kis/schools/hums/geog/water/occasionalpapers/home.html#trade).

Harkness, J.P. (1978). Factor abundance and comparative advantage, *The American Economic Review*, 68, December: 784–800.

Hoekstra, A.Y. (ed.) (2003). Virtual water trade: Proceedings of the international expert meeting on virtual water trade. *Value of Water Research Report Series*, 11. IHE, Delft, The Netherlands.

Hoekstra, A.Y. & Hung, P.Q. (2002). Virtual water trade: a quantification of virtual water flows between nations in relation to international crop trade. *Value of Water Research Report Series*, 11. IHE, Delft, The Netherlands.

IFPRI (1995). *Global food security*. IFPRI Report, October 1995. International Food Policy Research Institute, Washington, D.C. See (www.cgiar.org/ifpri).

IFPRI (1997). *China will remain a grain importer*. IFPRI Report, February 1997. International Food Policy Research Institute, Washington, D.C. See (www.cgiar.org/ifpri).

Krueger, A.O. (1977). *Liberalization attempts and consequences*. Ballinger Publishing Co., Cambridge, Massachusetts, USA.

Krugman, P.R. & Obstfeld, M. (1991). *International economics – theory and policy*. HarperCollins Publishers, New York, USA.

Lawler, K. & Seddighi, H. (2001). *International economics – theories, themes and debates*. Prentice Hall, London and New York.

Merrett, S. (2003). Virtual water and Occam's razor. *Water International*, 28(1): 1–3, 12–16.

Postel, S. (1999). *Pillar of sand*. Norton, New York, USA.

Postel, S. & Richter, B. (2003). *Rivers for life: managing water for people and nature*. Island Press, St. Louis, Washington University, USA.

Renault, D. (2003). Virtual water: principles and virtues. In: A.Y. Hoekstra, Virtual water trade: a quantification of virtual water flows between nations in relation to international crop trade. *Value of Water Research Report Series*, 12. IHE, Delft, The Netherlands: 77–91.

Rosegrant, M.; Agcaoli, M. & Perez, N.D. (1995). *Global food projections*. Diane Publishing Company, Washington, USA.

Rosegrant, M. & Cai, X. (2002). Global water demand and supply projections. Part 2: results and prospects to 2025. *Water International*, 27(2): 170–182.

Shiklomanov, I.A. (2000). Appraisal and assessment of world water resources. *Water International*, 25(1): 11–32.

Stauffer, T. (2004). *The economic limits of seawater desalination*. Unpublished draft.

Wakemagel, M. & Rees, W. (1996). *Our ecological footprint: reducing human impact on the Earth*. New Society Publishers, Gabriola Island, BC, Canada.

Weber, M. (1904 and 1917). *The methodology of the social sciences*. Translated and edited by Edward A. Shils and Henry A. Finch. Published in 1949. Free Press, New York, USA.

Wichelns, D. (2004). The policy relevance of virtual water can be enhanced by considering comparative advantages. *Agricultural Water Management*, 66: 49–63.

World Bank (2003). *World Bank water resources sector strategy*. World Bank, Washington, USA. See World Bank website.

World Bank-WRI (2003). *World Resources 2002–2004, decisions for the Earth: balance, voice, and power*. World Bank and World Resources Institute. United Nations Development Programme, United Nations Environment Programme, New York and Washington, USA.

World Commission on Dams (2000). *Dams and development*. Earthscan Publications, London, UK.

Zimmer, D. & Renault, D. (2003). Virtual water in food production and global trade: review of methodological issues and preliminary results. In: A.Y. Hoekstra, Virtual water trade: a quantification of virtual water flows between nations in relation to international crop trade. *Value of Water Research Report Series*, 12. IHE, Delft, The Netherlands: 93–109.

CHAPTER 9

# Virtual water – Part of an invisible synergy that ameliorates water scarcity: *commentary*

J. Ramirez-Vallejo
*Harvard University, Cambridge, Massachussets, USA*

ABSTRACT: Prof. Allan (this volume) argues that there is sufficient water in the world to meet current and future water needs, mainly because there are three processes in what he calls a threefold synergy: soil water, virtual water and the socioeconomic development that enables economies to trade their way to water security. In these comments, I argue that the virtual water argument does not hold true and that, although it is a useful concept to remind us of the water scarcity in some regions of the world, it is a concept that does not have significant implications on water and agricultural policy. These comments also include some reflections about the question regarding a possible water crisis in the future, the supply and demand of water resources, and the EU/USA agricultural subsidies. Finally, I present a hypothesis as why water resources development has not reach the top of the international agenda.

Keywords: *Virtual water, demand, agricultural subsidies, soil water*

## 1 INTRODUCTION

In his chapter, Prof. Allan (this volume) argues that there is sufficient water in the world to meet current and future water needs, mainly because there are three processes in what he calls a threefold synergy –soil water, virtual water and the socioeconomic development that enables economies to trade their way to water security. Therefore, according to Allan, the expected *water crisis* would eventually be solved. Then he develops each of these processes, some in more depth than others, and highlights several underlying fundamentals in the domain of water use and water policymaking.

The three processes are: (1) soil water, which according to him, is ignored by water budgets around the world; (2) virtual water, or water that is embedded in the international trade of agricultural products, mainly in meat and cereals; and (3) economic diversification or socio-economic development, which acts as a *hidden-hand* in enabling virtual water transactions. With these new processes in mind, Professor Allan, the father of the virtual water concept, presents an optimistic scenario of the future world water balance (Allan, 1996).

My comments are presented in the following order. First, I argue that the virtual water argument does not hold true and that, although it is a useful concept to remind us of the water scarcity in some regions of the world and the amount of water used in agriculture, it is a concept that in a world in which water subsidies and tariff and non-tariff measures will continue to be the norm, does not have significant implications on water and agricultural policy. In the second part of my comments, I will discuss the major question regarding a possible water crisis in the future. In particular, I try to explore the answer from the supply and demand side of water resources. Third, I will comment on the soil-water and socioeconomic development ameliorating processes identified and proposed by Prof. Allan, and will make a note on the EU/USA agricultural subsidies as a potential fourth process suggested by the author. Finally, I will present a hypothesis as why water resources development has not reach the top of the international agenda.

## 2   THE VIRTUAL WATER ARGUMENT

The *virtual water* concept comes from the idea that water should be treated as a production factor and is equivalent to the volume of water needed to produce a commodity or service. The *virtual water argument,* on the other hand, shows that the importation of agricultural products that require important amounts of water represents the importation of water into a water scarce country. Using Prof. Allan's words (this volume), "trade in commodities has already achieved a pivotal position in enabling the comparative advantages and disadvantages of regional water endowments to be balanced".

Under the virtual water argument, food trade then becomes an instrument to augment water supplies on the scale needed to meet the domestic food demand, which in trade-theory terms, is equivalent to saying that the *Heckscher-Ohlin* (H-O) theorem holds when applied to water as a key input in the agricultural production process.

Leontief (1953) first employed the factor content of trade approach in his well-known test of the H-O theorem, which states that a country exports products that use intensively a relatively abundant input. However, the proof of this theorem implies several assumptions such as factors can move without cost among industries within a country, but are completely immobile internationally, which might be the case of water as a production factor. Other more restricted assumptions are that the production functions for all countries (products) exhibit constant return to scale and each country has the same productive technology for each good, and consumers have the same utility functions, which clearly is not the case for the use of water in agriculture worldwide.

*Does the H-O theorem apply to water-dependent agricultural trade?* To answer that question Ramirez-Vallejo & Rogers (2004b) ran a test and found that the proposition of conformity in sign between water contents and excess water shares receives relatively little support. Using *Fisher's Exact Test,* the hypothesis of independence between sign of the water contents and of the excess water shares could not be rejected at the 95% level. They also tried to correct for differences in productivity of land and found similar results. This should not come as a surprise because this type of result has become the norm when testing the H-O hypothesis, Deardorff (1982), Ethier (1984) and Helpman (1984).

This finding again proves one more time, what Bhagwati (1964) states in his survey article: "Where the theory [H-O theorem] is deficient is in its not having been extended to a multi-country framework. The prediction holds when there are many commodities and two countries. However, when there are many countries and commodities it is not clear, in the absence of theoretical analysis, what conditions will suffice to make the Heckscher-Ohlin hypothesis true...".

The main reason for the failure of the H-O theorem is that under conditions sufficient to guarantee factor-price equalization there exist many efficient production configurations in the world consistent with the equilibrium factor prices and a given distribution of factor endowments among countries.

On the inability of the virtual water argument, Wichelns (2004) recognizes that the virtual water metaphor addresses resource endowments, but does not address production technologies or opportunity costs. More specifically, the price equalization hypothesis that underlies the H-O theorem rarely applies in the case of water as an agricultural input. Values of water differ significantly from one country to the next, and even within countries. The price for water that is actually paid by farmers and that is internalized in the farmer's decision process is distant from the true opportunity cost of water.

Under these circumstances, the farmer's choice of whether to plant water-demanding crops does not follow the water scarcity signal of the need to choose less-water-demanding crops. For instance, in the irrigation districts in Northern Mexico, farmers continue to cultivate cereals and water demanding crops (high levels of virtual water) because of the highly subsidized crop prices and low water fees administered by the government. In a scenario of no agricultural subsidies, the farmers in these districts should have switched from cereals to fruits and vegetables or to higher-value crops after NAFTA became effective to take advantage of the opportunities presented by the USA market.

*If water endowment does not explain the direction of agricultural trade or virtual water trade, what does?* The truth of the matter is that complex interactions among technology, policy, investment, environment, and human behavior influence the trade of virtual water. Indeed, any factor or condition that alters the demand or supply sides of the agricultural commodities markets impact agricultural trade pattern and therefore on the *Net Virtual Water Trade.* Among these factors are: trade agreements; economic and population growth; technological innovation; subsidies to agricultural products and inputs; international prices of agricultural products and inputs; macroeconomic policies in import and export countries; domestic microeconomic policies; political and economic crisis; degradation of natural resources; and cultural and religious changes and constraints (Rosegrant & Paisner, 2001; Ramirez-Vallejo & Rogers, 2004a).

On the other hand, the embodied virtual water in agricultural trade flows are explained by the variables considered by economic agents at the time production and trade decisions are made. These decisions are linked directly to price, product characteristics and real demand conditions. In other words, the decision taken by economic agents when trading agricultural products is directly related to attributes of the agricultural products, including their price, rather than to the water used for production alone. Rather water enters in the decision process as a major factor of production, but only indirectly.

Ramirez-Vallejo & Rogers (2004a) formally researched the determinants behind agricultural imports in order to understand the hidden factors that indirectly drive virtual water flows. They found, for example, a very strong relationship between income and virtual water imports with an income elasticity of 0.52 for the year 2000–2001. This means that in a scenario of a world two times richer than today, the world could experience an increase in virtual water flows of 50%, or close to 400 km$^3$. However, it is difficult to tell if this virtual flow would occur in the direction from water surplus countries to water deficit countries.

Income has a double role in virtual water. First, higher income and purchasing power holds out the possibility of being able to offset most of the constraints that prevent access to water in a particular place at a specific time. In addition, higher national income increases the possibility of importing food from other countries.

Ramirez-Vallejo & Rogers (2004a) also found that the level of support or protection to agriculture is also significant in explaining virtual water demand. Elasticity for the support

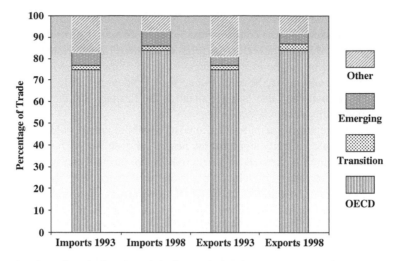

Figure 1.    Direction of Agricultural Trade before and after the Uruguay Round.

of agriculture was found to be $-0.9$. The higher the support for agriculture in one country compared to other countries, the less the amount of virtual water imports[1].

If one examines at who trades agricultural goods with whom, one finds that OECD countries trade mostly with each other. In 1998, 83% of OECD agricultural imports came from other OECD countries, while 85% of agricultural exports went to other OECD countries (Figure 1). There has been little penetration of OECD markets by less developed countries, whose combined share of OECD agricultural imports increased from 8% to 9% during the period of most significant agricultural liberalization around the world.

On the other hand, developing countries are sending an increasing share of their agricultural exports to other developing countries. One reason for this is that, on average, developing countries are growing more rapidly than developed ones. At the same time, the increasing share of non-bulk food trade in total food trade reinforces the reorientation of trade, since protection in developed countries is often higher for semi-processed and processed products (OECD, 2001).

In the case of emerging economies, they show a more diverse profile of trading relationships. It is interesting to note that Brazil sends half its exports to the European Union (EU), compared with just 27% in the case of Argentina. Nearly two-thirds of South African exports go to OECD countries, not least because of the underdevelopment of regional African markets.

The concept of virtual water is then appealing to educate public officials and society in general that water in some parts of the world is a scarce resource and that agriculture uses the great majority of water resources available on Earth. The argument also has an implicit lesson underscoring the importance of running efficiently irrigation districts so water could be allocated to other uses including ones benefiting the environment. However, the virtual water

---

[1] Other variables were found to have some explanatory power of the variance of virtual water imports. These are: (1) the average income (GDP); (2) population; (3) agriculture as value added (% of GDP); (4) irrigation (actual and potential); and (5) exports of goods and services (% of GDP). The sign found for each of these variables in the estimated equation were as expected, with the absolute value of elasticities ranging from 0.3 to 0.7.

argument, if applied improperly, could send the wrong message in terms of policy making in agriculture and water resources. For instance, a country could delay important investments now and decide instead to import foodgrains, or, it could choose not to remove price subsidies with the objective of saving water.

## 3   IS IT A DEMAND OR A SUPPLY PROBLEM?

### 3.1   *Supply or demand?*

As Professor Allan states, "optimists and pessimists interpret the underlying fundamentals differently". I consider myself a demand-driven pessimistic and a supply-driven optimistic unless some unlikely structural changes occur in policy.

Professor Allan argues in his chapter that there is sufficient freshwater and soil water in the world to meet current and future water needs if we accept the demographic predictions. I agree with Prof. Allan that the world does have the water resources to meet the needs of future populations, if one considers today's technology to mobilize it and to treat it in such a way that it will become available where it is needed.

Agriculture accounts for more than 70% of total water use in most countries. Therefore, the issue whether there is enough water is related to the question of whether there is enough water to feed the population and to end hunger and prevent recurrence of famine and starvation. IFPRI estimates that roughly 800 million people are chronically undernourished today –three times the population of the USA– and unfortunately nothing tells us that there will be a change in the trend in the near future (IFPRI, 2003).

The problem is not water supply when one could most probably induce the generation and use of sophisticated technologies to obtain enough water to satisfy water needs. With increasing pressure on use of natural freshwater in parts of the world, other non-conventional sources of water are growing in importance: (1) the production of freshwater by desalination of brackish or saltwater (mostly for domestic purposes); and (2) the reuse of urban or industrial wastewaters (with or without treatment), which increases the overall efficiency of use of water (extracted from primary sources), mostly in agriculture, but increasingly in the industrial and domestic sectors. This category also includes agricultural drainage water.

Although the unit cost of these technologies has been decreasing over time, they remain too expensive for most communities in the developing world. Thus, the real problem comes from the demand side as to whether there would be enough economic resources at the individual, regional or country levels to pay for the water/technology needed. The resources will be needed not only to cover the cost of water treatment but also its transportation to users. Table 1 shows how significant is the *per capita* annual water withdrawn by income groups for different uses in Africa. The water withdrawn *per capita* for the higher income group is on average three times larger than the amount of the lower income group.

### 3.2   *Induced innovation hypothesis*

If demand increases and local water supplies were scarce, then this situation would most likely induce a technology change. Hayami & Ruttan (1985) have presented various tests to their well-known *induced innovation hypothesis* in which they find that technical and institutional changes are induced through the responses of farmers, agribusiness entrepreneurs, scientists,

Table 1.    Sectoral water withdrawals by income group in Africa.

| Country income group | *Per capita* annual withdrawal (m$^3$) | | | |
| --- | --- | --- | --- | --- |
| | Total | Agriculture | Domestic | Industry |
| Low income | 385 | 351 (91%) | 15 (4%) | 19 (5%) |
| Middle income | 454 | 313 (69%) | 59 (13%) | 82 (18%) |
| High income | 1167 | 455 (39%) | 163 (14%) | 549 (47%) |

*Source*: World Bank (www.worldbank.org).

and public administrators to resource endowments and to changes in the supply and demand of factors and products.

Using their findings, one could say that the state of relative endowments and accumulation of the primary resources of land, labor, and water, would be a critical element in determining the pattern of technical change that will occur in water resources and agriculture in the future. In other words, technical change embodied in new and more productive inputs may be induced primarily either to save labor, to save land or to save water. If water were priced at its real opportunity cost, then technology changes would likely generate a lower-cost access to water resources. However, if water continues to be undervalued by society or highly subsidized then future technological changes will occur in other directions.

### 3.3    *Water and food security*

I agree with Prof. Allan that whether water will become a major constraint to the achievement of food security in many developing countries remains to be seen. However, I find it difficult to believe that by moving water resources virtually some countries would be able to solve their food problems regionally or at the national level. In his classic book *Transforming Traditional Agriculture*, T.W. Shultz (1964) suggested that significant growth in productivity cannot be brought about by the reallocation of resources in traditional agricultural systems. Significant opportunities for growth will become available only through changes in technology –new husbandry techniques, better seed varieties, more efficient sources of power, cheaper and more effective nutrients, and biotechnology.

It is difficult to disagree with Prof. Allan that economic diversification tends to reduce relative water scarcity. The economies of the globe usually adjust from the demand side and generate the required income for the population to effectively demand higher priced water. The example given by Prof. Allan in which job creation outside agriculture has generated incomes and revenues with which economies have been able to access virtual water, is a classic example of this type of adjustment of water demand through economic diversification.

## 4    COMMENTS ON THE OTHER TWO INVISIBLE PROCESSES: SOIL WATER AND SOCIOECONOMIC DEVELOPMENT

Prof. Allan introduces for the first time two processes that, in his opinion, have made possible for many communities around the world to access water without having to go to war. These processes substitute water in space and time when water is needed. One is the soil water about which he argues has not been taken into account when observing water supply at the country level. While the second process is the socioeconomic development process where diversification experienced by some economies has increased income and made possible buying intensive-water-agricultural products from other areas and therefore saving local water resources.

## 4.1    *Soil water*

Prof. Allan argues that soil water is not taken into account when computing national water endowments by international agencies. However, FAO, in its AQUASTAT information system, begins estimating the total amount of water from rainfall and then, after applying several assumptions, estimates the amount of surface and groundwater, external water resources and other uses. FAO defines available water resources as "water net balance in a given state of use and exploitation of the resources and not with a meaning of water offer. The *availability* may be: (1) equal to *resources* minus *withdrawal* at the local level of a subsystem, where a part of the water withdrawn cannot be returned into the system; or (2) equal to *resources* minus *final consumption* at a more regional scale (watershed, country), where the balance encompasses all the use systems" (FAO, 2002). If this methodology and these definitions are applied, then soilwater has always been considered as part of the water endowment or water availability of a country or region. This is seen even more clearly when looking at the water balance usually computed in hydrologic models used in watersheds.

For the sake of clarification, plants take only one part of soil water as a source of growth, the part denominated as capillary water. The other components are gravitational water and hygroscopic water which are not used by plants for a variety of reasons. Gravitational water drains quickly from the soil, and hygroscopic water remains adhered to the soil particles, therefore could be considered constant over time. The available water to the plant roots is the one located between the wilting point and the field capacity, and as explained above is part of the water balance used to estimate water endowments at the country and regional levels.

## 4.2    *The input and output substitution effects: a forth process?*

The input and output substitution effects is a process that could save water by obtaining more food per drop, or producing less water demanding products for the new preferences of consumers. In the case of inputs, the water required to produce food would be less if it were feasible to use other inputs that in some quantity may reduce water requirement. For example, a higher amount of fertilizer applied might increase yield with the same or less water requirements. In this case, fertilizer generates the concept of opportunity cost of virtual water. The same could be said for other inputs such as seeds, land, labor, machinery, and even financial resources. This is why if the prices are right, the correct signal is given to the innovation process to obtain high cost input saving technologies. In the case of output, a change in the daily intake diet preferences could change significantly in terms of water requirements. For instance, a switch from meat to fruits and vegetables would save significant amounts of water.

## 4.3    *EU/USA agricultural subsidies: another ameliorating forth (the forth force?)*

Prof. Allan points out the agricultural policies of the EU and the USA as a process that has led to the water scarcity since the middle of the 20th century. Based on his calculations, around

25% of total virtual water trade is a result of the lower world prices for grain produced by subsidies. This is true; however, the real effect could be overestimated because the increase in world prices of commodities in a world without subsidies could be significant[2] (Valdez & Zietz, 1995; OECD, 2000).

However, the level of protection to agricultural products via tariffs and duties and non-tariff instruments by all countries, developed and developing, has distorted the virtual water movement worldwide. Ramirez-Vallejo & Rogers (2004a) showed that, for example, using IFPRI's *IMPACT* model results, a scenario of full liberalization of agriculture compared to a baseline scenario would have a greater net effect of virtual water flows from the relocation of meat trade than from the adjustment in cereals' trade. When the net effect of the meat and cereals markets are added together, the two major contributors to the increase in virtual water trade would be the USA, which would increase its annual virtual water exports in about 86 km$^3$, and Latin America would have a similar increase: 89 km$^3$. These are the two water surplus regions in the world. The major changes in virtual water imports would occur in Asia in general (South Asia, Southeast Asia, East Asia) with an increase of 112 km$^3$, Sub-Saharan Africa with an increase of almost 40 km$^3$ and the former Soviet Union with an increase in water imports of 22 km$^3$, mostly because of an increase in meat imports. West Asia and North Africa together, on the other hand, would decrease the level of virtual water imports to about 7 km$^3$, but would remain as an important net importer of virtual water of about 176 km$^3$.

## 5   A FINAL HYPOTHESIS

Finally, I end my comments with a hypothesis on why water has not reached the top of the international agenda. Besides the lack of consensus on what the message on water development policy should be among water scientists[3], the challenge ahead is how to bring about significant shifts in perception of the role of water in poverty alleviation and competitiveness enhancement. As long as water is not considered important in alleviating poverty and increasing competitiveness, it will be very difficult to ask for something more from what Prof. Allan terms the political economy superstructure of water related decision making.

## REFERENCES

Allan, J.A. (1996). Water Use and Development in Arid Regions: Environment, Economic Development and Water Resource Politics and Policy. *Water Use and Development*, 5(2): 107–115.
Bhagwati, J. (1964). The Pure Theory of International Trade: A Survey. *Economic Journal*, LXXIV: 1–84.
Deardorff, A.V. (1982). The general validity of the Heckscher-Ohlin Theorem. *American Economic Review*, 72(4): 683–694.

---

[2] IFPRI found that full liberalization would have a significant effect on cereal prices by 2020 with rice increasing the most at 14%, followed closely by maize, wheat and other course grains. Meat prices would respond with even sharper price increases with beef alone being subject to an 18% increase.

[3] Some water specialists still argue in favor of more water subsidies in order to increase the supply of water resources infrastructure to the lower income segment of the population, while others, on the other hand, argue that water should be priced at the opportunity cost of the resource no matter how high this might be.

Ethier, W.J. (1984). Higher dimensional issues in trade theory. In: R.W. Jones & P.B. Kenen (eds.), *Handbook of International Economics*, Vol. I. Amsterdam, the Netherlands: 131–184.

FAO (2002). *Review of Water Resources by Country, Water Report 23*. Food and Agricultural Organization, United Nations.

Hayami, Y. & Ruttan, V. (1985). *Agricultural Development: An International Perspective*. Revised and Extended Edition. Jhon Hopkins University Press. Baltimore, Maryland, USA.

Helpman, E. (1984). The factor content of foreign trade. *Economic Journal*, 94: 84–94.

IFPRI (2003). *Annual Report 2002–2003, Trade Policies and Food Security*. International Food Policy Research Institute, Washington, D.C., USA.

Leontief, W.W. (1953). *Domestic Production and Foreign Trade: The American Capital Reposition Re-examined*. Reprinted in: R.E. Caves & H.G. Johnson (eds.), Readings in International Economics (London, Allen & Unwin, 1968).

OECD (2000). *Agricultural Policies in Emerging and Transition Economies*.

OECD (2001). *Agricultural Policies in OECD Countries: Monitoring and Evaluation*.

Ramirez-Vallejo, J. & Rogers, P. (2004a). Virtual Water Flows and Trade Liberalization. *Journal of Water Science and Technology*, 49(7): 25–32.

Ramirez-Vallejo, J. & Rogers, P. (2004b). Mexico: NAFTA, Virtual Water, and the Economic Value of Water (mimeo). DEAS, Harvard University, USA.

Rosegrant, M.W. & Paisner, M.S. (2001). *Global Food Projections to 2020*. International Food Policy Research Institute, Washington, D.C., USA.

Shultz, T.W. (1964). *Transforming Traditional Agriculture*. Yale University Press. New Haven, Connecticut, USA.

Valdez, A. & Zietz, J. (1995). Distortions in World Food Markets in the Wake of GATT: Evidence and Policy Implications. *World Development*, 23(6): 913–926.

Wichelns, D. (2004). The policy relevance of virtual water can be enhanced by considering comparative advantages. *Agricultural Water Management*, 66: 49–63.

V

Groundwater

CHAPTER 10

# Significance of the *Silent Revolution* of intensive groundwater use in world water policy

M.R. Llamas
*Royal Academy of Sciences, Madrid, Spain*

P. Martínez-Santos
*Complutense University of Madrid, Spain*

ABSTRACT:   A series of fairly new factors such as virtual water, the rise of desalination technologies and intensive groundwater use currently look as though they will exert a strong influence on future water policy. This chapter is concerned with the latter of the three: the *Silent Revolution* of intensive groundwater use in arid and semi-arid countries. Over the last half century, millions of farmers have independently drilled their own wells in the pursuit of the socio-economic advantages of groundwater irrigation. This has been due to fairly recent advances in well drilling and pumping, which together with the development of hydrogeology as a solid body of science, have made groundwater more widely available. The intrinsic benefits of groundwater irrigation in relation to traditional surface water systems, such as the ready availability of the resource or the resilience of aquifers against drought, constitute the main reason behind the spectacular groundwater development of many arid and semi-arid countries worldwide. Despite these undeniable benefits, certain problems (mainly related to groundwater quality degradation and water table depletion) have also arisen in some places. While no two cases are the same, a pattern of events can be observed in many of these regions, thus leading to the conceptualisation of this intensive groundwater-based development. Thus, five stages can be distinguished: *Hydroschizophrenia*, changes in water policy due to the *Silent Revolution*, *Farmer Lobbies*, *Conservation Lobbies* and *Social Conflict*. In any case, despite the significant role groundwater development is already playing in the eradication of poverty as well as towards fulfilling the United Nation's Millennium Goals, it cannot be seen as a panacea to solve all the world's water problems. These need to be dealt with on a case-by-case basis, in order to achieve an adequate conjunctive management of surface and groundwater resources. Finally, there is a real need to assess and correct the traditional imbalance in favour of conventional surface water systems that exists in most water agencies, and which is the main cause behind important social conflicts.

Keywords:   *Intensive groundwater use, silent revolution, irrigation efficiency, UN Millennium Goals*

## 1   INTRODUCTION

Currently, three issues pose an open challenge to the widely accepted paradigms of world water policy: the concept of virtual water, the improvements in desalination technology and the *Silent Revolution* of intensive groundwater use for irrigation. Abundant references to the first two already exist in scientific literature and have been widely discussed during the *Santander Workshop* (14–16 June 2004). This chapter is only concerned with the global implications of the latter.

In the last few decades, millions of farmers in arid and semi-arid countries have, at their own expense and risk, drilled their own wells in order to make use of the intrinsic advantages of groundwater-based irrigation. This phenomenon, the *Silent Revolution*, has thus been carried out with little or no planning on the part of governmental water agencies. The benefits of this *Silent Revolution* have been quite significant, even if some groundwater-related problems have also arisen in certain places. Precisely these have caused certain scholars and water policy makers to spread a series of *hydromyths*, which have led some to consider groundwater development *a pillar of sand* prone to collapse, or as a *bubble*, likely to burst (Postel, 1999). However, while these problems are sometimes real, it can be shown that they are many times pretended or exaggerated, or due to poor land use planning, rather than to intensive groundwater development.

The fact is that intensive groundwater use has been contributing for years to achieve two of the main goals set by the United Nations (2000) for the new Millennium: to halve by 2015 the number of people worldwide who suffer from malnutrition or do not have affordable access to drinking water. It must also be noted that fulfilling the another goal of the Millennium Declaration, i.e. to cut down by 50% the number of people living under the poverty threshold (US$ 1–2 per person per day), depends strongly on achieving the other two.

The aim of this paper is multiple. First, to provide a logically-phased overview of the *Silent Revolution* over time, paying attention to the key indicators involved; second, to outline the many socio-economic benefits groundwater irrigation has triggered in arid and semi-arid countries worldwide; third, to explain how the problems that arise from this *Silent Revolution* can be avoided or mitigated and sustainable groundwater development achieved; and fourth, to show how the lack of awareness on the reality of the *Silent Revolution* may induce a series of social and political conflicts (bridging the gap between scholars and decision makers, and those millions of poor farmers who fulfil their everyday needs with groundwater is an urgent and necessary action that can be implemented in a fast and economic manner, although it requires a political willingness).

It must be noted that irrigation nowadays amounts to about 70% of the total world water use (a figure which increases to about 80–95% in arid and semi-arid countries) and produces about half the food and fibres required by the human being (United Nations, 2003). Therefore, this chapter deals in particular with groundwater and agriculture, whilst those problems related to water in urban areas (or for industry or hydropower) are considered less important in the face of a potential water crisis, and thus remain beyond the scope of discussion.

## 2   THE *SILENT REVOLUTION* OF INTENSIVE GROUNDWATER USE FOR IRRIGATION

The last decades have witnessed a spectacular increase in the availability of new technologies such as the submersible pump, as well as significant improvements in the fields of drilling techniques and hydrogeology. Thus, in arid and semi-arid countries, groundwater irrigation has become much more profitable *de-facto* than surface water irrigation. As a consequence, a real *Silent Revolution* has taken place: during this period millions of independent (and usually poor) farmers in these countries have of their own accord implemented the necessary means to irrigate their land with groundwater. The participation of government agencies in the planning and control of these groundwater developments has usually been scarce.

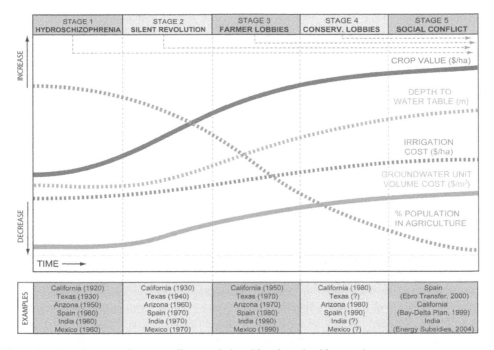

Figure 1.   Rough (ground)water policy trends in arid and semi-arid countries.

## 2.1   *Stages of the* silent revolution

Intensive groundwater development can be divided into a sequence of five stages that take place over time: *hydroschizophrenia*, *silent revolution*, *farmer lobbies*, *conservation lobbies* and *social conflict* (see Figure 1). It must be noted that while these stages typically span between 25 and 30 years (the approximate equivalent to one generation), this time might be longer in some cases, and some overlapping may also occur. For instance, in some countries where the last stage of social conflict has been reached (like Spain or India), it could be argued that *hydroschizophrenia* continues to take place.

### 2.1.1   *Stage 1: Hydroschizophrenia*

The first stage corresponds to *hydroschizophrenia*, a term mentioned by American hydrologist Raymond Nace in order to describe the attitude of many water decision makers who deal with surface and groundwater resources separately, often playing down the role of the latter. During this period, water policy is almost exclusively concerned with large surface water infrastructures heavily subsidised with public funds. At the same time the potential of groundwater resources tends to be overlooked.

Since hydrogeology is a fairly recent addition to the natural sciences, governmental water agencies have traditionally been managed by hydraulic engineers with a significant professional bias towards large surface water infrastructures. In fact, just about all major hydraulic works within the last one-hundred years have been under financial and operational control of government institutions. While this is understandable, unethical attitudes such as arrogance, negligence, vested interests, neglect and corruption have often played a role (Llamas &

Martinez-Santos, 2005). The practical consequence of this is that hydrological plans often consider surface waters as the only viable resource, even if competent studies prove local aquifers to be more advantageous, both from the economic and environmental point of view (Custodio, 2000, 2002). It is important to note that in certain situations, like in some Western USA states (e.g. California), *hydroschizophrenia* has been the consequence of legal constrains, rather than an *attitude* as such (the existing regulations were quite different in the case of surface water and groundwater).

Whilst it is not the aim of this paper to deny the interest of large surface water infrastructures, it is necessary to realise that these do not always constitute the unique, nor the best solution to cope with the need of a higher food output. In fact, it will be shown that groundwater irrigation can generally achieve more significant benefits than traditional surface water systems. It is also acknowledged that conjunctive use of surface and groundwater resources is in many cases the ideal solution. Many successful examples of conjunctive use exist in scientific literature, like the Barcelona water supply (Llamas, 1969), or others in Spain and the USA (Sahuquillo & Lluria, 2003). However, even if the technical solutions for the implementation of conjunctive use are well known, the main difficulties often arise from economic, legal and institutional constrains.

Some places, like the Punjab in Pakistan and India, or the California experience, have become an example of soil water logging due to over-irrigation from surface water systems. This problem, caused by the lack of a hydrogeological assessment, may be solved by depleting the water table through pumping (pumped groundwater can even be used for irrigation if quality allows). This is an example of poor conjunctive use.

### 2.1.2   *Stage 2: Silent revolution*

Wide availability (and affordability) of drilling and pumping techniques has led millions of farmers to carry out the *Silent Revolution* in many arid and semi-arid countries worldwide (Llamas & Martinez-Santos, in press; Deb Roy & Shah, 2003; Moench, 2003; Mukherji, 2003). While exceptions exist (California, Arizona and Texas, for instance), most of these farmers are poor and often illiterate. Thus, the intrinsic advantages of groundwater development, namely resilience against drought and ready availability of the resource on demand, constitute for them an important incentive for development. In fact, these result in the uncertainties about potential drought impacts being removed from the farmer's perception. Thus a new encouragement exists for them to invest in more efficient agricultural methods, including modern techniques such as sprinkler or drip irrigation.

Security against drought presents further effects. These are, for instance, described by Chadha (2004a), director of the Groundwater Central Board in India:

> "Development of groundwater has led to increased *drought proofing* of India's agricultural economy. The importance of this in the Indian context can be gleaned from the impact of droughts. In the 1960s groundwater was a relatively insignificant source of irrigation, particularly in Eastern India. In 1965–1966, rainfall (June to September), was 20% below normal, leading to drought conditions. National food grain production declined 19% over the previous year's level. In contrast, in 1987–1988 rainfall dropped almost 18% below normal, while food grain production only declined 2% over the previous year's level. Although the droughts are not directly comparable, the decline in production in 1987–1988 was significantly smaller than in 1965–1966, and much of it can be attributed to the spread of irrigation in general and of groundwater irrigation in particular.

Droughts have ripple effects throughout the Indian economy. Not only is there the direct loss of production; there are also numerous secondary effects. Vulnerable populations are particularly at risk, and are often forced to migrate in search for work. Public expenditures on drought irrigation and food distribution programs also increase substantially. The growth in India's irrigated areas, particularly the area irrigated with groundwater, has greatly reduced the economy's vulnerability to sharp reductions in rainfall, drought proofing and the rural economy in general and the crop sector in particular".

During this stage, the dramatic boost in pumpage may be coupled with a noticeable depletion of the water table, which in turn results in an increase in abstraction cost per unit volume. In contrast, the implementation of more efficient irrigation systems offsets this reality, slowing down the increase in the overall irrigation cost per hectare. Average crop value also begins to increase, due to the switch to cash crops. Finally, poor farmers become a strong middle class and are able to send their children to school and to college. As these are often trained for disciplines other than agriculture, a forward social transition begins to take place towards industry and the services sector (Moench, 2003).

### 2.1.3  *Stage 3: Farmer lobbies*

Ruthless groundwater development takes place in the pursuit of the aforementioned socio-economic benefits. Consequently, the trends described under the previous heading continue. Farmers have by then become wealthier and better educated, and as pumping costs increase with water table depletion, they begin to form strong pressure groups to lobby for *perverse* subsidies in the shape of cheap energy and/or water (dams and inter-basin transfers to be funded by all tax payers).

It is during this stage when groundwater related problems may begin to occur, thus accelerating the action of farmer lobbies. Such problems include water table depletion, groundwater quality degradation, land subsidence and ecological impacts upon wetlands and streamflows, and have been described in Llamas & Martinez-Santos (2005). Water table depletion is the most frequent, and thus there exists a widespread misconception that considers it the most serious of the lot. However, it may in many cases prove irrelevant (as will be shown further ahead). In fact, the authors are not aware of any cases where socio-economic havoc has been triggered by intensive groundwater exploitation of medium to large aquifer systems (surface over 500 km$^2$). Moreover, even in very small aquifers, like the case of Crevillente in Spain (90 km$^2$), abstraction takes place from a depth of about 500 m (see Table 2, in section 4.1).

Human-induced water quality degradation, or the *hydrocide*, is in fact the most significant one of the problems listed above, since it severely affects crop value and is likely to appear before water table depletion becomes an issue. However, it must be borne in mind that most groundwater-quality related problems are usually a consequence of poor land use. For instance, in the case of many countries in Central and Western Europe, groundwater quality degradation is not due to intensive pumping, but to the agrochemicals used in rain-fed agriculture.

### 2.1.4  *Stage 4: Conservation lobbies*

Large hydraulic infrastructures are often met with controversy, as they tend to demand large investments of public funds and to clash with the interests of different groups: for instance environmental organizations (WWF, Greenpeace) and other regional groups who consider these projects harmful to their rights and interests. These usually unite into a common front in order to oppose government authority and the economic power of large construction companies. Further hostility to these projects may come from scientific and economic-based analyses, usually carried out by scholars. For instance, the final report of the World Commission on Dams (WCD, 2000) notes that the economic advantages of many large surface water irrigation projects are doubtful.

### 2.1.5    *Stage 5: Social conflict*

The importance of water resources in the livelihoods of many people, as well as the added emotional factor which opposing parts may try to exploit in their own favour, can make these conflicts extend to wide sectors of society. The *Ebro River Transfer* in Spain, the CALFED controversy in the USA, or the social unrest due to the *Narmada Valley Project* provide current examples of these conflicts – which practically never in historical terms have become real water-wars (Asmal, 2000).

However, it often seems to be ignored that poor groundwater resources planning is at the very root of these conflicts. S.D. Limaye (July 2004, pers. comm.), vice-president of the IAH (International Association of Hydrogeologists) Asian Chapter, explained this situation in the following terms:

> "The attitude of major international funding agencies and other international organizations, towards groundwater as a resource and also towards countries making extensive use of ground-water, has always been skeptical. For example: World Bank and Asian Development Bank have sponsored several projects for surface water development. Their support for groundwater projects has been minimal, on the grounds that it is a hidden resource, difficult to explore and assess, energy is required for pumping, . . .
>
> In developing countries, politicians have always preferred surface water projects because they are grand, big-budget items with a great impact on the people in their constituencies and a greater opportunity for corruption".

As long as this situation continues, the current gap between the way millions of farmers worldwide attend to their everyday water needs and the perception of water policy decision makers seems likely to remain.

### 2.2    *Key indicators of intensive groundwater use*

In Figure 1, a series of social, economic and technical indicators (namely, crop value, abstraction cost, depth to water table, irrigation cost and % population in agriculture), constitute the key indicators to follow in the *Silent Revolution* of intensive groundwater development. A somere description of these is provided below.

(a) Groundwater abstraction cost per unit volume (US$/m$^3$): Groundwater abstraction cost increases with depth (because more energy is required for pumping and the need to drill deeper wells). Abstraction costs usually range between US$ 0.01/m$^3$ and US$ 0.20/m$^3$. Presently, the most significant water table depletion known to the authors has taken place in Crevillente (Spain), where groundwater is being pumped from a depth of 500 m at a cost of US$ 0.30/m$^3$. See Hamer (2002), for California; Chebaane *et al.* (2004), for Jordan; Hernández-Mora *et al.* (2001), for Spain; Deb Roy & Shah (2003), for Southeast Asia. An extreme case is the cost of groundwater abstraction carried out by one-dollar-a-day farmers by means of treadle pumps (Polak, 2004).

(b) Depth to water table (m): Intensive groundwater use causes a depletion rate in the water table often ranging between 0.5 and 5 m/yr (even more in some cases). However, this is seldom a deterrent issue except in shallow aquifers, which could reach physical exhaustion. In fact, even the increase in the cost of energy for pumping is a relatively small problem in comparison with potential groundwater quality degradation (like for instance in the case of seawater intrusion in coastal aquifers), and certain equity issues arising from the drying of shallow wells (usually owned by poor farmers who may not have the possibility to drill any further).

(c) Irrigation cost (US$/ha): Groundwater abstraction cost per unit volume may increase with time (up to ten-fold due to water table depletion). While irrigation cost also increases, it does so at a lower rate, as farmers begin to use more efficient agricultural and irrigation technology and switch to lower water-consuming crops (for instance, from corn or rice to grapes and olive trees). It is estimated that irrigation cost ranges between US$ 20/ha (e.g. in the case of cereals in good shallow aquifers) and US$ 1000/ha (in aquifers where water is pumped from a depth of hundreds of metres and crop water consumption is up to 3000 or 4000 $m^3$/ha/yr).

(d) Population in agriculture (%): This is a measure of the social transition triggered by groundwater development. Groundwater irrigation usually helps to improve the socio-economic status of poor farmers to a greater extent than traditional surface water systems (Deb Roy & Shah, 2003). This is a consequence of the entrepreneurial mentality of those farmers who decide to invest in their own wells, as well as to crop security due to the resilience of aquifers against drought. In the meantime, those farmers have been able to send their children to college to receive a better education. Some of these will be trained in agricultural disciplines, and then return to farming (albeit with a better grasp for technology). Others, however, will move on to become engineers, teachers, doctors, economists and so forth. As a consequence, within one or two generations, farmers have a chance to move from agriculture onto industry and the services sector. In Spain, for instance, the share of population in agriculture has decreased from 50% to below 6% in the last two generations. Similar examples can be found in many industrialised countries, where the main economic sector a century ago was probably agriculture (the case of the USA, a country where 3% of the population is able to feed the remaining 97%, and which still is the most important food-exporter in the world).

(e) Crop value (US$/ha): This is perhaps the most significant aspect of the *Silent Revolution*. As farmers become richer and more educated, they move from low-value to cash crops. This is mainly due to the ready availability of groundwater and the resilience of aquifers against drought, which also encourage farmers to make greater investments in order to improve their agricultural and irrigation technology and to grow cash crops. Crop value is not only related to crop type. In fact, the value of cash crops per hectare also depends largely on agricultural techniques applied and market conditions. Thus, it ranges widely: for instance, between US$ 500/ha (e.g. cereals) and over US$ 70,000/ha (tomatoes, cucumber and other greenhouse crops sold in other European states) in Spain, although these figures are probably similar in many other countries.

It must be noted that these indicators do not take into account the cost of externalities such as the degradation of aquatic ecosystems. There are two main reasons for it: one, the difficulty of assessing such cost, as described by the National Research Council (1997); and two, the fact that, according to Kuznets environmental curve, poverty is really the worst enemy of the environment (a good example of this is the situation in Southeast Asia). It must be borne in mind that concepts such as sustainability or the value of ecosystems are not easy to understand for governments nor for one-dollar-a-day farmers in developing countries. The situation is very different in richer countries, where the socio-economic setting is radically different, and a series of different mechanisms exist to protect the environment (for instance the Western USA, where an important tool for the defence of water environments is the *Endangered Species Act*).

## 3    GROUNDWATER AND HUMAN DEVELOPMENT: SOME BENEFITS OF THE *SILENT REVOLUTION*

### 3.1    *Socio-economic benefits of intensive groundwater use*

As stated before, the *Silent Revolution* is a market-driven phenomenon. In fact, the socio-economic efficiency of groundwater irrigation is between 3 and 10 times higher than that of surface water systems in terms of economic productivity (US$/$m^3$) and employment generation (jobs/$m^3$). Table 1 shows a socio-economic assessment of these figures in Andalusia, Spain. This analysis has been extended to other regions of Spain (Hernández-Mora & Llamas, 2001) with different climatic and social conditions, and the results are also similar. A recent paper (Vives, 2003) seems to confirm that the trend continues.

Data from countries such as India provide even more spectacular results. According to Dhawan (1995), yields in areas irrigated with groundwater are 1/3 to 1/2 higher than those in areas irrigated with surface resources. In a previous report, Dains & Pawar (1987) estimated that as much as 70–80% of India's agricultural output might be groundwater dependent. The Indian Water Resources Society (1999) published, among others, the following significant data:

– Groundwater is contributing at present 50% of irrigation surface, 80% of water for domestic use in rural areas, and 50% of water in urban and industrial areas.
– Groundwater abstraction structures had increased from 4 million in 1951 to nearly 17 million in 1997.
– In the same period the groundwater irrigated area had increased from 6 to 26 million ha, and it is estimated that this rapid pace of development is likely to continue and will reach 64 million ha in the year 2007.

By indirect calculation it may be estimated that in India the average amount of water applied in surface water irrigation is around 16,000 $m^3$/ha/yr; in groundwater irrigation this ratio is

Table 1.    Comparative socio-economic efficiency of surface water and groundwater irrigation in Andalusia, Spain.

| Indicator | Surface water | Groundwater | Total |
| --- | --- | --- | --- |
| Irrigated surface ($10^3$ ha) | 600 | 210 | 810 |
| Average economic production (€/ha) | 3300 | 8600 | 4600 |
| Average consumption at origin ($m^3$/ha/yr) | 7400 | 4000 | 6500 |
| Water productivity (€/$m^3$) | 0.42 | 2.16 | 0.72 |
| Employment generated (EAJ*/$10^6$ $m^3$) | 17 | 58 | 25 |

*EAJ: Equivalent annual jobs
*Source*: Hernández-Mora *et al.* (2001).

only 4000 m³/ha/yr. In other words, it seems that in India and on average, the economic yield in irrigation and by m³ is from 5 or 10 times higher when groundwater is used than when irrigation is carried out with surface water.

Some recent reports indicate that the spectacular increase of groundwater use in Southeast Asia continues. Chadha (2004a, 2004b), director of the Groundwater Central Board, considers that the renewable groundwater recharge in India is in the order of 450 km³/yr. In 1997, surface irrigated was about 46 million ha, and the total number of wells about 17 million. According to the same source, total pumping amounted to 150 km³/yr, and the total usable groundwater storage (down to 450 m in alluvial aquifers and 150 m in hard rock aquifers) to 11,000 km³. A few months after the previous reports, Shah (in press) points out that 60% of India's irrigated area is served by groundwater wells. An independent survey suggests that the figure may well be 75%, and even more if conjunctive uses are included. Shah estimates that India has been adding 0.8–1 million new tubewells every year since 1996, and that current pumping volume is in the order of 200 km³/yr (approximately twice the figure quoted by the Indian Water Resources Society).

Overall, most authors (almost without exception) conclude that groundwater irrigation is a major catalyst for rural development and that its productivity, even in the case of farmers who do not have a well and need to purchase water off others, is higher than that of surface water irrigation.

What are the reasons for the higher productivity of groundwater irrigation in relation to traditional surface water systems? The answer to this question, as previously stated, is multiple:

(a) In the first place are the physical advantages of groundwater in terms of drought resilience and ready availability of the resource on demand. These result in crop security, which in turn leads farmers to invest in better agricultural and irrigation technologies. Since abstraction cost (US$/m³) is higher than that of traditional surface water systems, wasteful over-irrigation is discouraged (however, when electrical energy is heavily subsidised, the incentive for efficient irrigation is smaller).

(b) Due to the above, farmers increasingly tend to a switch towards cash crops. A more efficient use ensues, thus maximising economic return per m³. In fact, while pumping cost increases with time (due to water table depletion), so does crop value, but at a much higher rate. Overall, it appears that crop value tends to be at least 20 times higher than abstraction cost.

The available data shows that the above seems to hold in arid and semi-arid regions worldwide. Therefore, there is a real need to carry out world-scale analysis of these figures in order to: (1) challenge those so-called world water crises; (2) give groundwater its rightful importance within the framework of global water policy; and (3) avoid problems to come if the current situation of chaotic groundwater development is to continue.

### 3.2 *Intensive groundwater use as a catalyst for social transition*

Since groundwater is a resource available and accessible to the poor, it has often acted as a catalyst for a quick social transition: often from a poor and illiterate society to a well-developed middle class, with a strong industry and/or services sector (Moench, 2003).

Several documented cases of this phenomenon exist in the scientific literature. For instance, Chebaane *et al.* (2004) cite groundwater-based irrigation development as the main driving force behind small villages becoming important provincial capitals in Jordan.

It must be acknowledged that a series of tragic examples exist in reference to suicides by Indian farmers (due to well failure in poor hard-rock aquifers), as well as to the re-location of small communities in Yemen – as a consequence of water table depletion, according to Moench (2003). While these are indeed sad cases (and cannot be overlooked), it is important to note that they correspond to very isolated situations under very specific hydrogeological settings. In fact, the authors of this chapter do not know of any cases of medium or large aquifers (surface area larger than 500 km$^2$) where socio-economic havoc has been triggered by intensive groundwater use.

### 3.3   *Groundwater developments are less prone to corruption than large surface water infrastructures*

Groundwater development is, by its own very nature, generally less prone to corruption than large surface water infrastructures. Indeed, the very nature of the latter (often large in size and investments) makes them a juicy prospect in the eyes of corrupt officials and corporate decision-makers. On the other hand, groundwater developments are usually carried out by individuals whose economic means are scarce.

Cases of corruption are not of an isolated nature. In a few international conferences and reports the practical importance of corruption and bribery in the water management has been emphasised. This was an issue that previously was very rarely mentioned in such conferences or in international reports. Perhaps the most important document on this topic is the book by OECD (2000) dealing with the *Convention on Combating Bribery of Foreign Public Officials in International Business Transactions*. This book was specifically quoted in the Report of the World Commission on Dams (WCD, 2000: 186–187, 249). This author has also mentioned the practical relevance of corruption as a frequent driving force in water policy in several papers (Delli Priscoli & Llamas, 2001; Llamas & Delli Priscoli, 2000; Llamas & Custodio, 2003; Llamas & Martínez-Santos, 2005). Postel (1999: 229) presents an interesting view of the role of consulting firms, construction companies and politicians in promoting large surface water irrigation projects, which is transcribed below:

> "The rules, especially for large schemes, often look something like this: a politician seizes the potential for bolstering political support by proposing an irrigation project in a strategic location. Engineering firms lobby the decision-makers in order to raise their chances of winning the project's construction contract. The politician, in collusion with colleagues who also want to please their constituents with pork-barrel projects, sees to it that the nation's taxpayers pay most of the bill. The farmers themselves pay only a small fraction of the project's cost; in return, they support the politician in the next election. With prices kept artificially low, farmers have little incentive to use water efficiently. The irrigation system never becomes financially self-sustaining because the meager fees collected from the farmers do not cover the system's operation and maintenance costs, much less its capital costs. National or state irrigation agencies keep the projects running with taxpayer funds. If budgets become tight and maintenance work is neglected, the systems fall into disrepair. Gradually, agricultural output and benefits begin to decline. Either the government saves the project by allocating more taxpayer money for expensive rehabilitation work, or the system deteriorates until farmers abandon it. Alternatively, international donors come to the rescue with funds provided by taxpayers in wealthier countries".

Another issue to be considered is the almost universal policy of public *perverse subsidies* for water and agriculture. According to Myers & Kent (1998), these subsidies are those which are noxious both for the economy and the environment. In most cases, the water users only pay a small fraction of the real cost of the water supplied. This is especially true in surface water for irrigation. Water policy all over the world has, during the past decades, focused on the management of the supply and not on the management of the demand. This has induced an almost universal wasteful use of water.

In most groundwater developments the situation may be quite different. Owners of wells usually pay for their construction, maintenance and operation. But they do not usually pay the external costs caused by the impact of groundwater abstraction.

## 4   NEGATIVE IMPACTS ARISING FROM INTENSIVE GROUNDWATER DEVELOPMENT

It would be unwise to advocate widespread groundwater use as a universal solution for the world's water problems. Uncontrolled groundwater development can lead to a series of effects including water table depletion (the most frequent), groundwater quality degradation, land subsidence as well as to other environmental impacts such as streamflow reductions and desiccation of wetlands. As a consequence, certain water officers (and often journalists) have voiced the *hydromyth* of groundwater as a fragile resource.

Out of these impacts, groundwater quality degradation, wisely termed *hydrocide* by Prof. Lundqvist, is the most significant one, albeit not the most frequent. In fact, the huge storage capacity of most aquifers suggests that it usually takes about 2–3 generations of heavy pumping before any of such negative effects becomes significant. In any case, it can be shown that most of these problems are due to poor groundwater governance and land use planning, rather than to intensive groundwater use.

### 4.1   *The controversy on groundwater mining*

Groundwater mining (pumping at a rate in excess of aquifer recharge) is often seen by many as a kind of ecological sin, opposed to the basic tenets of sustainable development. Nevertheless, a good number of authors disagree with this view, considering that under certain circumstances, groundwater mining can be a sensible alternative (Freeze & Cherry, 1979; Issar & Nativ, 1988; Llamas, 1992, 1999; Collin & Margat, 1993; Margat, 1994; Lloyd, 1997; Custodio, 2002; Price, 2002; Abderrahman, 2003).

Sustainability is often defined as "meeting the current needs without compromising those of future generations". In this regard, it is necessary to clarify that this definition is not comprehensive, and that sustainable development does not constitute an end in itself (Price, 2002). Indeed, questions such as "will future generations need it more than the present one?", do not find an answer in this definition.

Whether groundwater mining falls under the umbrella of sustainability has to be dealt with on a case-by-case basis. In most countries it is considered that groundwater abstraction should not exceed the renewable resources. In other countries – mainly in the most arid ones – it might be considered that groundwater mining is an acceptable policy, as long as available data assure that the groundwater development can be economically maintained for a long time, for example, more than fifty years and that the potential ecological costs and socio-economic benefits have been adequately evaluated.

In Saudi Arabia, according to Dabbagh & Abderrahman (1997), the main aquifers (within the first 300 m of depth) contain huge amount of fresh fossil water (a minimum of 2000 km$^3$) that is 10,000 to 30,000 years old. It is considered that these fossil aquifers can supply useful water for a minimum period of 150 years. Current abstraction seems to be around 15–20 km$^3$/yr. During a couple of decades the Saudi government has pumped several km$^3$/yr of non-renewable groundwater to grow low cost crops (mainly cereals) which were also heavily subsidised. The official aim of such activity was to help to transform nomadic groups into farmers. Apparently such *overdraft* has been a success. Now the amount of groundwater abstraction has been dramatically reduced and the farmer nomads have become high-tech farmers growing cash crops. Another example is the situation of the Nubian sandstone aquifer located below the Western desert of Egypt. According to Idriss & Nour (1990), the fresh groundwater reserves are higher than 200 km$^3$ and the maximum pumping projected is lower than 1 km$^3$/yr. Probably similar situations do exist in Libya and Algeria. Other examples of mining groundwater can be found in Llamas & Custodio (2003).

The Indian case has a special interest in this regard, not only because India is the number one country in the world when it comes to groundwater use, but also because a good number of authors have dismissed groundwater development as a *pillar of sand*, soon to collapse. A recent paper by the director of the Indian Central Groundwater Board (Chadha, 2004a) states that the number of *dark blocks* (by *blocks* meaning aquifers) in India has grown from some 250 in 1984–1985 to 450 in 1997–1998. However, this author puts a question mark on the World Bank's categorisation of blocks into *white, grey* and *dark*. There is not enough space here to discuss the manifold concept of overexploitation (see Custodio, 2002; Llamas & Custodio, 2003). Yet, it should suffice to say that such categorisation in blocks is based on the estimated difference between pumping and recharge, when it is well known that groundwater recharge can be substantially increased by water table depletion. It seems, in any case, that the accuracy of the data used in this assessment is rather low.

It is not easy to achieve a virtuous middle. As Collin & Margat (1993) state: "we move rapidly from one extreme to the other, and the tempting solutions put forward by zealots calling for *Malthusian underexploitation of groundwater* could prove just as damaging to the development of society as certain types of excessive *pumping*".

An extreme example of groundwater mining is the Crevillente aquifer, Spain (Table 2), a small (90 km$^2$) limestone formation where heavy pumping has taken place since the 1970s. As a consequence, the water table has been depleted at a rate of 10–15 m/yr, and required pumping elevation now reaches 500 m (wells are up to 700 m deep). The approximate abstraction cost

Table 2.   Groundwater development in the Crevillente aquifer, Spain: an extreme case.

| | |
|---|---|
| Aquifer settings (size, geology) | 90 km$^2$, limestones |
| Estimated recharge/abstraction | 2–16 Mm$^3$/yr |
| Initial pumping elevation (1970s) | 20–30 m |
| Current pumping elevation | 500 m |
| Groundwater abstraction cost | 0.30 €/m$^3$ |
| Irrigation cost (grapes) | 1000 €/ha/yr → 3300 €/ha/yr |
| Crop value | 25,000 €/ha → 15,000 €/ha |

*Source*: F. Corchón, of the Jucar Basin Authority (June 2004, pers. comm.).

is currently 0.30 €/m³. While these world-record figures are in place, the system has not yet collapsed.

However, crop value has gone down from 25,000 €/ha to about 15,000 €/ha, due to the lower fertility of the soil associated with an increase in groundwater salinity (not so much with the extra cost induced by groundwater depletion). This has become a pressing issue, and thus the Crevillente farmers' lobby, together with others (farmers who obtain their water from small nearby aquifers as well as urban water supply companies) have obtained the approval of a seemingly needless inter-basin transfer (a total of 90 Mm³/yr with a cost of 230 million €, to be funded 2/3 by public funds and 1/3 by the beneficiaries).

The idea that groundwater mining might be an ethically sound practice under certain circumstances was developed by Llamas (1999) in a UNESCO Conference held in Lybia in 1998. This thesis has been followed in the first report of the Chairman of the Subcommittee of the UNESCO World Commission on the Ethics of Science and Technology (Selborne, 2000; Delli Priscoli & Llamas, 2001).

### 4.2 *Water quality degradation: the* hydrocide

As stated above, water quality degradation is the main problem associated with intensive groundwater use. Groundwater abstraction can certainly cause, directly or indirectly, changes in groundwater quality. The change in hydraulic gradient due to groundwater abstraction might constitute cause of quality degradation, as it may result in intrusion from adjacent saline water bodies (other aquifers or the sea). This is a typical problem in many coastal regions in arid and semi-arid regions, although it is well known that coastal aquifers, like the ones in Israel or in Orange County, California, can be managed in a sustainable way. The relevance of the saline water intrusion not only depends on the total withdrawal in relation to the natural groundwater recharge, but also on the well field location and design, and on the geometry and hydrogeological parameters of the aquifer. Thus, in many cases the existing problems are due to uncontrolled and unplanned groundwater development and not to excessive pumping (Custodio & Bruggeman, 1982).

Quality degradation may not be related at all to excessive groundwater abstraction in relation to average natural recharge. Other causes may be responsible, such as return flow from surface water irrigation, leakage from urban sewers, infiltration ponds for wastewaters, septic tanks, urban solid waste landfills, abandoned wells, mine tailings and many other activities (Barraqué, 1997).

### 4.3 *Land subsidence*

Sedimentary formations are deposited at low density and large porosity. As subsequent layers are deposited the overburden compresses the underlying strata. The overburden is in static equilibrium with the intergranular stress and the pore water pressure. This equilibrium is quickly reached in coarse-granular layers, but in fine-grained layers with low permeability it may take a long time. The effect of this process is the natural progressive consolidation of sediments.

When an aquifer is pumped the water pore pressure is decreased and the aquifer solid matrix undergoes a greater mechanical stress. This greater stress may produce compaction of the existing fine-grained sediments (aquitards) if the stress due to the decrease in water pore pressure is greater than the so-called *preconsolidation* stress. This is a well-known situation which has occurred in some aquifers formed by young sediments, such as those in Mexico City, Venice, Bangkok and others (Poland, 1985). The Mexico City case is probably the most dramatic one, although since it is related to urban water supply and not to irrigation, it will not be discussed here.

Caves and other types of empty spaces may exist under the water table in karstic aquifers. When the water table is naturally depleted the mechanical stability of the *roof* of such empty spaces may be lost and the roof of the cave collapses. This is a natural process that gives rise to the classical *dolines and poljes* in the karstic landscape. When the water table depletion or oscillation is increased by groundwater abstraction, the frequency of karstic collapses can be also increased. The accurate prediction of such collapses is not easy (LaMoreaux & Newton, 1992).

In both cases the amount of subsidence or the probability of collapses is related to the decrease in pore water pressure which is related to the amount of groundwater withdrawal. Nevertheless the influence of other geotechnical factors may be more relevant that the amount of water abstracted in relation to the renewable groundwater resources of the aquifer.

### 4.4   *Environmental impacts on wetlands and streamflows*

The ecological cost of groundwater development should be compared with the socio-economic benefits produced (Barbier *et al.*, 1997; National Research Council, 1997). The evaluation of the ecological impacts is highly dependent on the social perception of ecological values in the corresponding region, a perception which is changing rapidly in most countries. For instance, the Framework Directive on Water of the European Union of the year 2000 paid attention to monitoring and conservation of aquatic ecosystems and especially to wetlands.

In arid and semi-arid regions, wetlands or oases are usually rare and frequently related to groundwater discharge zones. The development of groundwater for irrigation or other uses may often have a significant negative impact on the hydrological functioning of wetlands or oases (Fornés & Llamas, 1999). These impacts should be properly evaluated by decision-makers, bearing in mind that there is no blueprint solution. This issue, very much in relation with Kuznets environmental curve, is particularly relevant in the case of developing countries, where poverty constitutes the worst enemy of the environment, and where paradigms proper to wet industrialised countries are not likely to succeed. On the other hand, in industrialised countries such as Spain, conflicts between farmers and conservation groups are relevant (Brufao & Llamas, 2003; Llamas, 2003).

## 5   STRATEGIES FOR SUSTAINABLE GROUNDWATER USE IN ARID AND SEMI-ARID REGIONS

There exists a general consensus that, in order to avoid conflicts and to move from confrontation to cooperation, water development projects require the participation of the social groups affected by the projects, the stakeholders. The participation should begin in the early stages of the project and should be, as much as possible, bottom-up and not top-down. The first question is to define who the stakeholders are; the second, how, when and where they should intervene in the decision making processes.

The Spanish experience, in trying to implement groundwater management as a public dominion, indicates clearly that the active collaboration of *Groundwater Users Associations* is a key element (Hernández-Mora & Llamas, 2000, 2001). However, the process implementation demands sometime to switch from old to new paradigms (Lopez Gunn & Llamas, 2001).

In July 2001, the Spanish Parliament approved the *Law of the National Water Plan*. This Plan included several provisions that, if really enforced, would change the chaotic situation of groundwater development in Spain. Perhaps the most important article of this Law was the one which strictly demanded setting up of Groundwater Users Associations in every intensively developed aquifer and a thorough hydrogeological assessment of every aquifer which may be supposed to receive a surface water transfer from other catchments. The National Water Plan Law also stated that an intense and broad Water Education Programme has to be implemented. These provisions, together with the 1000 Mm$^3$/yr *Ebro River Transfer* to the *overexploited* aquifers of Southeastern Spain, have however been changed due to the government overturn of March 2004. The new Plan, subject to a heated economic and environmental debate, advocates the building of a series of desalination plants in the Mediterranean coast. As it can be easily seen, the paradigms from a century of *perverse subsidies* have not been overcome, and the existing administrative chaos in relation to groundwater seems not to be in the mind of Spanish water policy decision-makers.

Obviously, there is no blueprint for a universal solution. For example, in some arid and semi-arid developing countries, when dealing with correction of ecological impacts of over-exploitation, the influence of conservationists groups will probably be weak compared to the influence of farmers associations or urban water supply companies.

The necessary participation of the stakeholders demands that they are aware of the way the issue at hand will affect them directly or indirectly, and also a basic knowledge of the hydrogeological concepts involved in aquifer development. Probably in most countries there exist a good number of *hydromyths* or obsolete paradigms about the origin, movement and potential pollution of groundwater. In any stressed aquifer it is essential to organise different types of educational activities aimed at different groups: from school students and teachers to officials of Water Administrations, as described by McClurg & Sudman (2003), and other authors and institutions.

## 6   CONCLUSIONS

In the last few decades, a *Silent Revolution* of intensive groundwater use for irrigation has taken place in arid and semi-arid countries worldwide. This has been carried out by millions of independent farmers who, at their own expense and risk, have drilled wells in order to access the intrinsic advantages of groundwater. An analysis of the key social, economic and technical indicators involved, reveals that the guaranteed crop value is usually much higher that the pumping cost (at least 20 times, in most cases). Therefore, this *Silent Revolution* is market-driven, and thus unstoppable if business continues as usual.

The *Silent Revolution* has taken place with little or no control on the part of government authorities. As a consequence a series of problems, at times pretended and at times genuine, have arisen in some places. These have been often magnified, giving rise to the *hydromyth* of groundwater development as a *pillar of sand* (a fragile resource). However, to the knowledge of these authors, there are no documented cases where the *Silent Revolution* has induced socio-economic havoc. This may be due to the huge storage capacity of most aquifers, which allows for one or two generations of heavy pumping before serious impacts begin to be felt (it must be borne in mind that in the meantime a social transition away from agriculture has also taken place). However, it seems that this situation cannot last forever.

Therefore, three fairly cheap and concrete steps are suggested towards a sustainable use of groundwater resources:

(1)  Carry out a thorough and transparent assessment of the real situation of surface water and groundwater, together with their respective technical, economic and social efficiency. This assessment could probably carried out within a year in most cases.
(2)  Implement a massive water education campaign, paying attention to the special characteristics of groundwater, stressing the need for a *common pool resource* type development.
(3)  Governments should foster the creation of groundwater users associations in order to manage intensively developed aquifers.

REFERENCES

Abderrahman, W.A. (2003). Should intensive use of non-renewable groundwater resources always be rejected? In: M.R. Llamas & E. Custodio (eds.), *Intensive Use of Groundwater. Challenges and opportunities*. Balkema Publishers. Lisse, the Netherlands: 191–206.

Asmal, K. (2000). *Water: from casus belli to catalyst for peace*. Stockholm Water Week (address in the opening session, 14th August 2000).

Barbier, E.B.; Acreman, M. & Knowler, D. (1997). *Economic evaluation of wetlands: a guide for policy makers and planners*. Ramsar Convention Bureau. Gland, Switzerland: 127 pp.

Barraqué, B. (1997). Groundwater management in Europe; regulatory, organizational and institutional change. *Proceedings of the International Workshop: how to cope with degrading groundwater quality in Europe*. Stockholm, Sweden, 21–22 October: 16 pp (preprint).

Brufao, P. & Llamas, M.R. (eds.) (2003). *Conflictos entre el desarrollo de las aguas subterráneas y la conservación de humedales: aspectos legales, institucionales y económicos*. Fundación Marcelino Botín and Ediciones Mundi-Prensa. Madrid, Spain: 337 pp.

Chadha, D.K. (2004a). Status of groundwater development and management in India. In: *Resources Conservation and Food Security*. Concept Publishing Co., New Delhi, India. Vol. 1: 9–30.

Chadha, D.K. (2004b). Groundwater potential development and population: a critical review. In: Sundaram *et al.* (eds.), *National Resources Management and Livelihood Security*. Concept Publishing Co., New Delhi, India: 42–77.

Chebaane, M.; El-Naser, H.; Filch, J.; Hijazi, A. & Jabbarin, A. (2004). Participatory groundwater management in Jordan: development and analysis of options. *Hydrogeology Journal*, 12(1): 14–33.

Collin, J.J. & Margat, J. (1993). Overexploitation of water resources: overreaction or an economic reality? *Hydroplus*, 36: 26–37.

Custodio, E. (2000). *The complex concept of groundwater overexploitation*. Papeles Proyecto Aguas Subterráneas, A1. Fundación Marcelino Botín, Santander, Spain: 58 pp.

Custodio, E. (2002). Aquifer overexploitation: what does it mean? *Hydrogeology Journal*, 10: 254–277.

Custodio, E. & Bruggeman, G.E. (1982). *Groundwater problems in coastal areas*. Studies and Reports in Hydrology, 45. UNESCO, Paris, France: 650 pp.

Dabbagh, A.E. & Abderrahman, W.A. (1997). Management of groundwater resources under various irrigation water use scenarios in Saudi Arabia. *The Arabian Journal for Science and Engineering*, 22(IC): 47–64.

Dains, S.R. & Pawar, J.R. (1987). *Economic return to irrigation in India*. Report prepared by SDR Research Group Inc. for the U.S. Agency for International Development. New Delhi, India.

Deb Roy, A. & Shah, T. (2003). Socio-ecology of groundwater irrigation in India. In: M.R. Llamas & E. Custodio (eds.), *Intensive Use of Groundwater. Challenges and opportunities*. Balkema Publishers. Lisse, the Netherlands: 307–336.

Delli Priscoli, J. & Llamas, M.R. (2001). International perspective in ethical dilemmas in the water industry. In: C.K. Davis & R.E. McGinn (eds.), *Navigating in Rough Waters.* American Water Works Association. Denver, Colorado, USA: 41–64.

Dhawan, B.D. (1995). *Groundwater depletion, land degradation and irrigated agriculture in India.* Commonwealth Publisher. New Delhi, India.

Fornés, J. & Llamas, M.R. (1999). Conflicts between groundwater abstraction for irrigation and wetlands conservation: achieving sustainable development in La Mancha Húmeda Biosphere Reserve (Spain). In: C. Griebler *et al.* (eds.), *Groundwater Ecology. A Tool for Management of Water Resources.* European Commission. Environmental and Climate Programme: 227–236.

Freeze, R.A. & Cherry, J.A. (1979). *Groundwater.* Prentice-Hall. Englewood Cliffs. New Jersey, USA: 604 pp.

Hamer, G. (2002). The cost of water in Southern California. In: *Proceedings of California Groundwater Resources Association Conference* (September). Newport Beach. Power Point Presentation, 32 slides.

Hernández-Mora, N. & Llamas, M.R. (2000). *The role of user groups in Spain: participation and conflict in groundwater management.* CD-ROM of the X World Water Congress. Melbourne, Australia, 12–16 March. International Association of Water Resources: 9 pp.

Hernández-Mora, N. & Llamas, M.R. (eds.) (2001). *La Economía del Agua Subterránea y su Gestión Colectiva.* Fundación Marcelino Botín and Ediciones Mundi-Prensa. Madrid, Spain: 549 pp.

Hernández-Mora, N.; Llamas, M.R. & Martínez Cortina, L. (2001). Misconceptions in Aquifer Overexploitation. Implications for Water Policy in Southern Europe. In: C. Dosi (ed.), *Agricultural Use of Groundwater.* Towards Integration between Agricultural Policy and Water Resources Management. Kluwer Academic Publishers: 107–125.

Idriss, H. & Nour, S. (1990). Present groundwater status in Egypt and environmental impacts. *Environmental Geology and Water Sciences*, 16(3): 171–177.

Indian Water Resources Society (1999). *Water: Vision 2050.* New Delhi, India: 74 pp.

Issar, A.S. & Nativ, R. (1988). Water beneath the desert: keys to the past, a resource for the present. *Episodes*, 11(4): 256–262.

LaMoreaux, P.E. & Newton, J.G. (1992). Environmental effects of overexploitation in a karst terrain. In: I. Simmers, F. Villarroya & L.F. Rebollo (eds.), *Selected Papers on overexploitation.* International Association of Hydrogeologists. Selected Papers, 3. Heise, Hannover, Germany: 107–113.

Llamas, M.R. (1969). Conjunctive use of surface and groundwater in the water supply of Barcelona, Spain. *Bulletin of the International Association of Scientific Hydrology*, 14(3): 119–136.

Llamas, M.R. (1992). La surexploitation des aquifères: aspects techniques et institutionnels. *Hydrogeologie*, 4: 139–144. Orleans, France.

Llamas, M.R. (1999). *Consideration on Ethical Issues in Relation to Groundwater Development and/or Mining.* In: UNESCO International Conference on Regional Aquifer Systems in Arid Zones. Managing Non-Renewable Resources. Tripoli, Libya: 20–24. Also in: Technical Documents in Hydrology, V IHP, No. 42, UNESCO, Paris, France: 467–480.

Llamas, M.R. (2003). Epilogue. In: Special Issue on Water in the Iberian Peninsula. *Water International*, 28(3): 405–409.

Llamas, M.R. & Custodio, E. (eds.) (2003). *Intensive Use of Groundwater. Challenges and Opportunities.* Balkema Publishers. Lisse, the Netherlands: 478 pp.

Llamas, M.R. & Delli Priscoli, J. (2000). *Water and Ethics.* Papeles Proyecto Aguas Subterráneas, A5. Fundación Marcelino Botín. Santander, Spain: 56–99.

Llamas, M.R. & Martínez-Santos, P. (2005). Ethical issues in relation to intensive groundwater use. In: A. Sahuquillo, J. Capilla, L. Martínez Cortina & X. Sánchez Vila (eds.), *Selected Papers of the Symposium on Intensive Groundwater Use (SINEX).* IAH Selected Papers. Taylor and Francis Publishers, the Netherlands: 17–36.

Llamas, M.R. & Martínez-Santos, P. (in press). Intensive groundwater use: a silent revolution that cannot be ignored. In: *Proceedings of the 2004 Stockholm World Water Week.* Stockholm International Water Institute, Sweden.

Lloyd, J.W. (1997). The future use of aquifers in water resources management in arid areas. *The Arabian Journal for Science and Engineering*, 22(IC): 33–45.

López Gunn, E. & Llamas, M.R. (2001). New and old paradigms in Spain's Water Policy. In: Water security in the Third Millenium: Mediterranean countries towards a regional vision. *UNESCO Science for Peace Series*, Vol. 9: 271–293.

Margat, J. (1994). Groundwater operations and management. *Groundwater Ecology*. Academic Press: 505–522.

McClurg, S. & Sudman, R.S. (2003). Public and stakeholder education to improve groundwater. In: M.R. Llamas & E. Custodio (eds.), *Intensive Use of Groundwater. Challenges and Opportunities.* Balkema Publishers, Lisse, the Netherlands: 271–286.

Moench, M. (2003). Groundwater and Poverty: exploring the connections. In: M.R. Llamas & E. Custodio (eds.), *Intensive Use of Groundwater. Challenges and Opportunities.* Balkema Publishers, Lisse, the Netherlands: 441–456.

Mukherji, A. (2003). *Groundwater development and agrarian change in Eastern India.* Comment No. 9 in IWMI-TATA Water Policy Program (http://www.iwmi.org/iwmi-tata): 11 pp.

Myers, N. & Kent, J. (1998). *Perverse subsidies: their nature, scale and impacts.* International Institute for Sustainable Development. Winnipeg, Canada: 210 pp.

National Research Council (1997). *Valuing Ground Water.* National Academy Press, Washington D.C., USA: 189 pp.

OECD (Organisation for Economic Cooperation and Development) (2000). *No longer business as usual: fighting bribery and corruption.* Paris, France: 276 pp.

Polak, P. (2004). *Water and the other three revolutions needed to end rural poverty.* Invited paper in the World Water Week, 15–20 August, Stockholm, Sweden: 9 pp.

Poland, J.F. (1985). Guidebook to studies in land subsidence due to groundwater withdrawal. *Studies and Reports in Hydrology*, 40. UNESCO, Paris, France: 350 pp.

Postel, S. (1999). *The Pillar of Sand.* W.W. Norton and Co., New York, USA: 313 pp.

Price, M. (2002). Who needs sustainability? In: K.M. Hiscock, M.O. Rivett & R.M. Davison (eds.), *Sustainable Groundwater Development.* Balkema Publishers, Lisse, the Netherlands: 191–207.

Sahuquillo, A. & Lluria, M. (2003). Conjunctive use: a potential solution for stressed aquifers. Social constraints. In: M.R. Llamas & E. Custodio (eds.), *Intensive Use of Groundwater. Challenges and Opportunities.* Balkema Publishers, Lisse, the Netherlands: 157–177.

Selborne, J. (2000). *The Ethics of Freshwater Use: A Survey.* World Commission on the Ethics of Science and Technology. UNESCO, Paris, France: 58 pp.

Shah, T. (in press). Groundwater and human development: challenges and opportunities in livelihoods and the environment. In: *Proceedings of the Stockholm World Water Week.* Stockholm International Water Institute, Sweden.

United Nations (2000). *United Nations Millennium Declaration.* A/RES/55/2.

United Nations (2003). *Water for People, Water for Life.* UNESCO-WWAP. Paris, France.

Vives, R. (2003). Economic and social profitability of water use for irrigation in Andalusia. *Water International*, 28(3): 326–333.

WCD (World Commission on Dams) (2000). *Dams and developments. A new frame for decision-making.* Earthscan: 404 pp.

CHAPTER 11

# Is intensive use of groundwater a solution to world's water crisis?

A. Mukherji
*Department of Geography, University of Cambridge, UK*

ABSTRACT: This short chapter is a comment on the chapter titled "Significance of the *Silent Revolution* of intensive groundwater use in world water policy", by M.R. Llamas & P. Martínez-Santos (this volume). The main argument of Llamas & Martínez-Santos is that intensive use of groundwater offers an important window of opportunity for solving the so called *world water crisis*. In response to this, my comment is divided into three parts. In the first section, I try to ascertain the magnitude as well as the causes of water crisis in the world. In doing so, I particularly look at water poverty of nations. In the second section, I critically analyze the postulates of Llamas & Martínez-Santos and comment on how viable their recommendation of mitigating water crisis or meeting *Millennium Development Goals* (MDG) with intensive groundwater use is likely to be. In the third, and the final section, I present some of my thoughts on groundwater governance, which I believe will be vital, if groundwater is ever poised to tackle world's water problems.

Keywords: *Groundwater, governance, water poverty, water crisis*

## 1 WATER CRISIS: A MYTH OR A REALITY?

That the world is headed towards an impending water crisis is a cry heard with increasing intensity in the recent years and that too from different quarters, be it pro-privatisation neo-liberals or anti-globalisation groups, or be it advocates of large dams or opponents of it. Though the fact that there will be water crisis in very near future is more or less agreed upon, but there is disagreement about the precise nature of this crisis. Thus, one section of scholars whom I call the *neo-Malthusian pessimists* have contended that there will be actual physical scarcity of water because demand for water (due to population growth) will far outpace the supply for water. However, this view stands challenged in the face of past evidence which showed that human ingenuity in the face of scarcity is stupendous – more so in recent times because such ingenuity is supported by advances in science and technology. An apt example is the success of Green Revolution in India, which disproved prophesy of India's doom quite effectively. Again, there is enough evidence to show that as economies prosper, water use efficiency goes up and the nature of demand changes. Thus Gleick (2000) rightly points out that almost all water demand forecast done since 1970s have grossly over estimated water demand based on historical rates, while in reality, water use efficiency has gone up, thereby reducing water demand. The second group of scholars whom I called *cautious optimists* too have voiced their concerns about future water crisis, but they contend that water crisis is rarely a physical crisis, but more often a crisis of governance. Thus Seckler *et al.* (1998) predicted that by 2025, demand for water will outpace that of supply by anything from 57% to 25%. Cosgrove & Rijsberman (2000) made similar prognosis. Both suggested improving water use efficiency and constructing additional supply sources as possible remedy of such water crisis.

Again, the concept of water crisis, as simple it looks at the outset is in reality quite complex. Various definitions of water scarcity are in vogue, such as physical water scarcity or demographic water scarcity like the Falkenmark's index, IWMI's (Seckler *et al.*, 1998) concept of absolute and economic water scarcity and Wolfe & Brooks (2003) concepts of first order, second order and third order water scarcity. Similarly, measures to overcome water scarcity are just as diverse, ranging from supply augmentation to demand management to a judicious combination of both. One rather interesting work in this regard is that by Wolff & Gleick (2002), where they talk of *hard* and *soft* path of water management.

## 1.1   *Water poverty index and its determinants*

A very recent development in quantifying water scarcity and its multidimensional aspects has been the development of Water Poverty Index (WPI) by scientists at University of Keele and Centre for Ecology and Hydrology, Wallingford, UK (Sullivan, 2002; Lawrence *et al.*, 2002). This index provides relative ranking of 147 countries based on a composite index comprising of 5 sub indices, *viz.* water resource, access, capacity, use and environment. While WPI may be criticized over the choice of variables and whether all variables deserve inclusion in the first place and equal weight in the second place, yet it is a significant contribution because it puts the multi dimensionality of water poverty squarely into picture. In addition, by focusing on country wide situation, it presents a balanced view in that a country's water crisis is not exaggerated due to few basket cases, such as groundwater depletion in North Gujarat or arsenic contamination in Bangladesh. Third and most important, this index provides some food for thought in the form of understanding the relationship between physical water scarcity and water poverty of nations and effectively sheds light on what determines water poverty of a nation. Using simple graphs, I will make the following points:

(1) Water poverty of nation has very little to do with physical scarcity of water (Figure 1).
(2) And everything to do with Human Development Index (HDI) and per capita Gross National Product (GNP) (Figures 2a, 2b).
(3) Demand for water (depicted by water use index in WPI) has little or no relation with supply of water (depicted by resource index in the WPI) (Figure 3).
(4) On the other hand, water use index has very interesting relation with per capita GNP. As per capita GNP goes up, water use increases till a certain threshold value and after that it declines (Figure 4).
(5) Access to water, which is a component of WPI and indeed the true reflection of water poverty of nations has again very little relation with water resource availability (Figure 5).
(6) However, access to water is strongly and logarithmically correlated with per capita GNP (Figure 6).
(7) Finally water environment index, which is a sub index of WPI and is comprised of indices of water quality, water stress, environmental management, information capacity and bio diversity, is related to per capita GNP in an interesting way reminiscent of environmental Kuznets curve (Figure 7).

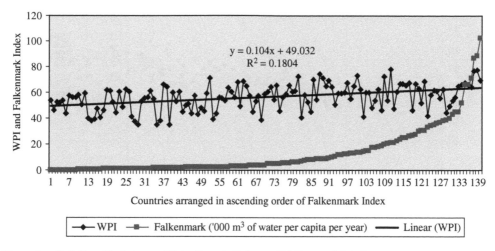

Figure 1.   Relationship between Water Poverty Index and Falkenmark's index.

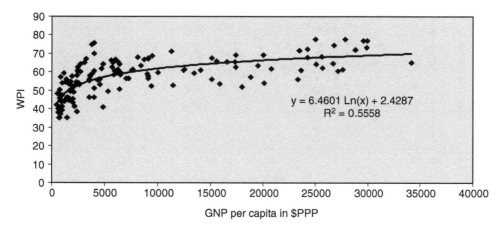

Figure 2a.   Relationship between Water Poverty Index and per capita Gross National Product ($PPP)[1].

Preliminary and unsophisticated as the above analysis is, yet it brings home several import-ant points. First, neither demand for water, nor access to water has much to do with resource endowments; instead both are strongly correlated to levels of income. While in case of access to water (which in my view is the true indicator of water poverty or the lack of it) is posi-tively and significantly related to per capita GNP, relationship of water demand with GNP is a bit more complicated and is best captured by a slightly inverted U shape curve. This means that as nations develop, water use efficiencies go up, a point corroborated by Gleick (2000) with help of examples from the USA. Second, water environment is a quadratic func-tion (U shape curve) of per capita GNP, signifying that in the initial stages, nations develop

[1] PPP means purchasing power parity. This is generally used to ensure comparability across different currencies.

184   *A. Mukherji*

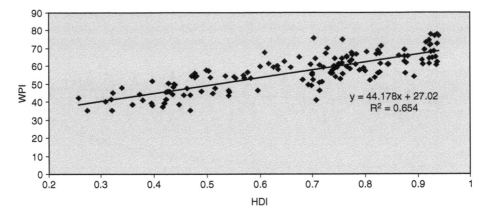

Figure 2b.   Relationship between Water Poverty Index and Human Development Index.

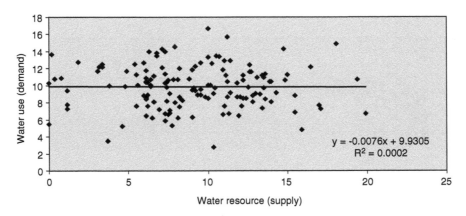

Figure 3.   Relationship between water demand and water supply.

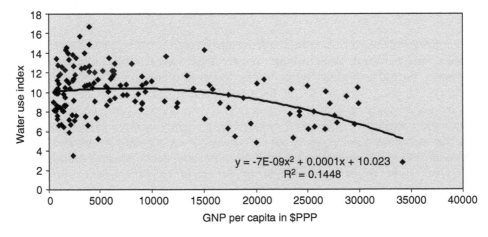

Figure 4.   Relationship between water demand and per capita Gross National Product.

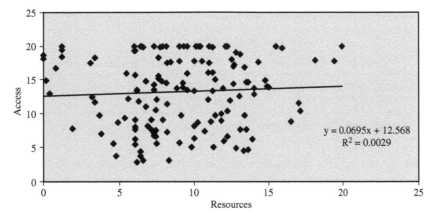

Figure 5.   Relationship between access to water and water resource availability.

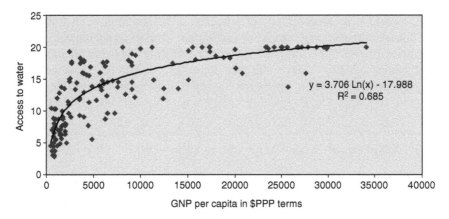

Figure 6.   Relationship between access to water and per capita Gross National Product.

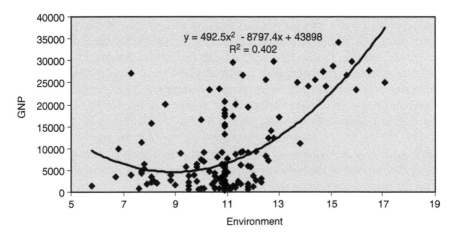

Figure 7.   Relationship between water environment and Gross National Product.

Table 1.    Determinants of water poverty of nations.

| Dependent variable | Intercept | Standardized beta coefficient | | | $R^2$ |
| | | Falkenmark index | HDI | GNP per capita in $PPP terms | |
| --- | --- | --- | --- | --- | --- |
| WPI | 27.119* | 0.262* | 0.781* | – | 0.722 |
| WPI | 50.002* | 0.282* | – | 0.606* | 0.482 |

* Denotes significance at .001 level for two tailed $t$ test.

at the cost of environment, but as development crosses a desirable threshold, demand for environmental goods and services increase and so does environmental quality. This is very similar to Kuznets environmental curve. Finally, water scarcity (or water poverty) is rarely a function of overall water resource availability in the country, but is more a function of level of economic and social development as reflected by HDI and GNP. In other words, water poverty simply reflects economic underdevelopment. The following two regression equations depict this.

In sum, therefore, physical scarcity of water has very little to do with water poverty of nations, while wealth generating capacity of a nation has. In effect, as economies prosper, water use efficiencies go up and this has much to do with availability of better technologies, which developed nations can afford. But at the same time, any effort to transpose such efficiency enhancing technology to a not so developed country is unlikely to succeed. Thus, in the long run, economic and human development is the best way to alleviate water poverty.

## 2    CAN INTENSIVE GROUNDWATER USE SOLVE WORLD'S WATER PROBLEMS?

Although, physical water scarcity is rarely a problem, yet it is undeniable that there exist severe problems regarding access to water. Thus, millions of people in the developing world do not have access to safe drinking water and sanitation – a number which United Nations MDG aims to half by the year 2015. In this context, Llamas & Martínez-Santos (this volume) contend that intensive and responsible use of groundwater has the potential of solving much of world water problems.

In support of their views, they present the following advantages of groundwater over surface water. First, using examples from India and Spain, they show that technical efficiency or water productivity ($m^3$/ha) of groundwater is several times higher than that of surface water. Second, in addition to technical efficiency, both economic ($/m^3$) and social (jobs/$m^3$) productivity of groundwater is higher than any other source of irrigation. Third, unlike centralized and bureaucratic development of most surface water systems, development of groundwater has been dispersed and *democratic* (also see Deb Roy & Shah, 2003; Shah *et al.*, 2003; Mukherji & Shah, 2005a, 2005b). This is what the authors appropriately call the *silent revolution*. In addition, unlike surface water schemes, groundwater extraction is rarely subsidized, so in effect farmers pay the full cost of groundwater extraction. Even in India, where electricity is highly subsidized, farmers using groundwater still pay anything between 3–5 times the price for water than do farmers in the canal command area. The fact that each unit of groundwater extraction entails additional cost, ensures that groundwater is used sparingly, resulting in higher water use efficiency as mentioned earlier. Since groundwater irrigation is under individual control, use of complimentary resources (*High Yielding Varieties* – HYV – seeds and fertilizers) is high, which again results in higher crop productivity (Shah, 1993; Kahnert & Levine, 1993). The authors also point out that groundwater irrigation involves less corruption, first because costs are much lower than even a small scale surface water scheme, and second because, in most cases, groundwater structures have been self-financed by the farmers. Though this is largely true, there have been allegations of corruption even in groundwater sector in India, especially when it involved any kind of subsidy scheme (Pant, 2003), but it is not nearly anywhere of those reported from surface water bureaucracy in India (Wade, 1984). Due to inherent advantages of groundwater use, it has important poverty alleviation implications in water abundant regions, a point noted by the World Bank as early as in 1989 (Kahnert & Levine, 1993). Quite often, a rather unexpected effect of intensive groundwater use has been the political empowerment of farmers. A case in point is the recent electoral upset in India, where it is claimed that farmers voted out a government which was doing exceedingly well in macro economic front (high growth in Gross Domestic Product, historically high levels of foreign exchange reserve, high foreign institutional investment and so on), but had failed to take into account farmers' demand in the form of lower input prices and higher output prices. An analysis of states where existing government was toppled correlates very well with the level of groundwater use. Thus, the agriculturally developed states like Punjab, Haryana, Tamil Nadu, Andhra Pradesh and Karnataka have all voted for a change in government and all these states are high users of groundwater. In fact, the issue of electricity reforms (mostly financed by either World Bank or Asian Development Bank and which entailed removal of electricity subsidy for agriculture) has been at the forefront of farmer's unrest in these states. So much so, the new Chief Minister of Andhra Pradesh has promised free electricity to the farmers on the very day he took oath of office, and the neighboring Chief Minister of Tamil Nadu has followed suit. Certainly these populist policies have deleterious effect, but the point I am making is that farmers who have enjoyed fruits of development (mostly due to groundwater led Green Revolution) also become politically vocal. But in other parts of India, such as eastern states of Bihar and West Bengal where level of agricultural development is still low (as is level of groundwater use), farmers lobby is either weak or is almost non-existence.

Llamas & Martínez-Santos (this volume) also point out certain misconceptions in vogue about groundwater use, which they refer to as *hydromyths*. Two such important *hydromyths* are that groundwater is a fragile resource and as such should be used as a last resort; and second, that rate of extraction of groundwater should be never more than rate of recharge. They go on to show that both these conceptions are erroneous and at best lead to confusion and apathy about the benefits inherent in groundwater use. On the whole, authors are optimistic and enthusiastic about the positive effect of intensive groundwater use. But to be fair, they also point out certain genuine drawbacks of groundwater, which stems mostly from mismanagement and hence can be corrected through improved management practices. However, this is easier said than done. They also note that the concept of overexploitation of groundwater often lacks scientific rigour and that there is rarely any aquifer in the world (above $1000\,km^2$) that has shown significant levels of dewatering. Perhaps true. This also implies that at a broad level, there is no groundwater resource crisis *per se* and whatever crisis remains can be adequately managed. But, although true at a macro level, this argument breaks down at the level of a farmer. For example, in the hard rock aquifers in India, farmers have been forced to commit suicide due to crop failure, which is very often, though not always related to well failure. It might very well be that the aquifer is not really dewatered as is often claimed in popular newspapers, but the fact remains that at the given level of economic status of the farmers, they could not deepen their wells further, neither could they grow crops which fetched them higher market value. In this circumstance, debating the exact definition of aquifer overexploitation will but seem irrelevant to affected farmers, though such definitional debate remains relevant for the water managers. Thus, the concept of aquifer overexploitation has as much social and economic implication as it has technical meaning. Having extolled the virtues of groundwater use and to a limited extent the possible negative externalities, the authors lay emphasis on awareness creation and stakeholder participation for fostering sustainable groundwater management.

This paper is a sterling contribution for various reasons. For one, it rightly stresses the positive benefits of groundwater, which is rarely emphasized in mainstream water resources discussions. Thus, while Cosgrove & Rijsberman (2000) and Seckler *et al.* (1998) discuss water future 2025, they do not pay any attention to the fact that today more than 50% of world's irrigated area is under groundwater irrigation and that water use efficiency in groundwater irrigated areas is anytime between 2 to 5 times more than surface water irrigation! Second, this paper is a welcome change in a sea of *pessimistic doomsday sayers* such as one that claims that almost 10% of world's food production is at the mercy of groundwater overdraft to the tune of around $200\,km^3$ and that such *pillar of sand* as it were is bound to collapse sooner or later (Postel, 1999). Third, the authors' attitude towards environmental value of groundwater is pragmatic, quite in contrast to the rather dogmatic views often imposed by western scholars on the developing world. They rightly comprehend the fact that utopian ideal of preserving groundwater for the future generations and for nature without providing any options for the present generation in poor countries is unlikely to succeed. They also effectively highlight the changing attitude to environmental values as society's make economic progress by citing examples from England, Spain and Turkey. Finally, by emphasizing such aspects as information sharing and stakeholder participation, they tend to demystify the hard and often incomprehensible science of hydrology for the benefit of common people.

However, the chief drawback of the chapter lies in its failure to take into account the negative aspects of groundwater use more thoroughly. In fact, negative externalities are more or less brushed aside in their optimistic *everything is all right with groundwater* attitude which is in stark contrast with the dominant pessimistic paradigm of groundwater and its ill effects. I claim that truth lies somewhere in between *pessimistic doomsday* saying and *optimistic every thing's all right* attitude. It is true that groundwater economy is indeed booming and has conferred huge socio-economic benefits in large parts of the developing world. But equally true is the fact that this boom is likely to burst without proactive governance and when it does, the ill effects will nullify or even surpass the social gains made so far (Mukherji & Shah, 2005a). This is precisely the issue I will deal with in the next section.

## 3   GROUNDWATER GOVERNANCE

In this section, I will lay down the conditions under which intensive use of groundwater is likely to confer benefits with minimum negative externality. To do this, I will claim that policies have to shift from *development* to *management* and finally to *governance* mode.

The present water crisis is "mainly a crisis of governance" – so declared the Global Water Partnership (GWP, 2000). This crisis of governance in the water sector is no where as visible as it is in groundwater. Figure 8 diagrammatically portrays the overall concept of governance.

The concept of governance has found acceptance among water professionals and GWP (2000) has defined water governance "as a range of political, social, economic and administrative systems that are in place to regulate the development and management of water resources and provision of water services at different levels of society". However, current discourse on water governance has not paid any special attention to groundwater *per se*, assuming that the essential elements of governance remain same across various water sectors, which is perhaps true at a certain conceptual level. However, equally true is the fact that governing groundwater is a greater challenge because: firstly, it is an invisible and fragile resource; and secondly, because intensive groundwater use of the scale we witness today is a recent phenomenon (of the last 30–50 years or so), many of the challenges encountered are new, and at times unexpected. A very good example of the unexpected nature of groundwater problem is the catastrophic arsenic poisoning in Bangladesh and West Bengal. Here, in order to provide safe and pure drinking water, World Bank and other donors encouraged extraction of groundwater through shallow tubewells little knowing that they were substituting the fear of surface water borne diseases with one of massive mass poisoning episode in human history (Centre for Science and Environment, 2002).

The first step in groundwater governance is adequate and good quality information, not only hydrogeological, but also socio-economic. In many instances, such information is missing or more possibly not accessible in the public domain due to unwillingness of the groundwater experts to share it with general public. While hydrological data is systematically gathered, if not disseminated, collection of social and economic data is few and far between. India undertakes reasonably detailed *Minor Irrigation Censuses* every five year, but results are published with great delay. Access to good quality data is perhaps less of an issue in developed nations, yet hardly any country so far has 50 years or so timeline hydrological data that some like Llamas (pers. comm.) contend is needed for thorough understanding of complex hydrological phenomena.

Secondly, what ails groundwater sector, very often, is the misallocation of roles and responsibilities assigned to various organizations that are in the business of managing groundwater. Thus, central and provincial governments are very often given responsibilities that are beyond their human and financial resources, while farmers and other stakeholders are asked

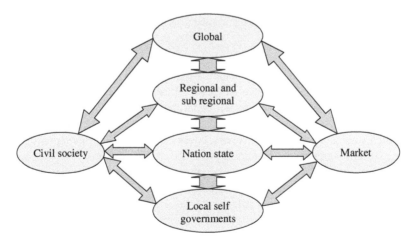

Figure 8.   The new and idealized concept of governance.

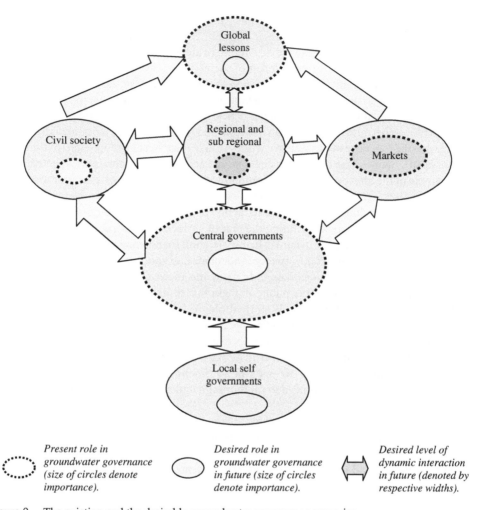

Figure 9.    The existing and the desirable groundwater governance scenarios.

to participate in aquifer management when their direct interests lie in non-participation. The problems inherent in groundwater governance are denoted in Figure 9. It shows the present state of governance and the desired governance structure and process that will possibly impart a modicum of order in anarchic groundwater economy. Developments in groundwater sector are guided very little by government but are a result of an interplay between many actors and factors in which state and government has much less central and dominant position compared to, say, in the surface water sector. But at the policy circles, there is very little recognition of the growing contribution of groundwater led economy, that too an economy that is developed and sustained by millions of small and marginal farmers. Thus, better groundwater governance it will mean recognizing greater role for the markets, the civil society and the local self governments and much less role for the central and provincial governments.

Finally, there is a need for attitude shift among the water policy makers and practitioners. Thankfully, in the field of groundwater, it has now come to be widely recognized that

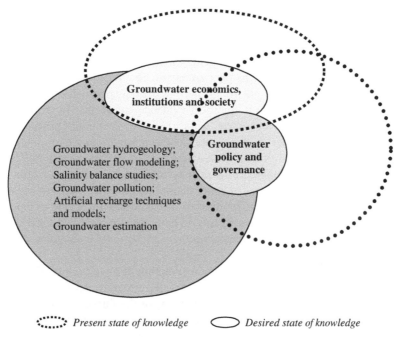

*Present state of knowledge*     *Desired state of knowledge*

Figure 10.   The knowledge development challenge of groundwater governance.

management is not the exclusive forte of hydrogeologists. In fact, in understanding the challenge of sustainable groundwater management anywhere in the world, it is critical to blend three distinct perspectives: (a) the characteristics and behavior of groundwater resource (resource perspective); (b) the characteristics and behavior of resource user communities (user perspective); and (c) the institutional frameworks under which the resource is appropriated and used (institutional perspective) (see Figure 10).

So far, global knowledge development as well as capacity building on groundwater use in agriculture has been dominated by the *resource perspective*; and a critical value-adding contribution is to be made in expanding global knowledge and capacity in user and institutional perspectives. Undergraduate and Masters' courses on hydrogeology in South Asia, for instance, spend less than 10% of teaching time on economics and sociology. Possibly same holds true in developed countries. When these students grow up as researchers and policy makers in groundwater sectors, their worldview tends naturally to get dominated by the resource perspective. And since economists and sociologists seldom foray into groundwater research, user and institutional perspectives have remained largely unexplored in the existing groundwater knowledge base.

## 4   CONCLUSIONS

From the foregoing sections, we saw that water crisis is very rarely a crisis of physical availability of water *per se*; it is much more a crisis of governance. More importantly, water crisis is but a symptom of the larger malaise of poverty that still plagues this world in this 21st century. Therefore, it follows the best way to tackle water crisis is to tackle poverty. However, it almost seems as if there is a global conspiracy aimed at keeping the poor people poor in perpetuity. This is manifested in such concepts as sustainable use of resource, which for instance talks of inter-generational equity, while forgetting that there exists a deep intra-generational gap – the gap between the rich and the poor countries. Second, associated with the concept of sustainability is the concept of environmental protectionism which often in the name of environment stalls development. But empirical evidence shows that in the initial years of development, environment takes a backseat, while as soon as economies prosper and demand for environmental goods and services increase, quality of environment too improves. Third, very often there is an attempt to impose techniques (e.g. of increasing water efficiency) that work well in the developed nations, to the poor countries with deleterious effect. Depending on the social and economic status of every nation, its priorities vary and rightly so. Any attempt to tinker with this is a formula for disaster, as has been seen in case of piped domestic water supply in most African countries. Finally, to deny development to poor countries in the name of environment is but a ploy to maintain the *status-quo* of inequitable distribution of wealth.

In this context of water crisis, which is but a symptom of poverty, role of groundwater is commendable. For one, it is relatively cheap to develop groundwater for irrigation; second, its development is decentralized and lies in the hands of the farmers; and third, because of inherent features of groundwater as a resource, it tends to be pro-poor. Thus, groundwater use and benefits thereof have deeply impacted humankind and it will not be exaggeration to say that groundwater use in agriculture has indeed revolutionized food production and created livelihood opportunity for millions of people. But, like every other natural resource, there are both costs and benefits and the prime aim in governing groundwater is to ensure that costs (or negative impacts) of intensive groundwater use does not exceed the benefits thereof. The costs or negative externalities hit the poor hardest and hence a word of caution is in place. Governing groundwater is a challenge, and if properly undertaken, it does have the potential of denting poverty, as it has already done in large parts of South Asia.

## REFERENCES

Centre for Science and Environment (2002). *Down to Earth*, Volume 11(1), May 31: 38.
Cosgrove, W.J. & Rijsberman, F.R. (2000). *World Water Vision: Making Water Everybody's Business.* Earthscan Publications, London, UK.
Deb Roy, A. & Shah, T. (2003). Socio-ecology of groundwater irrigation in India. In: M.R. Llamas & E. Custodio (eds.), *Intensive Use of Groundwater. Challenges and opportunities.* Balkema Publishers. Lisse, the Netherlands: 307–336.
Gleick, P.H. (2000). The changing water paradigm: a look at twenty-first century water resources development. *Water International*, 25(1), March: 127–138.
Global Water Partnership (2000). *Towards Water Security: a Framework for Action.* March.
Kahnert, F. & Levine, G. (eds.) (1993). *Groundwater irrigation and rural poor: options for development in the Gangetic basin.* World Bank, Washington, D.C., USA.
Lawrence, P., Meigh, J. & Sullivan, C. (2002). *The Water Poverty Index: an international comparison.* Keele Economics Research Papers, 2002/19. Keele University, UK: 24 pp.
Mukherji, A. & Shah, T. (2005a). Groundwater socio-ecology and governance: a review of institutions and policies in selected countries. *Hydrogeology Journal*, 13(1): 328–345.
Mukherji, A. & Shah, T. (2005b). Socio-ecology of groundwater irrigation in Asia: an overview of issues and evidence. In: A. Sahuquillo, J. Capilla, L. Martínez Cortina & X. Sánchez Vila (eds.), *Selected*

*Papers of the Symposium on Intensive Groundwater Use (SINEX)*. IAH Selected Papers. Taylor and Francis Publishers, the Netherlands.

Pant, N. (2003). *Key trends in groundwater irrigation in the eastern and western regions of Uttar Pradesh*. Paper submitted to IWMI-Tata Water Policy Program, Anand, Gujarat, India.

Postel, S. (1999). *The pillar of sand*. W.W. Norton & Co., New York, USA: 313 pp.

Seckler, D., Amarasinghe, U., Molden, D., de Silva, R. & Barker, R. (1998). *World water demand and supply, 1990–2025: scenarios and issues*. IWMI Research Report No. 19, International Water Management Institute, Colombo, Sri Lanka.

Shah, T. (1993). *Groundwater markets and irrigation development: political economy and practical policy*. Oxford University Press. Bombay, India.

Shah, T., Deb Roy, A., Qureshi, A.S. & Wang, J. (2003). Sustaining Asia's groundwater boom: an overview of issues and evidence. *Natural Resources Forum*, 27: 130–141.

Sullivan, C. (2002). Calculating a world poverty index. *World Development*, 30(7): 1195–1210.

Wade, R. (1984). *The market for public office: why the Indian state is not better at development*. Discussion paper 194, Institute of Development Studies, University of Sussex, Brighton, UK.

Wolfe, S. & Brooks, D.B. (2003). Water scarcity: an alternative view and its implications for policy and capacity building. *Natural Resources Forum*, 27(2), May: 99–107.

Wolff, G. & Gleick, P.H. (2002). The soft path for water. In: P.H. Gleick, *World's Water 2002*, Chapter 1.

# Water and Poverty Alleviation

CHAPTER 12

# Water and poverty alleviation: the role of investments and policy interventions

R. Bhatia
*Resources and Environment Group, New Delhi, India*

M. Bhatia
*Department of Agricultural and Resource Economics, University of California, Davis, USA*

ABSTRACT: This chapter shows that investments in water infrastructure (irrigation and hydropower dams, small check dams) provide direct and indirect benefits to the rural poor and agricultural labor that are higher than those obtained by large farmers and urban households. Improved management of water resources, through flexible allocation among sectors and users, provides benefits to the poor and can be an important instrument of poverty alleviation in developing countries. There is a need for monitoring the impact on the poor of community-managed rural water supply schemes where full cost recovery from the users is mandatory. Scaling-up of *islands of success* in community participation in urban sanitation faces constraints such as sources of public funds, cost recovery and financial viability, leveraging of community funds and private financial resources. To ensure protection, augmentation and equitable distribution of water resources, innovations will be required in water rights, pricing of water, targeted and cross-subsidies.

Keywords: *Impact of dams, flexible water management, community participation in water and sanitation, water and poverty, water rights, targeted subsidies*

## 1 INTRODUCTION

This chapter presents an overview of the role of policy interventions, investments in water infrastructure and improved management of water resources and services in poverty alleviation in developing countries. A number of case studies have been presented to show how water resources development and management can affect the rural and urban poor, both directly and indirectly, by providing livelihoods, food security, lower food prices and higher income and wages for all. Section 2 presents an overview of various dimensions of poverty and the Water Poverty Index (WPI) in the context of the sustainable livelihoods framework. Section 3 provides an overview of policy, investments and management interventions required in the water and related sectors for poverty alleviation/reduction. Section 4 presents empirical evidence to show that investments in irrigation and hydropower projects have considerable multiplier (direct and indirect) benefits to farmers and agricultural workers in terms of increased employment, real wages and income in agro-industries.

Section 5 shows how improved management of water resources can benefit the poor. Section 6 presents a review of experiences in community management of rural water supply services and community participation in urban sanitation in the context of the Millennium Development Goals of providing water and sanitation to millions of poor people in Asia and Africa. The chapter provides a summary of constraints to the *scaling-up* of these islands of success. In section 7, major policy issues are analyzed that affect investments in water resources development, protection, management and provision of services to the poor. These include legislation and regulation, water rights, pricing of water and financial sustainability of water investments, cross subsidies, and irrigation pricing in the context of subsidies provided by developed countries to their farmers. The last section presents some conclusions.

## 2    WATER AND POVERTY

### 2.1    *Dimensions of poverty*

Although politicians have always used poverty removal or reduction as a slogan, millions of people continue to live below the poverty line in Asia and Africa. The living condition of large populations, particularly in the slums, is a matter of shame for all those who enjoy the benefits of a higher standard of living. According to the *State of World Rural Poverty* produced by the International Fund for Agricultural Development (IFAD) in 1992, out of a population of some 4000 million people living in 114 developing countries, more than 2500 million live in rural areas, and of these approximately 1000 million live below the poverty line. More than half of this population lives in highly degraded lands. These people suffer from a lack of basic necessities like safe drinking water, adequate food and health care, which means that almost a third of the people in the developing world have a life expectancy of just 40 years. The IFAD report says that less than half the rural population had access to safe drinking water and even less to irrigation water to ensure sustained agricultural production.

It must be kept in mind that these indicators would vary from one country to another and from one social group to another and over time. For example, in the case of rural water supply in India, a community is classified as having safe drinking water supply if a source (government-owned) is available within a given distance (say 100 m) from the dwellings. Using this approach, it is claimed that 80% of rural households have *access* to safe drinking water. Such an approach is misleading because the indicator includes only public sources of supply (excludes private sources, which account for a high percentage in some states such as Punjab and Kerala). The method also does not take into account whether the *source* of water supply is functioning or not, whether water quantity is adequate for all the households (as per norms of 50 L/d *per capita*), and whether the water quality is good or bad (contaminated, saline or with arsenic or fluoride). Such measurements will have to be improved if these have to be used as indicators of meeting basic needs in the classification of poverty of regions or households.

#### 2.1.1    *Poverty and resource use*

As may be seen from any empirical estimates of poverty levels, the poor comprise heterogeneous groups in conflict with each other for resources, particularly for land, water and energy. There are a number of studies (Chopra *et al.*, 1990; Meinzen-Dick *et al.*, 1997a, 1997b; Bos & Bergkamp, 2001; Meinzen-Dick & Bakker, 2001; Soussan, 2001), that show that competition among the poor for these, usually common property resources, leads to overexploitation and degradation of resources and conflicts among users. Such degradation of resources results in reducing the *ability* of these resources (e.g. watersheds, wetlands, tanks, lakes, rivers) to provide livelihoods to a large number of poor people who do not own land or water resources.

Most development agencies and government institutions concerned with the implementation of poverty reduction programs, either from a supply-dominated approach or a demand-based approach, has to start by dealing with the issues related to quality and quantity of water resources. It is necessary to break the vicious circle that is created among the poor, who need water and other resources for their survival and livelihoods and overexploit these in their desperation.

Thus, starting from the supply-dominated approach we must distinguish between water service delivery problems among the rural poor and the urban poor. In many countries, it is the urban poor who face the serious health consequences of inadequate water supply, polluted water sources/supplies and high costs of provisioning and *coping strategies*. In many countries in Asia, urban populations are rising very fast and it is expected that almost 50% of total population in South Asia will live in urban areas. Majority of the people in Latin America are already in urban areas. Meeting the basic needs of water and sanitation services of urban poor at affordable prices is going to be a major challenge for the next two decades. The capital costs of providing basic services in congested urban areas are so high that innovative solutions will have to be found so that adequate quantities of safe water are available to the urban poor.

### 2.1.2   *Vulnerability of the poor*

The bulk of the world's poorest people, 800 million to 1000 million, live in arid areas and depend directly on natural resources including water for their livelihoods (UN, 1999, 2004). An important dimension of water and poverty is the vulnerability of the poor people to shocks or unexpected changes in water regimes (e.g. droughts and floods) and their consequences (e.g. decline in market prices of their outputs or rise in food prices). Millions of people, mainly the poor and disadvantaged, remain at risk from the lack of clean water, public health risks from inadequate sanitation affect some 50% of the world's population, and the number of people at risk from floods and drought continues to rise. At the same time risks from degraded ecosystems have increased inexorably, wetlands have been destroyed, over abstraction has lowered water tables and caused major rivers to cease flowing to the sea, and both ground and surface waters have been grossly polluted (Rees, 2001).

### 2.1.3   *Feminization of poverty*

Another important aspect of poverty relates to the feminization of poverty and its effects related to water and other resources. Available literature (UN, 2004) shows how poverty is related to ownership of property rights, especially land rights. From studies in India, it has been shown that the single most important economic factor affecting women's situation is the gender gap in command over property, an issue virtually neglected in research, policy and grassroots action. Based on a review of laws governing property and its inheritance, it has been pointed out that women's access to, and control over, landed property is crucial to their empowerment (Agarwal, 1997).

On a more basic subsistence level, restricting land ownership puts women at a disadvantage in providing for their families, gaining economic independence and enjoying personal security. It limits their production possibilities, such as cultivating crops, trees or small gardens and keeping livestock. Without these means to provide for themselves, women without land, as well as their children, have a significantly higher risk of absolute poverty. In virtually every legal system, gender inequality in land issues remains. For instance, some systems allow only small plots of land for women; others restrict the conditions under which women can inherit land and some are age discriminatory, prohibiting access to land for young women or the elderly. Women's empowerment in environmental decision-making and land management is well documented as crucial for sustainable human development. In recognition of this, the United Nations Development Program (UNDP), the United Nations Development Fund for Women (UNIFEM) and other UN agencies are working to help countries transform their legal and social structures and prioritize land rights for women[1].

### 2.1.4   *Ecological poverty*

According to Agarwal (1999), most rural people in the developing world live in an economy built on *natural capital*, which can also be described as a *biomass-based subsistence economy*. *Ecological poverty* is invariably the main cause of their impoverishment and an inability to meet their basic survival needs. Ecological poverty is defined as the lack of natural resources, both in quantity and quality, that are needed to sustain a productive and sustainable biomass-based economy. An excellent indicator of *ecological poverty* is the amount of time a rural household spends on collecting basic survival needs like water, fuel and fodder (for animals) from the local environment.

In conditions of high *ecological poverty*, this time can be excessive leading to enormous work pressures on the household. Since in most cultures, these activities are carried out by women, it is they who have to bear the maximum impact of *ecological poverty*. As pointed out by Agarwal & Narain (1980), degraded drylands and mountain sides can pose the biggest challenge for reducing female work burden. In a study carried out by the Center for Science and Environment (CSE), New Delhi, in a village of 213 people, the total work-hours spent in animal care, agriculture, fuel and fodder collection, animal grazing and other household and market-related activities during an entire year was 366,156, of which women contributed 58% of the labor and children another 26%. Men contributed a mere 15%. All this had a major impact on female literacy. Fortunately, the village had a school of its own. Most children went to school late but by about age 10 almost all children in the village were in school. But for girls education was very brief. No girl in the age group 15–20 was in school whereas 78% of the boys in this age group were still receiving education. No girl in the village had reached the high school level. In fact, less than 10% of the girls went beyond the primary stage.

As a contrast to the situation found in many states in India, women in Kerala (India) rarely spend more than an hour collecting basic needs like fuel, fodder and water and their work burden is far less than that of women in other parts of India. Kerala has often been highly praised by demographers for its low birth rate and high female literacy despite its low per capita incomes. But it may be useful to investigate whether the high availability of biomass and water in the tropical, humid environment of Kerala – the state is blessed with

---

[1] The Secretary General's report on improvement in the situation of women in rural areas (document A/52/326), underlines the critical importance of reconciling and strengthening the productive and reproductive roles of women farmers and entrepreneurs in improving their situation.

two monsoons a year – was an important factor in the success of the state's female literacy programme.

In rural areas where marginal farmers and landless laborers constitute 60% to 70% of population (e.g. India), a large majority of poor people do not have access to land and water for meeting their basic needs for survival and livelihoods. These poor groups then depend on government/community sources for meeting basic needs. As a result, a large number of women and girls spend a substantial part of their time in collecting water, fuel, fodder and food. According to Agarwal (1995, 1997), in India, the availability of natural resources to a large proportion of the poor rural population has been severely eroded over the past 20 years. The adverse class-gender effects include an increase in the time and energy that women and girls spend in fuel, fodder and water collection, a decrease in women's incomes from non-timber forest products and agriculture, an adverse effect on their health and nutrition, an erosion of women's social support networks, and a decline in their traditional knowledge of plants and species. The author maintains that the gender specificity of these effects stems from pre-existing gender inequalities in the division of labor, the intra-household distribution of subsistence resources, access to productive resources, other assets and income-earning opportunities, and participation in public decision-making forums. According to Agarwal (1995) rural women are worst off in regions where these three dimensions of disadvantage are strong and reinforce each other, as in many parts of Northern India, and especially Bihar. They are best off where all three types of disadvantages are weak, as in Southern and Northeast India, and especially Kerala.

## 2.2   *Water and livelihoods for the poor*

Water resources availability (rainfall and irrigation) affects the livelihoods of poor people by providing them with food (collected from community lands, forests, wetlands, tanks, fish ponds), employment and wages (both in kind and cash). *Livelihood thinking*, which developed in the 1980s as an alternative to *production thinking*, challenged beliefs about the neutrality of technology and the absolute ability of experts to promote optimal production systems. It also required a new professionalism to make resource management and technology serve small farmers. Livelihood thinking involves understanding water environments and technologies, understanding and working with the political processes through which local groups can question water assessment and allocation mechanisms, including *expert* solutions, and working directly with small farmers. This shift in orientation can foster local water-control initiatives that support users in negotiating their rights to water and livelihoods – within both water-basin and local water systems.

The dependence of poor people, particularly women, on common property resources, results in the overexploitation and degradations of land, water and other natural resources. Poverty shortens time horizons of the poor people leading them to overexploit resources. This means that poor people basic needs and livelihood considerations lead them to prefer consumption now over consumption in the future. The problem of *Tragedy of the Commons* gets exacerbated for a watershed or common water body (wetland, lake) when poor people have to eke out a livelihood from the limited natural resources and meet their basic needs of fuel, fodder and water. The situation worsens for the poor when their hand pumps and wells become dry because rich farmers pump water for irrigation, where private benefit of extracting groundwater becomes more important than preserving a groundwater reservoir (belonging to a community) and using it in a sustainable manner. This leads to a *killing* of the resource before its time. There comes a time when the resource is gone (or groundwater table has declined) and is not there for anybody's use.

It may be emphasized that the poor comprise a heterogeneous group that may be in conflict with each other for the access to natural resources, such as land, water and energy; so it is important to have a broad social participation to find the proper solutions. Further, there has to be a growing awareness of the need to shift the design of programs and projects targeted to reduce poverty (by increasing the water supply) into projects that have greater emphasis on the vital connection that exists between land use, secure water accessibility and poverty. Until now the health impacts of improvement of service delivery – either the supply-dominated approach or the demand-based approach – were seen as important. However, it has now become increasingly evident that the quality and quantity of water is a fundamental requirement in order to escape the vicious circle that is created among the poor, who need water and other resources for their survival and livelihoods and overexploit these in their desperation.

## 2.3   The Water Poverty Index (WPI)

The Water Poverty Index (WPI) (Sullivan & Meigh, 2003; Sullivan *et al.*, 2003) is an interdisciplinary tool that integrates the key issues relating to water resources, combining physical, social, economic and environmental information associated with people's ability to get access to water and to use water for productive purposes. It is most relevant at the community or sub-basin scales. The WPI can best be applied in practice to generate useful data that may be used to generate benefits, especially for poor people who suffer from inadequate access to water. WPI values would need to be generated over wide areas, and this would require substantial institutional development. This may require the use of existing census procedures and simplified data collection, including widespread data collection through schools. A number of technical issues relating to implementation of the WPI are:

(1) How the different spatial scales inter-relate; (2) how the assessment of the physical resource and the collection of social and economic data may be made compatible; and (3) how the WPI value can be used in practice, and what are the issues and problems that this presents.

The WPI concept is closely linked to the ideas of sustainable livelihoods and WPI is a way of measuring water status focusing on poverty and the livelihoods components of the poor. The five key components of WPI – Resources, Access, Capacity, Use and Environment – are closely analogous to the livelihood capitals (infrastructure, physical capital, social capital, and financial capital for investment. There is no one to one equivalence, but the WPI and sustainable livelihoods concepts fit together (Sullivan *et al.*, 2003: 193). In WPI, it is considered that a lack of water is consistent with a lack of one of the basic prerequisites to an effective life, but lack of water will have many additional repercussions. For example, low quantities of water can be shown to have a direct relation to health, as personal and food hygiene will be less effectively carried out. Further more, there are a number of illnesses that can result from poor water quality or contaminated water. In terms of productivity, water is usually a factor, even in subsistence households. For many poor households, their incomes from livestock, fishing, critically depends on availability of water in canals, ponds, tanks and wetlands. Thus, inadequate access to it (water) will impact on economic performance, and – obvious but often forgotten – time spent collecting water will not be available for other activities (Sullivan *et al.*, 2003). To redress any kind of poverty, access to water and other natural resources (along with availability of the livelihood capitals) must be redistributed more equitably.

## 3 POLICY, INVESTMENT AND MANAGEMENT INTERVENTIONS FOR POVERTY ALLEVIATION/REDUCTION

To meet the objective of poverty alleviation/reduction, a number of actions will be required at the policy and institutional levels (including legal and regulatory issues) in the water sector and related sectors. Substantial investments will be required in water infrastructure such as check dams, hydro dams, multipurpose dams, canal systems and groundwater pumping. New paradigms in management of water resources will be required that will ensure flexibility in allocation of water among sectors, users and uses. Community involvement in the management of water resources and in the provision of water and sanitation services can address the problems of the poor and slum dwellers.

It may be pointed out that the poor benefit from a number of interventions, both directly and indirectly. Some of these interventions, discussed in detail in the next sections, are:

- Investments in water infrastructure (e.g. check dams, multi-purpose dams, groundwater projects) provide direct and indirect benefits to the poor, *inter alia*, in terms of higher employment, higher real wages, increased non-farm employment and lower food prices.
- Improved management of resources (e.g. flexibility in sectoral allocation of water, water conservation, improved water quality).
- Policies, legislation (e.g. water rights, water markets) and regulations may be modified to benefit the poor directly or indirectly by providing livelihood assets.
- Community management of water and other natural resources may ensure that benefits from the use of water resources are equitably distributed and the poor benefit from better management of resources.
- Community management of water delivery services in rural and urban areas will ensure the participation of communities in planning and management of delivery systems based on *demand responsive approaches* (DRA) and ensure their financial and institutional sustainability.
- Targeted subsidies and programs of service provisioning directly help the poor in securing needed water supplies at prices they can afford. More efficient service delivery would *free-up* financial resources that could be deployed for meeting the needs of partially served and unserved poor.

## 4 INVESTMENTS IN WATER INFRASTRUCTURE AND POVERTY ALLEVIATION

Investment in water projects has provided livelihood, employment and incomes to the poor people in developing countries. There are considerable multiplier benefits from irrigation and power projects to farmers and agricultural workers in terms of increased employment and real wages and income in agro-industries. Although this may not be true in some countries (e.g. China, Spain) major water development projects in Brazil, India, Malaysia and the USA show large direct benefits (from irrigation and hydropower) but with indirect benefits typically twice as large (World Bank, 2003). In many cases the poor benefit enormously from this economic activity. In Petrolina, in Northeast Brazil, for example, water infrastructure has been the basis for the development of a dynamic rural economy. This has meant the creation of a large number of high-quality, permanent agricultural jobs (40% held by women). And for every job in agriculture, two jobs have been created in the supporting commercial and industrial sectors. These opportunities have meant a reversal in the historic pattern of out-migration, with the benefiting districts growing at twice the state average. Similarly in India, water infrastructure has evened out the seasonal demand for labor, resulting in major gains for the poor. Irrigation projects benefited, both directly and indirectly, by raising the demand for labor, a large number of poor agricultural laborers who had no assets or training of any kind. For example, during its operation over 30 years, the Bhakra dam in Northern India has transformed the lives of millions of people in the states of Punjab, Haryana and Delhi, and provided seasonal employment to hundreds of thousands of poor migrant workers from Bihar and Eastern Uttar Pradesh.

## 4.1   Poverty reduction impacts of Bhakra dam, India

The results of a multi-country World Bank study (for details: Bhatia et al., 2004) show that investments in irrigation and hydropower projects have very significant indirect economic impacts that could be as high as 90% to 100% of direct economic impacts.

The multi-sectoral, economy-wide models used for multiplier analysis also provide quantitative estimates of income distribution and poverty impacts of dams. The *Social Accounting Matrix* (SAM)-based, fixed price, multiplier models provide income and consumption estimates under *With Project* and *Without Project* situation for various household groups (rural self employed, agricultural labor, marginal farmers, household quintiles in rural and urban areas). These data have been used to compute *gains* or *losses* associated with the dam for each category of household. The benefits to the poorest people (e.g. agricultural labor) from irrigation and hydropower projects are sometimes higher than the benefits to other households. Thus, investments in large water projects help significantly in the reduction of poverty in the regions and beyond. For example, in the case of the Bhakra dam, it has been estimated that agricultural labor gained a 65% increase in income as compared to a rural average increase of 38% under the *With Project* scenario compared with a hypothetical situation where the project had not been undertaken (Figure 1)[2].

One of the major benefits of irrigation projects for the urban and rural poor is a substantial reduction in prices of foodgrains[3]. Given the high crop yields and substantial marketed surplus of foodgrains resulting from surface and groundwater pumping (helped by recharge from irrigation canals and power from the Bhakra dam), only two states (Punjab and Haryana) provide around 70% of the all-India requirements of foodgrains distributed by the government in fair-price-shops to the urban poor all over India (at relatively low food prices). The urban poor have also benefited from surplus foodgrains from Bhakra distributed all over the country through *fair-price* shops. Bhakra dam contributed 30 million tons or 60% of total foodgrains procured by public distribution agencies in 2000–01. Remittances (around Rs 3548 million

---

[2] Does not include the income of the 1 million workers who migrate seasonally from Bihar and Eastern Uttar Pradesh each year.

[3] This is evident from food prices in areas where crop yields are high (Bhatia, 2003). Higher marketed surplus in irrigated areas provides additional foodgrains to the public distribution system that provides foodgrains to the urban poor at prices they can afford.

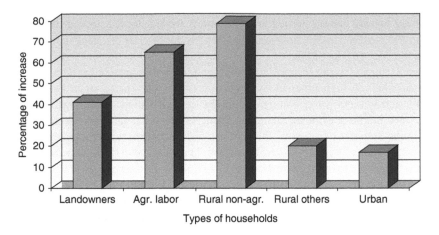

Figure 1.  Percentage change in income of different types of households *With* Bhakra dam.

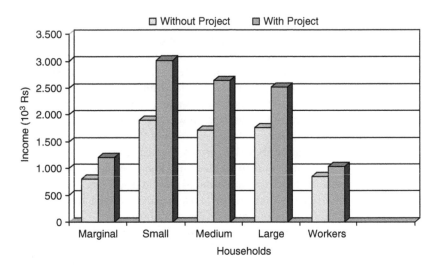

Figure 2.  Income of households *With* and *Without Project*[4].

or US$ 75 million in 1995–96) sent by migrant labor working in the Bhakra command have benefited millions of poor in the villages in Bihar and Uttar Pradesh, with resulting multiplier and downstream effects.

### 4.2  *Income distribution impacts of small check dams in Bunga village, India*

In a recent case study (Malik & Bhatia, 2004), an attempt has been made to analyze the direct, indirect and multiplier effects of the check dams constructed in Bunga village in Haryana. Bunga has been a successful example of how community management of water and natural

---

[4] Rs: Indian rupees. Rs 1 = US$ 0.02161 (September 1, 2004).

resources can benefit people, particularly the poor. In the Bunga village, a small village of 178 families (1100 persons), near Chandigarh in Haryana, India, the two check dams are providing irrigation to about 276 ha of land.

The results show that the gain for land holding households (except for large farmers) from the dam was relatively higher than the average difference of 48% between village-level incomes under with and without project situations. In the case of Bunga check dam, the household categories are landless workers, marginal farmers, small farmers and large farmers. The increase in the incomes of marginal farmers was 50% as compared with the average increase of 48% under *With* and *Without Project* situations. The increase in the income of small farmers was 59% as compared to the average increase of 48% under *With* and *Without Project* situations.

Further, the worker households did not benefit much from the dam because the demand for increased labor from irrigated crops was met from family labor by most of the farm families. Due to relatively small farm holdings, the farm households did not increase their demand for hired labor that would have benefited landless households. However, the worker households benefited indirectly from the dam in terms of higher incomes from milk production and shops.

### 4.3    *Poverty and irrigation*

Based on an empirical study (Bhatia, 2003), it was found that irrigation was positively associated with reductions in poverty levels in different states of India. The proportion of persons above the poverty line is higher in states/regions where a higher percentage of cropped area receives irrigation (e.g. Punjab, Haryana and West Uttar Pradesh). The empirical results based on the analysis of data for 17 states for selected years and 81 agro-climatic regions (for 1984–85) show that: (1) irrigation is positively associated with the level of per capita consumption expenditure; (2) variations in real wages alone account for 60% to 70% of the inter-state variation in the proportion of population above the poverty line. Real wages, in turn, are influenced by the level of irrigation in a state; (3) crop yields of rice and wheat are significantly higher in states with higher proportion of area irrigated; and (4) wheat prices are significantly lower in the states where the proportion of gross irrigated area to gross cropped area is relatively high.

Multiple uses of irrigation canals and tanks, in many countries, have provided opportunities for the poor in terms of additional employment in crop production and income from livestock and fishing. For example, in the *Kirindi Oya Irrigation and Settlement Project* (KOISP) in Sri Lanka, significant benefits have been identified from crop production, fishing, tourism, livestock grazing, fuelwood collection and shell mining (Meinzen-Dick & Bakker, 2001).

Despite many direct and indirect benefits to a large number of people (including the poor and disadvantaged), a number of large water projects have resulted in adverse social and environmental consequences. Some of these relate to sufferings of people (particularly the tribals, native populations) due to absence of, or delays in, relief and rehabilitation of project affected people (WCD, 2000). Environmental consequences relate to salinity and water logging in large tracts in Pakistan and Western India, reducing cultivable land and affecting drinking water supplies for people and animals.

### 4.4    *Irrigation and returns to investment in education*

Pritchett (2001) has found that for India, irrigation infrastructure has a major impact on returns to investments in education. Returns to investing in five years of primary schooling compared to no schooling in Indian districts where agricultural conditions were conducive to adoption of *Green Revolution* technologies, was as high as 32%. However, in districts where conditions were not conducive to such irrigated agriculture, estimated returns to such schooling were negative. At the same time, it has been found that in unirrigated districts in India 69% of people were poor, while in irrigated districts this share drops to 26%. [It should be noted, however, that information about these respective poverty shares before irrigation investments took place, were not known].

## 5   POVERTY IMPLICATIONS OF IMPROVED MANAGEMENT OF WATER RESOURCES

Improved management of water resources can help the poor people, both directly and indirectly, in a variety of situations. Here we describe a few such situations where water resources management can help in poverty alleviation/reduction.

### 5.1   *Flexible water allocation and management benefit the poor*

In a recent study (World Bank, 2004) for the Indian state of Tamil Nadu, it has been shown that flexible water allocation among sectors provides substantial benefits in terms of higher incomes for almost all the income groups. For example, the increase in overall income at the state level is about 20% compared with a scenario of fixed water allocation. Some of the details of the study on income distribution impacts of improved water management practices are given below.

This analysis of water allocation in the state of Tamil Nadu for the year 2020 has been done under two alternative scenarios:

*S-1. Fixed allocation*
Business as Usual Scenario (BAU): Fixed sectoral water allocation based on proportion of water allocated to agriculture, industry etc. as estimated in the year 2000. Additional supplies of sustainable groundwater will also be allocated in the same proportion as in 2000.

*S-2. Flexible water allocation*
Volume of water available in this scenario is the same as in *S*-1 but water is allocated to all sectors based on the estimated economic value of water in each sector (as defined by the Willingness to Pay – WTP – for water in each sector). It is assumed that control structures for water distribution among various users will be in place and incentives to transfer water from one user to another will be present. This may require water rights, including rights to transfer/sell water to other users as well as institutional interventions such as the *water users associations* (WUAs). This will also include needed interventions in terms of transfer and delivery mechanisms including the needed secondary and tertiary system improvement options such as lining and *on-farm development* (OFD) works for effective water control. A number of simulations have been carried out for this scenario by taking into account additional water available from the Godavari link, water conservation in agriculture and industry and removing the subsidy on the price of electricity used for water pumping.

The analysis is carried out in two steps (for details: World Bank, 2004). Estimation of direct and indirect impacts of alternative strategies of water allocation on the economy has been assessed through a two stage inter-related modeling process. In the first stage an *Optimization Model* has been formulated to assess the direct impact of the increase in water demand on intra and inter-sectoral allocation of water, and the direct impact such reallocation of water has on these water using sectors. The second stage uses a *Social Accounting Matrix* (SAM)-based, fixed price multiplier model of the Tamil Nadu economy to quantify the direct and indirect impacts on other sectors of the economy and on the poverty levels under alternative scenarios. The SAM-based model takes (as input to the model) the estimated output levels of sectors directly affected by alternative scenarios of water allocation (e.g. optimum levels of output of various crops, industries from the *Linear Programming Model*).

### 5.1.1    *Income distribution aspects of water allocation*

The SAM-based multiplier model disaggregates households into nine categories. Based on the survey done in 1997, these nine income groups represent the following occupational classification:

Rural household categories: (R1) rural self employed in non-agriculture; (R2) rural agriculture labor; (R3) rural other labor; (R4) rural self employed in agriculture; (R5) rural other households.

Urban household categories: (U1) urban self employed; (U2) urban regular wage salary; (U3) urban casual labor; (U4) urban other households.

In rural areas, R2 rural agriculture labor constitutes 45% of total rural population of 34 million people. Urban casual labor or U3 accounts for 19% of urban population, while U2, urban regular wage earners, constitute the largest group at 43% of the total urban population of 27 million.

It may be noted that under flexible water allocation (*S*-2), all income groups receive higher incomes (except R2), when comparing with incomes received under fixed water allocation scenario (*S*-1). On average, rural population gains 14% while R3 (rural other non-agriculture labor households) gain 21% and rural non-agriculture households (R1) gain 19%. This is because all these households gain indirectly from higher incomes generated in services, construction, transport and communications and ancillary manufacturing. As a result of reduction in direct incomes from farming, rural self-employed (farmers) gain only 3% while agricultural labor households have incomes under solution *S*-2 that are marginally (2%) lower than incomes under solution *S*-1.

It may be useful to compare per capita incomes across income groups for the years 2000 and 2020 (Figure 3) under *Fixed* and *Flexible* allocation. Over the two decades (2020 over 2000), under the flexible water allocation strategy, overall per capita income increases by 500%. The incomes of rural population increase by 280%, while those of urban people increase by 580%. The incomes of the lowest income group in rural areas also increase (by 120%), although much lower than that for other rural or urban groups. The lowest income group in urban areas, however, gains 625% in *per capita* income, even higher than the average for urban population. The income groups in the Figure 3, represent the following occupational classification as in the survey done in 1997: (R1) rural self employed in non-agriculture activities; (R2) rural agriculture labor; (R3) rural other labor; (R4) rural self employed in agriculture; (R5) rural other households; (U1) urban self employed; (U2) urban regular wage salary; (U3) urban casual labor; (U4) urban other households.

### 5.2    *Community-based management of resources and poverty alleviation in India*

Two villages in India (Sukhomajri and Ralegan Siddhi) have shown that it is possible to reduce poverty by bringing about community-managed ecological regeneration (for details: Agarwal,

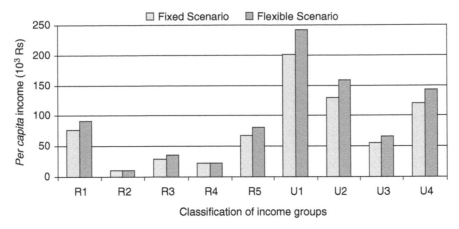

Figure 3. *Per capita* income of different groups under *Fixed* allocation and *Flexible* allocation (year 2020).

1999). In both villages, development of transparent and participatory community-based decision-making institutions, and establishment of community property rights over the local natural resource base was critical. These institutions decided natural resource management priorities, resolved conflicts within the communities, and determined burden- and benefit-sharing rules. Technologically, the starting point was rainwater harvesting – centuries-old tradition of Indian villages – which slowly led to the regeneration of the entire *village ecosystem* and the associated rural economy.

The village of Sukhomajri, near the city of Chandigarh, has been widely hailed in India for its pioneering efforts in microwatershed development. In 1979, when the nation was facing a severe drought, the villagers built a small tank to capture the monsoon runoff and agreed to protect their watershed, in order to ensure that their tank did not get silted up. Since then the villagers have built a few more tanks and have protected the heavily degraded forest that lies within and around the catchment of its minor irrigation tanks. The tanks have helped to increase crop production by nearly three times, and the protection of the forest area has greatly increased grass and tree fodder availability. This, in turn, has increased milk production. With growing prosperity, Sukhomajri's economy has undergone a change. The villagers have replaced their thatch-and-mud dwelling with brick-and-cement houses, and most of the houses boast of radio sets, electric fans, sewing machines and television sets. A survey conducted in 1998 revealed that the income distribution in Sukhomajri, situated in an agriculturally-depressed region, today matches the income distribution of rural Haryana which is one of the most agriculturally prosperous states of India[5].

In 1975, Ralegan Siddhi was a village stricken by poverty (Hazare, 1997). It had hardly one acre of irrigated land per family (1 acre = 0.4047 ha). Yield was less than 0.75 tons/ha. Food production was only 30% of the village requirements, some 15%–20% of the families were

---

[5] One of the most impressive savings resulting from the project is in the reduced cost of desilting the Sukhna lake which supplies water to the downstream city of Chandigarh. The inflow of sediment has come down by over 90%. This saves the government Rs 7.65 million (US$ 165,000) each year in dredging and other costs.

undernourished and most men migrated each year to look for work. In 1979, a social worker began work in the village by constructing storage ponds, reservoirs and gully plugs. Due to the steady percolation of water, the groundwater table began to rise. Because of the increased availability of irrigation water, land that was lying fallow came under cultivation and the total area under farming increased from 630 ha to 950 ha. The average yields of millets, sorghum and onion increased substantially. Every effort was made in the village to ensure equitable access to the resources generated. This was particularly important when the conserved water was still small in compared to potential demand. Water is distributed equitably. As cultivation of sugarcane – common in the command areas of state-managed irrigation systems in the state – requires a large quantity of water, it was forbidden in the early years in Ralegan Siddhi to ensure that the limited amount of water available was distributed equitably to all farming households. Only low water-consuming crops were allowed. All families got water in turn. No farmer would get a second turn of irrigation until all families had been served. Even where individuals had private dug wells, they were persuaded to share water with others. Water conservation efforts resulted in increased availability of groundwater that in turn facilitated the development of community wells. Water from these wells, supplied at a moderate price, has enabled farmers to grow two to three crops a year including fruits and crops, some of which are exported. A 1998 survey revealed that the monthly income distribution in Ralegan Siddhi is much higher than the income distribution estimates of rural Maharashtra.

## 5.3    *Community management of watershed and natural resources*

There are interesting cases of cooperative watershed management (World Bank, 2003), stimulated by the recognition of dam owners that upper catchment management is imperative for maintaining the value of their assets. Thus, for example, the Nam Theun 2 hydropower project in Laos provides support for communities to improve the management of the catchment. And the private companies which operate the water concessions in Manila are similarly investing heavily in catchment management to preserve the quality and quantity of the water on which the city depends.

Another interesting case relates to the efforts to manage the watershed of the river La Quebrada Chocho, Colombia, with the creation of an *Association of Users* to buy the land around the watershed, reforest it, and preserve it in an integrated manner. Participatory diagnostic raised awareness of problems and built consensus on solutions such as tree planting and the need for a solid management team supported by *Participatory Action Research* (PAR) trainers. Training was useful to bring a common vision on watershed management, to solve conflicts, to build a management team. Decentralized community management brings ownership of integrated water management.

In another case involving native populations in Mexico relates to civil society initiatives to restore the traditional aztecan *chinampas* at the Xochimilco watershed. This case shows the implementation of a common agreed project among different water users for a complete hydrological regeneration of the *chinampa* area. It also gives examples of a large-scale experience of conflict resolution on watershed management, in an urban environment to devise a plan of action, providing us with their experience of how they have blended *traditional* and *modern* technology into a civil society's initiative.

The NGO-led Shivalik hills watershed management project in India, financed by the World Bank (2003), seeks to scale up the lessons from many successful NGO-led watershed management projects. The project aims at simultaneously reducing erosion, increasing ground-water recharge, and improving the livelihood of the poor. The major investments are in terracing, establishing small check structures in eroded ravines, planting vegetative cover on denuded hills, building small dams, and digging wells and canals which can make better use of the preserved water resources.

## 6   WATER AND SANITATION SERVICES FOR THE POOR

During the last two decades or so, *access* to safe drinking water and sanitation facilities has been provided to millions of poor people through government programs and international efforts of multilateral and bilateral agencies. The extent of water and sanitation services actually available to the poor has yet to be evaluated (as discussed elsewhere in this chapter). Still, millions of poor people in urban and rural areas have yet to receive basic facilities of safe drinking water and sanitation. For example, South Asia accounts for 500 million of the world's poorest people with an income below one-dollar *per capita* per day. Of the total population of 1313 million (1998), 300 million people do not have access to safe water today, millions more get much less water than considered necessary for basic needs and personal hygiene. If adequate investments are not made by 2025, the additional population of more than 500 million people will not get enough safe water, particularly in slums in mega cities. In the case of sanitation, in South Asia alone, 920 million do not have access to sanitation; if investments are not made, 1450 million will be without sanitation, causing health problems and loss of output. Hence, new investments in the water sector are critical to the survival, well-being and poverty alleviation of 1000 million people in South Asia.

The estimated investments in the water sector in India (Bhatia & Malik, 1999) would be of the order of US\$ 420,700 million over the next 25 years (undiscounted, at 1997 prices). It may be noted that the bulk of the investments (39%) will be required for providing urban water supply, urban sanitation and urban wastewater treatment. Treatment of industrial effluents would require an investment of US\$ 68,800 million over the next 25 years. A little over a quarter of the capital cost will be required for food security investments including surface irrigation, ground-water pumping, drainage, watershed development and flood control. Investments of the order of US\$ 76,000 million (18%) would be required for rural water supply and sanitation services.

Although considerable progress has been made in increasing *coverage* levels, the sustainability of such approaches and technologies have been questioned (Hansen & Bhatia, 2004). It has been argued that financial and institutional sustainability of government sponsored systems cannot be ensured due to lack of cost recovery and accountability. These government-sponsored systems are characterized by supply/technology fixes and do not take into account what services people want and what they are willing to pay for them. Lack of any notions of opportunity cost of water in alternative uses has resulted in tariffs that are too low for domestic and industrial users. Non-targeted subsidies have resulted in the rich people getting the benefits of water services and subsidies in urban areas. Even for the relatively better-off sections of society, uncertain and inadequate water supplies have resulted in high cost *coping strategies* to meet their demands for water.

In many cities in the developing world, millions of the poor people living in unauthorized colonies do not have access to adequate water supplies and sanitation services and continue to be partially served or unserved. In the absence of adequate provision by the government agencies, people have to do with inadequate and contaminated water supplies. The poor get intermittent and unreliable water supplies from public standposts, while the rich, who have house connections, use water lavishly and pay subsidized rates, much below the cost of supplying water (for details for Delhi: Zerah, 2000). As a consequence, many poor people have to pay high prices (almost ten times the prices paid by the rich) for getting water to meet their basic needs of drinking and cooking and personal hygiene. Millions of poor people in rural areas do not get adequate water supply and have to walk miles to get water for meeting their basic needs. Such an arrangement is not sustainable and significant changes will be required in the approaches to provision of water services to meet the objectives of poverty reduction.

Poor men and women do not have access to resources to lift them out of poverty. They could do better if the resources were made available and they could control them to upgrade their environments (Esrey & Andersson, 1999). If local communities own their systems, they will be more responsive to demand from community members and be able to share and recover costs. It is also important that when responsibilities for managing water systems are given to local communities, financial resources and technical assistance should also be made available to the local leadership. For example, in Tamil Nadu, India, in the initial stages, although responsibilities for water supply to rural communities was transferred to *panchayats*, they did not get the budgeted money for maintenance of these systems, and engineering staff was not transferred to work under the *panchayats*. In addition, micro-credit facilities should be made available to community organizations, NGOs and others to provide loans and guarantees. Small-scale private sector organizations and groups (such as vendors, truckers) can provide cost-effective services to urban poor and need to be helped by changing regulations and financial assistance policies of banks and micro-credit institutions (deLucia and Associates, 2000).

## 6.1    *Community managed rural water supply schemes – Equity and sustainability issues in Swajal water supply scheme in Uttar Pradesh, India ( for details: deLucia and Associates, 2001)*

In an attempt to improve rural water supply and sanitation services in the state, the Government of Uttar Pradesh piloted an alternative service delivery model-*Swajal* (Own Water). *Swajal* is part of the *Uttar Pradesh Rural Water Supply and Environmental Sanitation Project*. The Project aims at delivering sustainable health and hygiene benefits to the rural population through improved water supply and environmental sanitation services.

The *Swajal* project envisages a decentralized decision making and demand responsive institutional framework. The model consists of three organizations: the Village Water and Sanitation Committees (VWSC) at the community level, the Support Organizations (usually NGOs) at the district level, and the Project Management Unit at the centre. Selection of a village to be covered under the project is done with the consent of the village community represented by the VWSC. The VWSC and the village *panchayats* (body of elected representatives) need not necessarily be the same entity. In some villages the two institutions could be the same, while in others they could be different.

After the launch of the scheme, much is not known about the water supplied through the standposts in these villages – the only source from which poor non-member households can draw water. What has been the impact in terms of number of standposts now available, the quantity of water that is made available through these standposts, the number of users per standpost, the time spent by households in collecting water from these standposts, and the implicit tariff paid by these households *vis-à-vis* those having a household connection. Have the poor households relying on standposts for their water needs become worst off in terms of availability of water, time spent in collecting water as also the implicit tariff paid for a unit of water both in terms of what they were paying prior to the scheme, as well as now in comparison to the households having a piped water connection? Based on a short field visit to two villages (near Dehradun, in June 2000), it seems that in the pumping scheme where fuel costs are high, financial viability has been achieved at the cost of equity, and the poor are paying for inadequate and unreliable water supply, while the rich are enjoying the benefits of adequate water supply through house connections provided under the community management. There are significant trade-offs between objectives of meeting basic needs of the poor and financial viability of small community-managed water delivery systems. For example, in the case of *Swajal* community schemes in Uttar Pradesh, India, cost recovery for a diesel-based pumping system in a village required that each poor household pays Rs 10 (US$ 0.22) per household for getting water from a public standpost (compared with Rs 50 – US$ 1.08 – per month for a house connection). However, given the objective of maximizing cash revenues, the Village Water Committee, over time, increased the private connections and reduced public standposts from 10 to 5 for about 50 poor households in the village. As a consequence, the poor households have to spend a lot more time in procuring water from the standposts that is generally inadequate for their needs.

## 6.2   *Community participation in urban sanitation*

For obtaining the health benefits of reduced infection from faecal-oral diseases, particularly for the children, it is necessary to provide *adequate* sanitation, that includes hygienic, well-maintained, easily accessed toilets that are used by all family members, and safe and convenient disposal of wastewater. In 2000, the total number of urban dwellers lacking *adequate* sanitation was between 850 to 1130 million, much larger than the 400 million lacking *improved* provision. This means that the funds required to meet *Millennium Development Goals* for urban sanitation are likely to be more than double of what has been estimated so far.

During the last two decades, the participation of communities and NGOs has resulted in the provision of toilet and bath facilities to about 4 million urban slum dwellers through community toilet complexes in India. In addition, community participation has helped in the construction of private toilet facilities for about 1.5 million urban people in several countries (for details: Bhatia & Bhatia, 2004; Bhatia & Hansen, 2004; Hansen & Bhatia, 2004) including India, Bangladesh, Pakistan, Angola, Burkina Faso, Ghana, Brazil and Bolivia. Involvement of communities provides incentives for selection of low-cost technologies, ensures efficient management of services and leverages a part of the required capital funds.

## 6.3   *Constraints to scaling-up of islands of success*

### 6.3.1   *Capital costs are substantial for scaling-up*
However, the efforts of community participation have to be scaled-up by a factor of 100 times if adequate sanitation services are to be provided to about 500 million urban poor and slum dwellers over the next decade or so. Taking a conservative figure of US$ 20 *per capita* for providing sanitation services in urban areas, the estimated capital requirements are around US$ 10,000 million for the poorest in urban areas in Asia, Africa and Latin America. Besides, such a *scaling-up* would require actions at a number of levels in the international organizations and central and local governments in developing countries.

### 6.3.2    Public funds for capital costs constrain scaling-up

Although NGOs play an important role in organizing communities, the institutional structure of NGOs does not enable them to raise capital funds from financing institutions. Further, NGOs are not interested in obtaining loans to construct facilities and take financial risks. Invariably, the efforts of the NGOs (as in India) in providing community toilet facilities are constrained by the total funds available to the municipalities or local bodies for providing sanitation services. For investments in private toilets, subsidies (20% to 60% of capital costs in Ouagadougou, Burkina Faso) and loans (in Kumasi, Ghana) with the exception of *Orangi Pilot Project* in Karachi, Pakistan are typical instruments.

Governments, local bodies, international lending institutions and bilateral aid agencies have to understand and accept that far greater funds are required for meeting the *Millennium Development Goals* of providing adequate sanitation than the resources available in the past. Governments and local bodies should consider additional taxes on water use (e.g. in Burkina Faso) and/or property taxes that are specifically marked and allocated to provisioning of sanitation services to the poor. Financing for sanitation will require special funding provisions as well as providing sufficient funds through *poverty reduction support credits*.

Additional funds have to be raised through financing mechanisms for leveraging resources through private sector participation and/or through greater community resources. Innovative institutional arrangements will be required so that NGOs can set-up companies that can raise capital funds and manage financial risks. The design of financing mechanisms for leveraging resources would require linking with domestic credit markets; emphasis on risk management; pre-investment funds for project preparation and financial resources for efficient management.

### 6.3.3    Financial sustainability requires targeted subsidies to cover O&M costs

Given the limited ability of the poor to pay for sanitation services, the greatest challenge for community sanitation systems has been low tariffs, generating revenues that are too low for cost recovery and financial sustainability of existing or new services (Bhatia, 2004). For a long-term sustainability of investments in urban sanitation, the community schemes have to raise revenues from users who are too poor to contribute enough to cover even the Operation and Maintenance (O&M) costs. Other sources of revenues have to be found to meet O&M costs and a part of the capital costs. For example, in India, NGOs assisting communities can raise revenues from their toilet complexes by renting space or selling biogas and thus reduce their dependence on government funds and subsidies. However, this will require changes in policies and regulations for NGOs and funding organizations such as the municipalities or international donors/banks.

### 6.3.4    Private toilet programs do not cover the poorest

In some community projects, most of the households that have invested in on-site sanitation systems seem to belong to the middle class or poor but relatively better-off households (e.g. the participants in the *Orangi Pilot Project* in Karachi, Pakistan; in Ouagadougou, Burkina Faso; in Luanda, Angola; and in many towns in Brazil). For extending these programs, the subsidies need to be targeted specifically towards the poorest sections, who will not be able to afford the systems, even the low cost options. Such subsidies are in the nature of capital cost sharing through subsidized loans. Cross-subsidization or direct support from the local governments may be helpful in some cases. From the experience of many projects (e.g. Condominial system in Brazil), it is important to recognize that cost recovery and subsidy rules must be set in a clear and transparent manner.

## 7  POLICY INTERVENTIONS FOR POVERTY ALLEVIATION

### 7.1  *Legislation, water rights and water markets*

Water legislation provides the basis for government intervention and action, and establishes the context and framework for action by non-governmental entities; hence it is an important element within the enabling environment. Specific water laws have been enacted in a considerable number of countries, but some still lack a water resources law *per se*. Although references to water resources may be found in the national legislation, these are often dispersed in a multitude of sectorally oriented laws and may be contradictory or inconsistent on some aspects of water resource usage (Solanes & Gonzalez-Villarreal, 1999; GWP-TAC, 2000a, 2000b).

Water rights can affect the poor in a variety of ways including: (1) granting of special water rights to the poor that are not linked to land-ownership; (2) protection of traditional and customary water rights that are available to the poor, tribal and native populations; and (3) protection of water rights of native populations by ensuring that federal or state governments do not give away those water rights to companies using the water resources for irrigation/hydropower/households/industry.

In Sukhomajri village in Haryana, India, the villagers created a new institutional arrangement in 1981 in the form of a water users' association evolving a commonly acceptable method of distributing water (Chopra *et al.*, 1990). A village society was formed with all households from the village as members. It was laid down that the landless or those whose lands did not get water due to high terrain, etc., also had a right to water and could sell it to those whose demands for water exceeded their share. Water rights were established for all members of the society. In case water distribution pipelines did not reach the land of any member, water-selling rights were also established. On an average, about two water discharges were made per household during *Rabi* (winter) season. While water storage dams, pumps and other equipment were maintained out of water charges ranging from Rs 6 to 10 per hour (US$ 0.13 to 0.22 per hour) of water discharge, the surplus of the society was used in laying additional pipelines, etc. While the society does not make any profit in the distribution of irrigation water, the creation of equal water rights is its prime institutional contribution. The scope of the village water users' association, meant initially for the distribution of irrigation water, was extended to the management of a large number of common property resources (e.g. contracting forest land from the government, managing fishing in the irrigation tanks).

Changes in policies and legislation have a substantial impact on opportunities for the poor. For example (World Bank, 2003), in 1992 Mexico passed a new water law, which introduced radical changes in the way in which water is managed. Of greatest importance in these changes were giving the users much greater say in the management of water, and introducing tradable water rights. In some areas the effects have been dramatic, with substantial reductions in the (unsustainable) pumping of aquifers, and with water moving from traditional low-value crops to new high-value crops. Each drop of water now generates much higher economic returns and each hectare of land and each drop of water now generate a direct demand for more than twice as much agricultural labor (and therefore opportunities for the poor).

### 7.2  *Targeted subsidies, water tariffs and pollution charges*

According to the principle of managing water as an economic and social good, the recovery of costs should be the goal for all water uses, unless compelling reasons indicate otherwise.

Yet, this principle entails inherent difficulties: how can principles of equitable access to water used for basic human needs be taken into account at the same time? At a minimum, full supply costs should be recovered in order to ensure sustainability of investment and the viability of service providers. However, in many situations, even the achievement of this objective may require direct subsidies for years to come.

Poverty alleviation policies might be incompatible with abrupt implementation of full supply cost recovery in, for instance, some surface irrigation systems. In the provision of municipal and rural water supply there are well-established practices of cross-subsidization from better-off water users to the poor. The use of cross-subsidies does not necessarily compromise the financial sustainability of utilities but they distort prices and patterns of demand. For management purposes such subsidies should be made in a transparent manner and, where possible, direct subsidies are the preferred option to reduce distortions in the system. Under normal circumstances industries should meet at least the full economic cost of the supplied water (Bhatia & Falkenmark, 1993; Bhatia et al., 1994; Bhatia et al., 2000).

Chile has been able to implement a well-working system of focal subsidies in the water and sanitation sector (Briscoe et al., 1997; GWP-TAC, 2000a). The success of the system depends on the concerted effort and institutional capabilities of the national government, the municipalities and the water companies. Other countries in Latin America have attempted to replicate the very successful Chilean experience. However, the funds available did not match the needs of the users, neither did the institutional capability of governments match the monitoring requirements of system implementation and enforcement. For this reason some countries, such as Argentina, have resorted to traditional cross-subsidies, despite the obvious drawbacks of the system. The lesson is that before suggesting either focal or cross-subsidies, countries and financing institutions should ensure not only financial and economic viability, but also that institutional structures do allow efficacious implementation.

### 7.3    Special programs for financing connection costs

Maylinad Water Services Inc. in Metro Manila has a program called Bayan Tubig (Water for the Community) under which the company provides individual connections to poor households. Although water tariffs are the same for all households, the increasing block tariffs mean low tariffs for households using 10 m$^3$ per month or less. For poor households there are schemes for payment of connection charges (US$ 60) in 12 monthly installments. In squatter settlements, where space is a problem, water lines are run through neighborhoods and connected to a battery of water with each meter assigned to a nearby home. From the meter, each homeowner makes his own plastic connection, aboveground. Around 60,000 such new connections (50% of total new connections) have been made under this scheme supplying water to more than 400,000 poor people.

### 7.4    Water tariffs

Water tariffs provide little incentives for the sustainable use of water if charged at a flat rate independent of the amount used. In such cases, setting the right fee structure, imposing progressively higher unit cost prices on high-volume users, may induce the more judicious use of the resource, although the level of demand reduction will depend upon the nature of the high-volume users. Such a structure also contributes to the financial sustainability of water authorities and to covering the cost of administering water resources management. (GWP-TAC, 2000a).

It has been argued (Bhatia *et al.*, 1994; Bhatia & Kijne, 1994; Rogers *et al.*, 1998; Bhatia *et al.*, 1999) that sustainable and efficient use of water require the tariff to match not only costs of supply (i.e. O&M and capital costs), but also opportunity costs, economic externality costs, and environmental externality costs. Very often the tariffs do not even meet the full supply costs, and sometimes the value of water is lower than the cost of supply! The evidence is presented from the Subernarekha and Yamuna basins in India. For example, a comparison of costs and values associated with urban and industrial water use in the Subernarekha river basin is given in Rogers *et al.* (2002). The full cost of water to be US$ 0.47/m$^3$. The value in industrial use was US$ 2.60/m$^3$; the value in urban use was US$ 0.25/m$^3$. The industrial tariff is more than double the urban tariff, but still incredibly low compared to the cost and value of water. So the urbanites over-consume and value the resource less. In this case there is an industrial to domestic to agriculture subsidy. In each case the agricultural tariffs are much smaller than cost of supply in both basins.

## 8   CONCLUSIONS

To recapitulate:
- There is ample evidence to show that investments in water infrastructure (irrigation and hydropower projects, small check dams) have considerable multiplier (direct and indirect) benefits to poor farmers and agricultural labor in terms of increased output, employment and real wages.
- Improved management of water resources (through flexible allocation among sectors and users) provides benefits to the poor and can be an important instrument of poverty alleviation/reduction in developing countries.
- A review of experiences in community management of rural water supply services points to the need for careful monitoring of the impact of some community schemes on the poor, particularly those where it is mandatory to recover, from the users, the high cost of pumping schemes.
- Community participation in urban sanitation in the context of the Millennium Development Goals should be encouraged. However, there are significant constraints in scaling up these *islands of success* in terms of sources of funds, cost recovery and financial viability. Public funds will be required to provide targeted subsidies to the millions of poor who can not afford to pay for water and sanitation services. Leveraging of community funds and private financial resources will require innovative financing mechanisms.
- To ensure protection and augmentation of water resources and their equitable distribution, innovative practices will be required in legislation and regulation, water rights, pricing of water and cross-subsidies. Financial sustainability of water investments will be necessary so that the poor get the water and sanitation services they need and deserve.

## REFERENCES

Agarwal, A. (1999). *Population and Sustainable Development: Some Exploratory Relationships*. Centre for Science and Environment, New Delhi, India.

Agarwal, A. & Narain, S. (1980). *Strategies for the involvement of the landless and women in afforestation: five case studies from India* (mimeo). World Employment Programme, International Labour Office, Geneva, Switzerland.

Agarwal, B. (1995). Gender, Environment and Poverty interlinks in rural India: regional variations and temporal shifts, 1971–1991. *UNRISD Social Development News*, 13.

Agarwal, B. (1997). *Gender and Legal Rights in Landed Property in India*. Institute of Economic Growth, Delhi, India. Draft.

Bhatia, M. (2004). *Financing and Financial Viability of Community Sanitation Projects: A Review of Sulabh and SPARC Programs in India*. Draft.

Bhatia, R. (2003). *Poverty and Irrigation in India: Some Empirical Findings*. Research Report, Resources and Environment Group, New Delhi, India.

Bhatia, R. & Bhatia, M. (2004). *Financing and Financial Viability of Community Sanitation Projects: a review of Sulabh and SPARC Programs in India*. Resources and Environment Group, India. Draft.

Bhatia, R. & Falkenmark, M. (1993). *Water Resource Policies and the Urban Poor: Innovative Approaches and Policy Imperatives in Water and Sanitation*. CURRENTS, World Bank. (Paper presented at the Dublin Conference on Water and Environment).

Bhatia, R. & Hansen, S. (2004). *Millennium Development Goals Challenges: What Can We Learn From the Experience of Community Participation in Urban Sanitation Programs?* Paper prepared for the Norwegian Ministry of Environment.

Bhatia, R. & Kijne, J. (1994). Conflicts in Water Use: Sustainability of Irrigated Agriculture in Developing Countries. *Proceedings of the Stockholm Water Symposium*, August 9–12, 1994, Stockholm, Sweden.

Bhatia, R. & Malik, R.P.S. (1999). *Financing Needs of the Water Sector in India*. Resources and Environment Group, New Delhi, India. Draft.

Bhatia, R.; Rogers, P.; Briscoe, J. & Sinha, B. (1994). *Water Conservation and Pollution Control in Indian Industries: the Role of Regulation, Water Tariffs and Fiscal Policies in India, Water and Sanitation*. CURRENTS, World Bank.

Bhatia, R.; Rogers, P. & de Silva, R. (1999). *Water is an economic good: how to use Prices to Promote Equity, Efficiency, and Sustainability*. Paper presented at the Stockholm Water Symposium.

Bhatia, R.; Kumar, R.; Misra, S. & Robins, N. (2000). *Full Cost Pricing of Water – Options and Impacts: A Case Study of the Impacts of Moving to Full Cost Pricing on Freshwater Demand, Recycling and Conservation at the Tata Steel Company, India*. UNIDO-IIED. Draft.

Bhatia, R.; Scatasta, M. & Cestti, R. (2004). *Indirect Economic Impacts of Dams: Methodological Issues and Summary Results of Case Studies in Brazil, India and Egypt*. Vol. I–II, Forthcoming, World Bank, Washington, D.C., USA.

Bos, E. & Bergkamp, G. (2001). *Overcoming Water Scarcity and Quality Constraints*. Water and the Environment, IFPRI, Washington, D.C., USA.

Briscoe, J.; Salas, P.A. & Peña, T.H. (1997). *Managing water as an economic resource: Reflections on the Chilean experience*. World Bank, Washington, D.C., USA. Draft.

Chopra, K.; Kadekodi, G. & Murthy, M.N. (1990). *Participatory Development: People and Common Property Resources*. Sage Publications, New Delhi, India.

deLucia and Associates (2000, 2001). *Best Practice Policy and Financing Approaches for Small-Scale Infrastructure Service Providers in South Asia*. Report prepared for the World Bank, October 2000 and March 2001.

Esrey, S. & Andersson, I. (1999). *Poverty-Environment Interventions in Water and Sanitation: Key Issues and Policies*. UNDP-UNICEF.

GWP-TAC (2000a). *Integrated Water Resources Management*. Global Water Partnership, Technical Advisory Committee. TAC Background Paper 4. 71 pp.

GWP-TAC (2000b). *Letter to my Minister*. By Ivan Chèret. Global Water Partnership, Technical Advisory Committee. TAC Background Paper 5. 30 pp.

Hansen, S. & Bhatia, R. (2004). *Water and Poverty in a macro-economic context*. Paper commissioned by the Royal Norwegian Ministry of the Environment.

Hazare, A. (1997). *Ralegan Siddhi: a veritable transformation*. Ralegan Siddhi Pariwar Publications, Ralegan Siddhi, India.

Malik, R.P.S. & Bhatia, M. (2004). Indirect Economic Impacts of Check Dams in the Bunga Village, India. In: R. Bhatia, M. Scatasta & R. Cestti, *Indirect Economic Impacts of Dams: Methodological Issues and Summary Results of Case Studies in Brazil, India and Egypt*. Vol. II, Chapter 3. Forthcoming, World Bank, Washington, D.C., USA.

Meinzen-Dick, R. & Bakker, M. (2001). Irrigation systems as multiple-use commons: water use in Kirindi Oya, Sri Lanka. *Irrigation and Drainage Systems*, 15(2): 129–148. Also published as EPTD Discussion, Paper 59. International Food Policy Research Institute (IFPRI), Washington, D.C., USA.

Meinzen-Dick, R.; Swallow, B.M.; Jackson, L.A.; Williams, T.O. & White, T.A. (1997a). *Multiple Functions of Common Property Regimes*. IFPRI Environment and Production Technology Workshop. Summary Paper 5. Washington, D.C., USA.

Meinzen-Dick, R.; Brown, L.; Feldstein, H. & Quisumbing, A. (1997b). Gender, Property Rights, and Natural Resources. *World Development*, 25(8): 1303–1315.

Pritchett, L. (2001). Where has all the Education gone? *World Bank Economic Review*, 15: 367–391. Oxford University Press.

Rees, J. (2001). *The Risk and Integrated Water Resources Management*. Paper prepared for the 9th International Conference on the Conservation and Management of Lakes. Lake Biwa, Japan, November, 8–16.

Rogers, P.; Bhatia, R. & Huber, A. (1998). *Water as a social and economic good: how to put the principle into practice*. Global Water Partnership, Technical Advisory Committee, Paper 2.

Rogers, P.; Bhatia, R. & de Silva, R. (2002). Water is an economic good: how to use prices to promote equity, efficiency, and sustainability. *Water Policy*, 4: 1–17.

Solanes, M. & González-Villarreal, F. (1999). *The Dublin Principles for water as reflected in a comparative assessment of institutional and legal arrangements for integrated water resources management*. Global Water Partnership, Technical Advisory Committee, Paper 3.

Soussan, J. (2001). *Poverty & Water Security*. Paper prepared for the Water and Poverty Initiative, Global Water Partnership, Stockholm, Sweden, December 2001. Draft.

Sullivan, C. & Meigh, J. (2003). Considering the Water Poverty Index in the context of poverty alleviation. *Water Policy*, 5(5): 513–528.

Sullivan, C.A.; Meigh, J.R. & Giacomello, A.M. (2003). The Water Poverty Index: Development and application at the community scale. *Natural Resources Forum*, 27(3): 189–199.

UN (United Nations) (1999). *The State of world population*. United Nations Population Fund.

UN (United Nations) (2004). *Interim Report of the Task Force 7*. United Nations Millennium Project.

WCD (World Commission on Dams) (2000). *Dams and Development: a new framework for decision-making*. The Report of the World Commission on Dams. Earthscan Publications, UK.

World Bank (2003). *Water Resources Sector Strategy – Strategic Directions for World Bank engagement*. Washington, D.C., USA. Draft.

World Bank (2004). *Tamil Nadu Water, Economy and Poverty Study*. A Draft Report by the Study Team, World Bank, New Delhi, India.

Zerah, M. (2000). *Water – Unreliable Supply in Delhi*. French Research Institutes in India.

CHAPTER 13

# Do investments and policy interventions reach the poorest of the poor?

C.A. Sullivan
*Centre for Ecology and Hydrology, Wallingford, UK*

ABSTRACT:   Macroeconomic policy instruments are often held to be the most effective tool for poverty alleviation. In the water sector, various approaches have been used. These include flexible water allocation schemes, water trading and transferable rights. Illustrations of the benefits of cross-subsidization schemes are cited, and the need for private sector involvement in water distribution is frequently stressed. In contrast to this, the case is made in this chapter that in addressing the needs of the poorest of the poor, such measures are far from ideal. Weaknesses are highlighted both in terms of theoretical frameworks and in terms of procedural failure, and it is due to these, it is argued, that poverty continues to be ubiquitous throughout the world. How development is measured, and how benefits from water infrastructure are assessed, are brought into question, presenting an alternative explanation of why poverty is such a difficult ill to cure.

Keywords:   *Poverty alleviation, Water Poverty Index, investment, water policy*

## 1   INTRODUCTION

The chapter *Water and poverty alleviation: the role of investments and policy interventions* (Bhatia & Bhatia, this volume) provides an excellent review of water projects, highlighting potential approaches that could be used as best practice tools for poverty alleviation. Many examples are cited, from India and other parts of the world, demonstrating the practical benefits from community participation, flexible water allocation schemes, property rights, water trading, etc. The benefits of community water management are discussed on a number of levels, with examples from many places to illustrate how it can work in practice. Finally, the chapter concludes that while clearly investment in infrastructure will have a positive and more than proportionate benefit for poor people, there is need for more involvement by the private sector to finance it, as one of the problems of reaching the very poor is the fact that they simply do not have the cash to support any kind of water provision system in a sustainable way.

The flexible allocation models described in the chapter do demonstrate how cross-subsidization can take place at all scales. If one sector bears some extra costs as a result of reallocations, it may mean that other sectors gain, and supplies are maintained to the poor, or for domestic use. The additional cost to some sectors is compensated for, in *pareto* terms, by the marginal benefits of extra water access accruing to the poor, and other, possibly more productive sectors. This shows how the distributional impacts of cross-subsidization can be spread widely throughout the economy, along with the potential to use water in a more cost effective way, assuming allocations are made on the basis of efficient returns to water use.

Of course, it must be said that the determination of what is most efficient in any location will be determined by social, cultural and political values, so representative consultation would be essential if equitable decisions and financial instruments were to be put in place.

The study from Tamil Nadu state in India does suggest however that under a flexible allocation strategy, although overall incomes would increase significantly, more benefits accrue to urban communities than to the rural population, who make up some 34 million people in that state. Of these, 45% are classified as rural labour, and according to the analysis presented, these are the group who benefit least of all from this approach, actually losing 2% of their household incomes under a flexible allocation scheme. In reality therefore, this means that while overall per capita income is estimated to increase by 500% between 2000 and 2020, for those who do not manage to move to other forms of employment, the rural labouring poor experience a less than proportionate rise in incomes, effectively subsidizing the benefits gained by others.

While this study does provide an interesting insight into both the methodologies of livelihood analysis, and of how water management may impact on different groups, it does also highlight some of the weaknesses of certain widely used techniques. In the analysis presented, "water is allocated to sectors on the basis of the estimated economic value of water in each sector, as defined by *Willingness to Pay* (WTP) for water in each sector". There is much literature on the issue of valuation of environmental goods and services (Farber & Costanza, 1987; El Serafy, 1991), and the work by Costanza *et al.* (1997) on ecosystems highlighted the uncertainty associated with such values. The WTP methodology itself is fraught with pitfalls, and to some extent at least, is an unreliable numeraire for policy formulation (Friend, 1993; Hueting *et al.*, 1998). The whole debate of how water is valued (Briscoe, 1996; Faucheux & O'Connor, 1998; Spash, 1998), what it is used for, and who are the beneficiaries, are recurring issues in water management, and the rest of this chapter is used to raise questions about how we judge development impacts, and how the benefits of water allocation are actually assessed.

## 2    ADDRESSING WATER NEEDS

Water is a fundamental component of every ecological system. The development of policies and tools which recognize the role of water as part of our life support system is essential for the achievement of global security. In Africa, Asia and Latin America, some 70% of human populations currently live in fragile ecosystems, where significant disruption to ecological services could bring about what effectively would be irreversible consequences (Raskin *et al.*, 1997). This can already be seen globally in terms of increased rates of desertification, soil salinization and deforestation, and in many areas of the world, such as the Sahel, Amazonia, the Aral Sea, etc., where the impact of human activities has already begun to disrupt the life support functions of those systems (UNESCO, 2003).

At the human scale, this situation is even more pressing, as currently, at least 1100 million people lack access to an improved water source, and 2400 million lack access to basic sanitation. As a result, every day some 6000 children in developing countries die for want of a clean water supply and sanitation (WHO-UNICEF, 2000), and thousands more suffer from the impacts of ill-health. In general, water scarcity negatively impacts on food security and livelihood choices, and consequently, economic progress is stifled. Furthermore, educational opportunities for poor families are often missed, and women continue to bear an unjust burden of water provision, particularly in Africa and Asia.

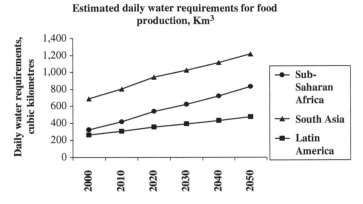

Figure 1.    Increasing demand for water as human populations rise.

The impact of human population growth is a major issue when considering the future challenges for water management (Falkenmark, 1990). If we examine the effect that this will have on demand for water for both domestic use and food production, we can see that expected rises in demand for these uses are significant (Postel, 1998). This is illustrated by Figure 1, which shows the likely increase in water for food production over the next 50 years. This is based on a water requirement for food production of 1400 L/d of water per person, with population growth rates maintained at the 1999 level. It is important to note however that these figures make no attempt to incorporate any increases which will also occur as a result of rising standards of living.

Since the overall amount of water available for human use is limited by time and space, it is reasonable to assume that as demand increases following population growth, per capita resource availability shrinks, generating conflicts over use. Gleick (2000) has examined many aspects of water resources and entitlements, especially with respect to global security, and today, the issues of both poverty and water are now attracting considerable attention from a security point of view. The widespread publication of global disparities in water accessibility in such meetings as the *World Summit on Sustainable Development* in Johannesburg in 2002 and the *3rd World Water Forum* in Kyoto in 2003 have emphasized the need to address the problem of water management more effectively, both at a local and international scale.

## 3    RECONSIDERING THE LINKS BETWEEN DEVELOPMENT AND ECONOMIC GROWTH

In the past, economic development has in most cases taken precedence over all other issues (Lumby, 1979), but more recently, attention has been drawn to the need to expand the way in which we view the development process (Lipton, 1988; Sen, 1995). Numerous examples can be found where ecological disruption has resulted from development projects designed to increase agricultural or industrial production. These have occurred because knowledge of the complexities of ecosystems is limited, and values of the relevant environmental attributes have been ignored (Jacobs, 1997). Compounded by a scientific approach which has been focused rather than holistic, to some extent at least, this has led to unrepresentative theories of growth economics. These theories, on which many development projects are founded, are based on understandings which:

(a)  suggest that man-made and natural capital can infinitely be substituted, and
(b)  ignore the constraints on production from the laws of thermodynamics (Daly, 1999).

Clearly, while it can be shown that some measure of economic capital is generated from the depletion of natural resources, it can also be easily shown that certain natural resources cannot be reproduced by utilization of financial or physical capital (Kaufmann, 1995). This refutes the concept of *perfect substitutability of factors of production*, which is a basic assumption underlying the positions held by eminent economists such as Julian Simon (Simon & Khan, 1984) and Wildfred Beckerman (1995). Furthermore, the fact that money generated by exploitation of natural capital is accounted for in terms of *income streams* rather than *capital depletion*, brings about an inevitable undervaluation of such resources, and consequent policy failure.

The existence of entropy, as explained by the laws of thermodynamics, means that even the most efficient production system must produce waste, thus making it clear that the idea of infinite resource recycling and substitution is physically impossible. The failure of early growth theories to take account of these real world conditions is one of the reasons why many water projects developed in the past have failed to live up to expectations, and why numerous examples exist of inequitable, and perhaps unexpected, development outcomes.

## 4   THE RELATIONSHIP BETWEEN WATER USE AND ECONOMIC DEVELOPMENT

At a national level, it can be seen that countries which have higher levels of income tend to have a higher level of water use, as can be shown in the examples in Table 1.

If however we look more closely at the returns from water use by sector, we can see that in some countries, different sectors play different roles in economic performance, and the use of water in these sectors may be a determining factor in the progress of an economy. Table 2 shows sectoral correlation coefficients for a selection of rich and poor countries currently facing some degree of water shortage. In this example, these countries have been identified

Table 1.   Water use and national income.

| | GDP per capita, US$ (1990) | Annual water withdrawals per capita, m³ (1970–1987) | | |
| --- | --- | --- | --- | --- |
| | | Domestic | Industrial and agricultural | Total |
| Tanzania | 110 | 8 | 28 | 36 |
| Sri Lanka | 470 | 10 | 493 | 503 |
| South Africa | 2530 | 65 | 339 | 404 |
| United Kingdom | 16,100 | 101 | 406 | 507 |
| Sweden | 23,660 | 172 | 307 | 479 |
| United States | 21,790 | 259 | 1903 | 2162 |

Source: World Bank (1992: *Development and the Environment*, Tables 1, 33).

as having a per capita *Gross Domestic Product* (GDP) of either less than, or more than US$ 10,000 (on purchasing power parity rates), and the degree of water shortage is indicated by having less than 0.5 km$^3$/yr per capita freshwater withdrawals.

From this information, we can see that in the agriculture sector in poorer countries, there is some correlation between agricultural water use and the contribution of agricultural outputs to GDP. For industry, in poor countries, higher levels of industrial water use tend to generate greater contributions to GDP, whereas there is no such correlation in rich countries. While these correlations provide no suggestion of a causal relationship, they do suggest that in poor countries, when water is used for industry, the economic returns are of greater importance to the nation than in richer countries. It also reflects the fact that in higher income countries, the service sector (which is not a heavy water user) is much more important than it is in poor countries. These figures also indicate that in poor countries, there is a high negative correlation between the contribution of agricultural activity to the economy, and the country's score on the *Human Development Index* (HDI), reflecting the fact that a high level of agricultural dependence may be associated with lower levels of economic and social wellbeing. This confirms the fact that developing economies heavily dependent on agriculture tend to be price takers on world commodity markets, often getting very little return for the use of any of their resources, including water.

From this sample of countries, there also appears to be a stronger link in the higher income group between per capita water use and per capita GDP, than in the lower income group, again suggesting that the richer countries are in a better position to get higher benefits from their water use than do poor ones. While this may reflect the fact that they have better control of other factors of production, this is clearly an area of research which should be explored much further, and while these figures can only be taken as illustrative, they do suggest that how water is managed within an economy can have a direct impact on the economic welfare of society. It is important to note, however, that these figures take no account of rainfed farming systems that provide significant amounts of food throughout the world, much of which may be unaccounted for in national accounts due to home consumption. Furthermore, no account is taken in these figures of environmental damage associated with any form of water use or allocation.

Table 2. The relationship between economic performance and water use.

| | Correlation between agricultural water use (%) and contribution to value added to GDP from agriculture | Correlation between industrial water use (%) and contribution to value added to GDP from industry | Correlation between value added to GDP from agriculture and HDI score | Correlation between water withdrawals and HDI score | Correlation between water withdrawals and GDP per capita value |
|---|---|---|---|---|---|
| Countries with high GDP per capita | −0.03 | −0.07 | −0.06 | 0.51 | 0.47 |
| Countries with low GDP per capita | 0.21 | 0.69 | −0.51 | 0.11 | 0.27 |

*Note*: Both groups of countries suffer from some degree of water shortage.

## 5   THE NATURE AND SCALE OF POVERTY

Poverty is a multidimensional concept, and numerous definitions of it have been provided in the vast literature that has been generated on the subject. Recurring themes in the literature include:

– Poverty in the context of development (van der Gaag, 1988; Sen, 1995; UNDP, 2000);
– Poverty measurement (Lipton, 1988; Desai, 1995; World Bank, 1996);
– Poverty thresholds (Orshansky, 1969);
– Poverty and gender (Rosenhouse, 1989);
– Poverty and welfare (World Bank, 1998);
– Poverty and food (Malseed, 1990);
– Poverty and politics (Uvin, 1994);
– Poverty and health (WHO, 1992);
– Poverty and vulnerability (Sullivan *et al.*, 2002; Meigh & Sullivan, 2004).

One consensus that certainly exists in all this literature is firstly that there are too many poor people, and often they find themselves trapped in poverty by circumstances beyond their control.

## 6   HOW ECONOMISTS MEASURE POVERTY

Methods currently in use to assess poverty need to be considered when discussions of poverty alleviation take place. There are a number of approaches used to quantify poverty levels, including the *Poverty Line*, the *Headcount Index*, and the *Poverty Gap*. All of these approaches are based on national income figures, and as averages, are not very representative of regional variations. As a result, they often fail to accurately represent the levels of poverty experienced in different communities. Importantly, measures of per capita income are recognized to be inadequate to represent human wellbeing. While money measures may provide some means of comparison of economic activity, they take no account of non-monetary attributes of human wellbeing, nor of the value of women's household labor, nor indeed of depreciation of natural capital.

## 7   ABSOLUTE POVERTY

In absolute terms, there is always a minimum level of certain resources that people need in order to survive, and without these, a person will find himself in a state of poverty. Access to water resources is likely to be a major determinant of absolute poverty. Absolute poverty describes poverty "without reference to social context or norms and is usually defined in terms of simple physical subsistence needs but not social needs". Furthermore "absolute definitions of poverty tend to be prescriptive definitions based on the *assertions* of experts about people's minimum needs" (Gordon & Spicker, 1999). The World Bank employs an absolute definition of poverty in the form of a poverty line. The poverty line, or minimum income required, is deemed as US$ 1 per day, and more than 1000 million people in the world today fall below this level. While such a definition does serve to quantify the extent of poverty in a monetary sense, it does not encompass important social factors, such as being *free of obligations to make*

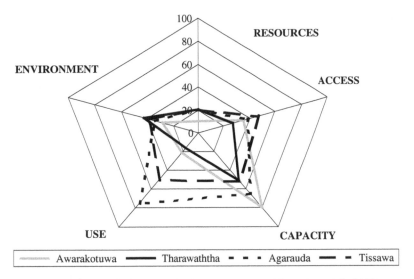

Figure 2. Displaying findings from the *Water Poverty Index* (Sullivan & Meigh, 2003).

*children work for others*, or *the ability to decline demeaning jobs*. These are two examples of the elements of poverty that are often identified by the poor themselves. As a result of these omissions, many measures may fail to accurately indicate the levels of poverty experienced in different communities. Importantly, measures of per capita income are now recognized to be inadequate to accurately represent human wellbeing (UNCED, 1992; Scoones, 1998).

## 8  LINKING WATER AND POVERTY

In an attempt to address the inadequacy of many different water evaluation techniques, new approaches have been developed, and one of these, the *Water Poverty Index* (Sullivan, 2002; Sullivan *et al.*, 2003), tries to capture the diversity of the many facets of water, and its links to wellbeing. This is an indicator based approach, and Figure 2 illustrates how the results of the *Water Poverty Index* (WPI) can be displayed, delivering complex information to policy makers in a simplified way. Using this approach, it is not only possible to identify the spatial variability of water provision, but also the variation in component values is shown, highlighting strengths and weaknesses in particular locations. The use of this approach does facilitate the inclusion of a greater variety of issues, and increases the accountability of water management decisions. Using this approach, decision makers are able to take account of many aspects of water management, to distinguish between locations, and prioritise investments.

## 9  ACCOUNTABILITY AND UNACCOUNTABILITY

Accounting for water is an important challenge underlying efficient and equitable management. *Unaccounted-for water* is a term often heard in discussions with water professionals, representing as it usually does, an apparently unrecoverable cost. This has long been an area of attention for executives of water companies, and management is encouraged to reduce its level by whatever appropriate means. While this may be admirable from a business efficiency point of view, it is often not equitable, as in many instances, those currently benefiting from *unaccounted-for water* are the poorest in society, and least able to pay.

In economic analysis, there are often many activities which are *unaccounted-for*. These would include obvious things like crime and the proceeds from it, but also other things like charity work, and unpaid housework. Often, another *unaccounted-for* aspect of economic analysis is the environment. Environmental costs associated with production are almost always *unaccounted-for*, and there is much need to reassess the processes by which we evaluate development options when these *unaccounted-for* dimensions create such uncertainty (Hannon, 1991; Jacobs, 1994; Daly, 1999). Furthermore, accounting systems are designed on the basis of standards established in developed economies. Sometimes this creates problems, as activities and values particularly relevant in some places are not relevant in others (Opschoor, 1998). The question of whose values should be counted is particularly important in considering water provision.

This leads us to the point of *unaccounted-for* people. These could include the millions of people who over the last five decades have been ousted from their homes in the interest of policy implementation and water management (Dreze *et al.*, 1997). It could include the millions who live in the squatter settlements fringing almost all the world's large cities. In the case of these people, they are often *unaccounted-for*, since they lack any property rights or tenured access. *Unaccounted-for* people would also include those millions who are simply not registered in any formal way in the population records. Even in the most advanced places, data is often missing. So in megacities across the world, many people are *unaccounted-for*, simply because that is what they want to be. This failure to account accurately for human numbers is another reason why many attempts to reduce poverty are often flawed.

Another barrier to the achievement of better water access is due to the widespread implementation of water pricing. While there is much logic in this development, there is little doubt that large numbers of the world's poor are simply not in any position to pay anything for water at all. At the same time, it is known that in many countries, the poor pay more for water than the rich, (as shown in Table 3), and this is often given as justification of why more streamlined pricing should be adopted.

While some examples of small scale cross-subsidization are to be found in many places, there is little doubt that much more should be done to implement more such schemes to address this. It could be argued that as a requirement in all urban water development projects, there

Table 3.   Water price variation in selected developing cities.

| Water source | Water prices, US cents per 1000 litres | | | |
| --- | --- | --- | --- | --- |
| | Lima | Kampala | Bandung | Dar es Salam |
| Utility provision | 0.5 | – | 0.5 | – |
| Kiosk | – | 1.0 | – | – |
| Water truck | 2.3 | – | – | – |
| Water vendor | – | 8.7 | 3.5 | – |
| Handcarts | – | – | – | 1.6 |
| Stand pipes | – | – | – | 5.0 |

should be some provision for an open access water point to provide for the needs of those who otherwise may *slip through the net*.

One way in which this the problem of subsidization can be considered on a much larger scale is in the case of large water projects. In many countries, both irrigation and hydropower schemes often benefit from high degrees of government subsidy, although often these are not made clear. In such cases, the real costs associated with water consumption for these purposes are to some degree *unaccounted-for*, and it may be that the beneficiaries are not really those in the greatest need. Examples of this can be seen in numerous countries where dams and reservoirs which are built are actually used largely by private commercial farmers, or power companies, rather than the poorest of the poor. One way of overcoming these problems would be through the development of more transparent systems of subsidization, more effective prioritization of investments, and the development of better identification of those who are in the greatest need.

In the overcrowded cities of the world, millions of people may be within a short distance of a water supply point, but in practice, this may not be available to them. It could be that they do not have legal access to this water, or that they are prevented from its use by caste or other social constraints. In addition, in most urban settings, there is a huge proportion of household time spent queuing to get access to the water points, and single points may be shared by thousands of households. This may be simply due to the huge number involved, or due to low pressure reaching outlying shantytowns where so many have to make their homes. This is one of the reasons why some argue that the data generated by the *Joint Monitoring Programme* (WHO-UNICEF, 2000) does not provide an accurate picture of water deprivation, as that information is simply based on physical distance from a water source. The additional time spent in queuing to have access to water points represents another example of the *unaccounted-for* aspects of water management.

Today, in many cases, poverty is increased by ecosystem degradation, and in its turn, poverty may create more degradation (Kaufmann, 1995). At the same time, when people are empowered to manage their own resources, this can be extremely beneficial, and can bring about a real change in the way resources are used for different purposes. There is little doubt that there is much need for more effective monitoring of the links between water and poverty, and it is important to identify those places where the need for improvement is greatest, thus enabling the prioritization of action. By identifying and tracking the physical, economic and social drivers which link water and poverty, decision-makers will be able to identify *who needs water, where*, thus enabling action to be taken on the ground to meet these pressing needs.

## 10   CONCLUSION

This chapter has tried to argue that while policy interventions and investments have a significant role to play in poverty alleviation, they often fail to account for the most vulnerable groups of the poor. It is recognized in most countries that the activities of a sizeable sector of the population are not included in national income statistics. These are the *unaccounted-for* millions: often very poor people who participate in activities which are not formally recognized and recorded. People classed as the *rural poor* are often in this category, since their economic activities are largely not accounted for. The majority of rural people are likely to generate their own food, consuming it directly, without recourse to a market. In addition, they may gather products from nature such as fuelwood, along with forest foods and medicines. In societies where capitals are tied up in livestock, people are reluctant to reveal the extent of their livestock holdings, and it may be the case that the ecological costs associated with their activities are also *unaccounted-for.*

There is no doubt that there is a vital role to be played by investments and policy interventions in the water sector. Nevertheless, overcoming inequities in water allocation is one of the objectives of improving stakeholder participation and better water governance, and there is a clear need for more inclusive and accurate assessments of access to water, and the way it is used. While agriculture in general uses around 70% of all water used for human consumption, and the level of domestic water needed to meet all human health needs is less than 5% of total withdrawals, it is clear that there could be a small amount of water taken from agriculture (through improvements in water use efficiency), to provide for better domestic provision. While this suggestion may be said to be simplistic, there is no doubt that in a number of countries, great improvements could be made to the quality of life of millions of people if more account was taken of what was *unaccounted-for*, and if more attention was paid to building the political will necessary to achieve this.

## REFERENCES

Beckerman, W. (1995). *Small is stupid: Blowing the whistle on the Greens.* Duckworth, London, UK.

Briscoe, J. (1996). *Water as an economic good.* Paper presented at the World Congress of International Commission on Irrigation and Drainage. Cairo, Egypt.

Costanza, R.; d'Arge, R.; de Groot, R.; Farber, S.; Grasso, M.; Hannon, B.; Limburg, K.; Naeem, S.; O'Neill, R.V.; Paruelo, J.; Raskinet, R.G.; Sutton, P. & van den Belt, M. (1997). The value of the world's ecosystem services and natural capital. *Nature*, 387: 253–260.

Daly, H. (1999). *Ecological Economics and the Ecology of Economics.* Edward Elgar, Cheltenham, UK.

Desai, M. (1995). *Poverty, Famine and Economic Development.* Edward Elgar, Aldershot, UK.

Dreze, J.; Samson, M. & Singh, S. (1997). *The Dam and the Nation: Displacement and Resettlement in the Narmada Valley.* Oxford University Press, New Delhi & Oxford.

El Serafy, S. (1991). The environment as capital. In: R. Costanza (ed.), *Ecological Economics: the Science and Management of Sustainability.* Columbia University Press, New York, USA.

Falkenmark, M. (1990). Rapid population growth and water scarcity: the predicament of tomorrow's Africa. In: K. Davis & M.S. Bernstam (eds.), *Resources, environment and population.* Supplement to *Population and Development Review*, 16: 81–94.

Farber, S. & Costanza, R. (1987). The economic value of wetland systems. *Journal of Environmental Management*, 24: 41–51.

Faucheux, S. & O'Connor, M. (1998). *Valuation for Sustainable Development.* Edward Elgar, Cheltenham, UK.

Friend, A.M. (1993). Feasibility of environmental and resource accounting in developing countries. *Environmental Accounting: A Review of the Current Debate.* UNEP, Environmental Economics Series, 8. Nairobi, Kenya.

Gleick, P. (2000). *The World's Water 2000–2001.* Island Press, London, UK.

Gordon, D. & Spicker, D. (eds.) (1999). *The International Glossary on Poverty.* Zed Books, London, UK.

Hannon, B. (1991). Accounting in ecological systems. In: R. Costanza (ed.), *Ecological Economics: the Science and Management of Sustainability.* Columbia University Press, New York, USA.

Hueting, R.; Reijnders, B.; de Boer, B.; Lambooy, J. & Jansen, H. (1998). The concept of environmental function and its valuation. *Ecological Economics*, 25: 31–37.

Jacobs, M. (1994). The limits to neoclassicism. Towards an institutional environmental economics. In: M. Redclift & T. Benton (eds.), *Social Theory and the Global Environment.* Routledge, London, UK.

Jacobs, M. (1997). Deliberative Democracy. In J. Foster (ed.), *Valuing Nature. Ethics, Economics and the Environment.* Routledge, London, UK.

Kaufmann, R.K. (1995). The economic multiplier of environmental life support: can capital substitute for a degraded environment? *Ecological Economics,* 12: 67–79.

Lipton, M. (1988). *The Poor and the Poorest.* World Bank, Washington, D.C., USA.

Lumby, S. (1979). *Investment Appraisal and Financing Decisions.* Chapman and Hall, London, UK.

Malseed, J. (1990). *Bread without dough: understanding food poverty.* Horton, Bradford, UK.

Meigh, J.R. & Sullivan, C.A. (in press). The Impact of Climate Variations on Water Resources: an Index-Based Approach to Assess Human Vulnerability. *Climate Change* (Special Issue).

Opschoor, J.B. (1998). The value of ecosystem services: whose values? *Ecological Economics,* 25: 41–43.

Orshansky, M. (1969). How poverty is measured. *Monthly Labor Review,* 92(2): 37–41.

Postel, S.L. (1998). Water for food production: will there be enough in 2025? *Biosciences,* 28: 629–637.

Raskin, P.; Gleick, P.; Kirshen, P.; Pontius, G. & Strezepek, K. (1997). Water Futures: Assessment of Long-Range Patterns and Problems. In: *Comprehensive Assessment of Freshwater Resources of the World* (background document for Chapter 3). Stockholm Environment Institute, Boston, Massachusetts, USA.

Rosenhouse, M. (1989). *Identifying the poor: is headship a useful concept?* LSMS Working Paper, 58. World Bank, Washington, D.C., USA.

Scoones, I. (1998). *Sustainable Rural Livelihoods: a Framework for Analysis.* IDS Working Paper, 72. IDS, Brighton, UK.

Sen, A. (1995). *Mortality as an indicator of economic success and failure.* Discussion paper, 66. London School of Economics and Political Science, UK.

Simon, J. & Khan, H. (eds.) (1984). *The Resourceful Earth.* Blackwell, Oxford, UK.

Spash, C.L. (1998). Environmental values and wetland ecosystems: CVM, ethics and attitudes. In: *Social processes for environmental valuation.* The VALSE project final report. Versailles, France.

Sullivan, C.A. (2002). Calculating a Water Poverty Index. *World Development,* 30: 1195–1210.

Sullivan, C.A. & Meigh, J.R. (2003). The Water Poverty Index: its role in the context of poverty alleviation. *Water Policy,* 5: 5. October.

Sullivan, C.A.; Meigh, J.R. & Acreman, M.C. (2002). *Scoping Study on the Identification of Hot Spots – Areas of high vulnerability to climatic variability and change identified using a Climate Vulnerability Index.* Report to Dialogue on Water and Climate, Centre for Ecology & Hydrology, Wallingford, UK.

Sullivan, C.A.; Meigh, J.R.; Giacomello, A.M.; Fediw, T.; Lawrence, P.; Samad, M.; Mlote, S.; Hutton, C.; Allan, J.A.; Schulze, R.E.; Dlamini, D.J.M.; Cosgrove, W.; Delli Priscoli, J.; Gleick, P.; Smout, I.; Cobbing, J.; Calow, R.; Hunt, C.; Hussain, A.; Acreman, M.C.; King, J.; Malomo, S.; Tate, E.L.; O'Regan, D.; Milner, S. & Steyl, I. (2003). The Water Poverty Index: development and application at the community scale. *Natural Resources Forum,* 27: 1–11.

UNCED (1992). *Report of the United Nations Conference on Environment and Development.* United Nations, New York, USA.

UNDP (2000). *Overcoming Human Poverty.* United Nations Development Programme, New York, USA.

UNESCO (United Nations Educational Scientific and Cultural Organization) (2003). *World Water Development Report.* UNESCO, Paris, France.

Uvin, P. (1994). *The International Organisation of Hunger.* Kegan Paul, London, UK.

van der Gaag, J. (1988). *Confronting Poverty in Developing Countries: Definitions, Information and Policies.* LSMS Working Paper, 48. World Bank, Washington, D.C., USA.

WHO (1992). *Health for All Database.* World Health Organization, Geneva, Switzerland.

WHO-UNICEF (2000). *Global Water Supply and Sanitation Assessment 2000.* Joint Monitoring Programme for Water Supply and Sanitation. World Health Organization and UNICEF, Geneva, Switzerland.

World Bank (1992). *World Development Report.* Oxford University Press, New York, USA.

World Bank (1996). *Poverty Monitoring and Analysis Systems.* World Bank, Washington, D.C., USA.

World Bank (1998). *Standardised Welfare Indicators.* World Bank, Washington, D.C., USA.

# Water and Nature

CHAPTER 14

# Water and nature. The berth of life

F. García Novo
*Estación de Ecología Acuática, University of Seville, Spain*

F. García Bouzas
*Junta de Andalucía, Seville, Spain*

ABSTRACT: Life is presented as an historic phenomenon arising on the planet in close connection to water environments and the water cycle. Early stages of life on Earth are underlined, pointing to some relevant roles of water in cell physiology. The breaking up of water during photosynthesis is focused as a far reaching step toward the development of life.

The importance of water in energy exchange and circulation of the atmosphere is discussed and climates are characterised in terms of water availability, regime and temperature. The role of water in the alteration of minerals and in transport processes of continents are discussed in terms of geomorphology and soil evolution under each climatic type. As a summary, a classification of environments is presented, showing how living organisms have succeeded in coping with varied conditions during the evolutionary process.

Communities are introduced as the ecological basis of life in ecological systems presenting a few examples of regulatory mechanisms involving water functions. Wetlands are presented in some detail and the relationships of basic ecological traits such as community structure, diversity and productivity in relation to water availability are discussed.

The advent of cultural man started a cascade of ecological phenomena that impaired the primeval functioning of ecosystems. Environmental consequences of river impoundment, water diversion, irrigation, and drainage of wetlands are dealt with. Eutrophication, contamination and salination of natural waters are discussed. Exploitation of aquatic organisms such as fish and the overall effects of disturbance on biodiversity are also shown. The introduction of aquatic species to new biological areas and the side effects of introductions on natural systems, rivers and cultivated areas and the invasion of urban aquatic environments by wild organisms is presented in this perspective of human induced biodiversity changes.

Some examples of subtle human alterations of aquatic systems are presented suggesting new technical developments were needed to attain sustainability development policies.

Keywords: *Water cycle, ecology of water, impacts on natural waters, water life, wetlands, water cultures*

## 1 LIFE ORIGINS

The origin of Life on our planet is still uncertain. The earliest evidence dates back some 3860 million years (Mojzsis *et al.*, 1996), this evidence consisting of isotopically light carbonaceous inclusions within grains of apatite (basic calcium phosphate), only known to build up within living cells. The deposits belong to a sedimentary formation.

Early fossils were aquatic organisms. Terrestrial forms appeared much later in the timeline. Life access to continents was mediated through aquatic forms invading estuaries, swamps and areas with plentiful water. The expansion of living organisms from aquatic to aerial environments allowed for the colonisation of continents and was followed by a large increase in biomass and species diversity.

All known life forms basically are tight arrangements of (organic) vesicles containing water solutions and dispersions, their composition being controlled by permeability and transport functions of vesicles walls and membranes. The description of molecular mechanisms involved in water transport across membrane pores represents a recent advance in biochemistry.

Life appeared in water environments and has evolved for most of its span as aquatic organisms. Life *performs* in water.

## 2   ECOLOGY

For minute living forms, such as micro organisms, the diffusion process is sufficient to exchange materials with the surrounding water environment. More efficient mechanisms involve the incorporation of water droplets to the cell in the form of vesicles, letting organic particles be exploited as trophic resources. The opening of fine ducts in the organism letting water solutions flow through the body largely increased the surface available for exchange of ions and gases or the delivery of particles. It had profound consequences in evolution: circulatory systems allowed for the integration of a single organism composed of specialised organs. The opening of outlets connecting ducts to the outer medium lies in the base of diversification of digestion, excretion, respiration and reproduction. Multicellular organisms and finally Metazoa, expanded as physiological water roles multiplied.

The appearance of contracting organelles led to pulsing structures in life forms, making possible the flow of digestive tracts or circulatory systems. The formation of flagella let organisms move, avoiding unfavourable environments and gave them the means to tap distant resources. Dispersal to new areas or to patchy resources in aquatic media became feasible. When sexual systems evolved, it was movement coupled to sensory mechanisms what allowed individuals or complementary sexes to couple. Sessile organisms, unable to move, retained a free mobile stage in water as gametes, larvae, or juveniles. Contracting cells arranged into muscles added a longer range of movement for animals, letting them commute between distant environments and to exploit resources, such as other organisms, on a more efficient scale.

The development of Life has exploited the water environment in a growing number of ways. Water as source or sink for gases, ions, salts and particles has been mentioned. However, diffusion flow proceeds slowly in still water as opposed to turbulent diffusion. Turbulence, or energy dissipation inducing it, has been exploited by some life forms which adapted to the littoral, to strong current sites, or thermal vents where turbulent diffusion occurs. At every scale seas show currents that have been integrated by many organisms in their life cycles: from tide or daily local currents to oceanic giro, all scales have been encountered and exploited as an energy source (for transport), by one group of organisms or other. And evolutionary sequences often document in organisms the appearance of convergent body forms and organs imposed by water properties.

For displacement through water, a fusiform body provided with a round head and a thin tail is the more efficient morphology. Propulsion organs should be in the rear. Fish-like forms are shared by the different lineages of fish, amphibians, and cetaceans among mammals. Other fast swimmers such as penguins, sea lions, and seals, have heads with big jaws or beaks and palmed legs instead a long tail, altering their silhouettes, but the body shape sticks to the fusiform pattern imposed by water viscosity.

As opposed to this pattern, planktonic species may evolve to bigger sizes and longer life cycles. Size implies, according to Stokes' Law, a faster precipitation. Vertical movements in the aquatic column may displace the organism from the favourable environment. To impair precipitation many plankton species possess chetae, appendages or specialised ramified organs with a large surface.

Turbulence favours diffusion in the aquatic environment supplying sessile organisms with nutrients, oxygen, and organic particles. The easy access to supplies cannot be separated from the transport of mineral particles abrading epidermis or clogging ducts and respiratory organs. Intense turbulence only occurs in certain locations where large energies are released to the water such as the littoral where waves break or in steep river channels. In each case the energy stored as wave oscillatory motion or the potential energy of the river flow is released, turning into kinetic energy of the water.

A few morphological patterns have been adopted by those organisms surviving in turbulent waters. For plants that demand light intercepting surfaces, laminar extensions shape into narrow bands forming belts, ribbons, narrow flat branches and threads. Brown algae (*Feophyta*) living in littoral waters often exhibit a thallus with a strong anchoring organ, a stem and a leaf-like expansion In *Laminariales* that may be laciniate such as *Himantalia*, or *Chorda*, digitated as in many *Laminaria*, flat or thread-like branches as in *Cistoserira* or *Fucus* species. It has been documented for *Laminaria* that the tensile shear induced by wave action elongates the *stem* of the plant generating heat and favouring metabolic activity. Red algae (*Rodophyta*) exhibit similar morphological patterns standing better water turbulence. It is worth noting that many aquatic vascular plants from flowing waters adopt comparable morphologies with roots and rhizomes fixing them to the substrate and strong laciniate leaves standing turbulence. *Zostera* and *Posidonia* live on shallow marine waters; species of *Ceratophyllum*, *Myriophyllum*, *Potamogeton* and many others are present in rivers. An extreme adaptation to intense water turbulence is exhibited by species from *Podostemonaceae*, a tropical family living in cascades, also presenting laciniate forms.

Animals do not need exposing a large body surface to radiation and confront water turbulence with a strong skeleton and an adequate organ for fixation to a solid substrate. The development of organs resistant to water shear and erosion, let animals exploit environments with very high turbulence colonising cliff rock surfaces exposed to intense wave action. The best hydrodynamic shape in response to water turbulence is a cone fixed to the ground. A strong conical skeleton protects the animal soft organs inside. A foot secures the cone to the substrate and an upper opening, often protected with a lid, keeps the communication to the environment. Or else, the cone apex is turned outside and the base is attached to the substrate. From early Palaeozoic era some coastal marine invertebrates from unrelated groups have evolved to produce cone-shaped individuals: *Archeocyatids* (*Porifera*), *Brachiopoda* or *Madreporaria*, or the Mesozoic *Rudista* molluscs, or the contemporary limpets (*Patella*), are a few examples of shape convergence in response to water turbulence.

Sessile animals build complex colonies to exploit a moderate turbulence: the numerous coral-like forms where individual animals trap organic particles or minute organism brought in by water currents. In some cases, Red algae *Criptonemiales*, may approach this pattern the thallus presenting calcareous deposits and the algae assuming articulate segments (*Corallina, Halimeda*), or shaping into crusts in a lichen like form (*Lithothamnium*) or in coral like patterns (*Porolithon oncodes*).

Vertebrates originated in aquatic media terrestrial forms having their origin in the fish-pulmonate fish-amphibian-reptile sequence. It is intriguing that some of the terrestrial lineages turned back to continental waters or to the ocean to exploit trophic resources or to live more or less permanently in aquatic media. The extinct reptiles Ichtiosaurs and Plesiosaurs, the numerous marine and aquatic turtles; among mammals, the cetaceans, seals and manatees, beaver, otters, capybara (*Hydrochoerus capybara*); among birds, penguins, albatross, and the large number of waterfowl are but a few examples of the process. As an old trait of their terrestrial lineage, these vertebrates use lungs to breathe atmospheric air and need surfacing from time to time. Many groups also demand terrestrial ground for reproduction: nesting, egg hutching, rearing, delivering or mating. Breeding colonies regulate offspring numbers, avoiding overpopulation bursts in the species.

Morphology usually conforms to aquatic habitat restrictions, but in several cases, *terrestrial* morphologies are capable of efficiently exploiting aquatic resources. The morphologies of the marine iguana of Galápagos (*Amblyrhynchus cristatus*), the dippers (*Cinclus cinclus, C. mexicanus*), or the osprey (*Pandion haliaetus*) do not suggest their ability to dive, forage or catch underwater. Brown and black bears (*Ursus arctos, U. americanus*) are great salmon catchers in fast rivers. Examples are plentiful to show how water surfaces are crossed by some species to exploit resources on the other side. These examples belong to this category and water fowl, at large, is also included.

In a few cases some aquatic animals exploit terrestrial organisms: fish species preying on insects flying above the water surface throwing a jet of water (*Arawana*) or reaching for the fruits in branches of trees flooded in the high water, *várzea*, of the Amazon basin (*Colossoma macropomum*). Some fish, such as the European eel (*Anguilla anguilla*) leave the water and creep through meadows over long distances, behaving like snakes. *Perioftalmus* is a remarkable fish from mangroves that abandons water to *run*, with the help of fore fins, on the muddy shore searching for food. Large aquatic predators such as alligators and crocodiles often attack terrestrial animals approaching the wetland or living in it. Piranha fish (*Serrasalmus, Pygocentrus*), are capable of vicious collective attacks to terrestrial vertebrates venturing into their waters.

The distinction between aquatic and terrestrial environments is blurred after intense rains or under wet climates. Soil surface easily form pockets of water that may last for a few hours or days, long enough for a temporary aquatic ecosystem, to develop. *Anphipods* are a group of crustaceans that often dominate temporary ponds in mid latitudes. In tropical climates some plant groups, notably *Bromeliaceae*, store rainwater among the sheath of upper leaves. These minute water bodies present a distinct planktonic composition according to light intensity, plant size and the bromeliad species involved. *Sarracenia* and *Nepentes* are genera of carnivorous plants with modified leaves that store water. The decomposition of insects falling into the water trap releases some nutrients usable for the plant. But the trap offers a durable water body that allows for the development of amphibian toads. Bromelias, nepentes, and many other plants serve as temporary nurseries for toads. Time and again water plays multiple ecological roles.

Water in the soil profile plays some important roles in the processes: rainfall infiltration through the soil dissolves minerals, bringing their elements to the soil solution that in turn behaves like a chemostat in a dynamic equilibrium adjusting its concentration of ions, salts and gases to the reactive sites of soil minerals, and to local hydraulic pressure. Vertical water flow redistributes the elements from one soil horizon to the next. The slow water circulation to deeper levels completely modifies lithosphere composition.

Water dissolution also creates pores and ducts in the rock matrix. Water movement in films through soil structures carries away particles. The action of ice or the swelling and shrinking of some clay minerals and organic matter deposits, according to their water content, also favours mechanical movements in the soil all caused by water action.

With the expansion of plants to the terrestrial media, the water environment was restricted to the soil matrix where the finer roots reach the water films among edaphic particles. Extraction of nutrients from the soil is largely mediated by the soil solution and is achieved by a concentration process carried out by permeases, implying (metabolic) energy expenditure.

The strongly negative water potential in the atmosphere, as compared to that of organisms, menaces terrestrial life forms with desiccation. If impermeable coats or thick epidermis developed to prevent water losses, the abstraction of $O_2$ or $CO_2$ from the atmosphere was impaired. Free communication with the atmosphere implied the occurrence of transpiration for terrestrial organisms. Water economy became a key trait of terrestrial organisms exhibiting powerful absorption organs, a stringent regulation of water losses, water recovery processes and an efficient use of water in their metabolism. Water circulation and water loss opened new evolutionary arenas to terrestrial plants and animals.

Vascular plants have developed the xylem, a vascular system made of bundles of fine pipes with rigid walls capable of withstanding strong pressures. Vascular plant leaves couple negative water tension in leaf parenchyma to water columns contained in xylem tissue bringing negative pressures to root level. Soil water is easily absorbed by the plant, accomplishing two relevant functions: to compensate for transpiration losses and to bring fresh soil solution to the roots, supplying them with fresh nutrients. It should be noted that the energy needed for root water absorption and transport to leaves is largely supplied by the atmosphere as a gradient of water potential from air to soil. Together with photosynthesis, this mechanism represents a net input of energy to plant metabolism.

For terrestrial animals, desiccation has selected peculiar ecological traits such as the impermeable cuticles of many insects, the loss of flying wings (i.e. *Tenebrionidae* beetles), the recovery of water from faeces and urine (i.e. reptiles of very dry habitats), the storage of water inside the body (such as in the toad *Pelobates cultripes*), and others.

Many diurnal animals benefit from sun radiation, warming up to increase activity. It is a well known condition for lizards and many other reptiles, including aquatic turtles, alligators and crocodiles. Birds and mammals have evolved homothermal physiological mechanisms where activity from muscles and other organs releases heat that is lost through thermal conductivity through the skin and specialised organs. Transpiration can play an important role here for homotherms as water evaporation from skin absorbs a large heat load letting the body regulate temperature.

Soil horizons are somehow protected from desiccation; consequently, edaphic organisms often benefit from moist environments and small surfaces of water films. On a minute scale, soil structure with long connected pores among granules of minerals, small rock fragments and organic matter particles, can compare to a network of tunnels with water pockets.

The soil surface receives fresh organic matter from vegetation, less so from animals. Mineral particles are brought by wind or incorporated into rainfall together with dissolved materials.

Organic matter contains the energy supply for decomposers, a large fraction of soil organisms that help structure the soil by breaking down the organic compounds, finally producing humic and fulvic acids and releasing some of the mineral ions to the soil solution. Clay minerals often arise from secondary formation within the soil profile. The large exchange capacity of humus favours the storing of nutrients in the soil profile. Water storage (field capacity) also benefits from the accumulation of organic matter, fine particles and clay minerals in this profile.

Vascular plants are favoured by the expanded soil retention of nutrients and larger water reserves, showing a larger primary productivity that, in turn, sustains a higher secondary productivity. At the same time vegetation intercepts some of the precipitation before reaching the surface and pumps out water reserves in transpiration, reducing soil moisture.

The building up of a soil horizon reduces runoff and stores water in the profile. It is through these mechanisms that terrestrial vegetation plays a significant role in the soil water balance and the partitioning of precipitation among evapotranspiration, infiltration/aquifer recharge and runoff. Runoff depends on precipitation/infiltration rate, surface slope and vegetation development. Living communities actively control transpiration to the atmosphere, and in some cases induce higher precipitation by means of interception of low clouds crossing through the canopies in cloud forests. In wet tropical climates it is remarkable the variety of plant epiphytes growing on leaves, stems or trunks of perennial species, taking advantage of air moisture replenished by cloud circulation and precipitation. In some vegetation types such as *Laurisilva* it is the aerial structure of vegetation that favours precipitation inside the forest as low clouds move through it.

Should the forest be suppressed, precipitation would be much reduced, completely altering infiltration and runoff. Soil horizons would lose most of their organic matter and their field and exchange capacities would be much reduced. Logging operations under the above circumstances often result in irretrievable devastation to ecosystems.

On Hierro Island (Canary Islands, Spain), it was the mighty Garoe tree (*Ocotea foetens*) that supplied water to meet the islanders' needs before Europeans settle the island. A hurricane uprooted the tree by 1610, and the population, deprived of water, fled from the island.

A few animal species have also developed some mechanisms to tape water from atmosphere. The fog basking beetle (*Onymacris unguicularis*) from Namibia desert climbs to dune crests where it adopts a head down stance with the dorsal surface exposed to the fog bearing winds. Droplets slip towards the head and the beetle can gain a 12% weight in water in a single night (Cloudsey-Thompson, 1991).

## 3   CLIMATE AND THE WATER CYCLE

The water mass of the planet has been estimated to be 1600 million $km^3$, and 97% belongs to the oceans including water and ice. Annual evaporation from ocean represents 448,000 $km^3$ and another 72,000 $km^3$ come from the continents[1]. 100 $km^3$ daily water flows into the sea, 37,000 $km^3$ annually. Precipitation on continents averages 730 mm/yr supplying vegetation, refilling wetlands, keeping drainage network flowing and favouring circulation of biogeochemical cycles. Water scarcity in dry lands or water in ice rigorously limit life development also bringing rock alteration and sediment transport to a reduced state.

---

[1] Water balance figures change slightly with published source.

Water plays important roles in the circulation of energy through the atmosphere. The major energetic circulation takes place as radiation. Infrared Radiation (IR) exchanges between the lower troposphere, and the ocean and continent surfaces are about four times as important as Photosynthetic Active Radiation (PAR) received from the sun at the continent surface level. At IR wave lengths water vapour exhibits important absorption bands (4–7 and over 18 micron). It intercepts a fraction of solar radiation and more strongly the Earth irradiation fluxes that largely correspond to a 4–20 micron interval. Cloud water droplets are efficient absorbers but also serve to reflect radiation at both visible and invisible wave lengths. From outer space, our *Blue Planet* exhibits strong white patterns of cloud formation against a blue background of oceans and shades of vegetation blurred by atmosphere.

Energy flows in the troposphere are also associated with mass air displacements. The radiative energy absorbed by soil and ocean surfaces, heats these outer layers that in turn transfer some of the energy to the base of air column raising its temperature (sensitive heat). Expanded air masses become unstable and tend to rise in the troposphere until they find a new equilibrium point inducing convective and advective air movements. A second mechanism for energy transfer (about 5 times more) is water evaporation from ocean or continent surface to the air. The amount of water involved approaches 36 mg/m²/s and the associated energy flux is about 82 W/m², averaging the flow for the planet surface. Plants behave as evaporating surfaces or intermediaries between atmosphere and soil in the transpiration process.

Average residence time for water in the atmosphere is short: 3 to 4 days, but the relevant process is that water enters the atmosphere as vapour and leaves it as water or ice in rain, hail or snow. The energy difference per gram of water mass in the cycle varies between 600 and 700 cal. The water cycle thus appears as a powerful energy transfer mechanism from oceans and continents to atmosphere. Energy budget is dissipated through winds that mix atmospheric layers favouring the vertical transfer of gases and air colloids and fine particles. Horizontal winds transfer air masses from source areas, redistributing water and energy, lowering the temperature and precipitation at low latitudes and increasing it at high latitudes. Some wind energy is also transferred back to the ocean inducing sea currents and waves.

Temperature and precipitation patterns at surface level largely define climate. Mineral alteration of rocks and life development are both favoured by high temperatures and water availability. Cool climates, where water remains ice or in arid climates where water is scarce and easily evaporates, show limited mineral alteration and a limitation of living forms as well as low productivity for continents. For oceans, life is loosely associated with climate because temperature plays a modest role and precipitation has little or no importance. Instead, it is PAR radiation, having to do with latitude, and nutrient availability, that play major roles.

### 3.1 *Continents: the slope/river continuum*

Rainfall is a key character to the continents of our planet. Every year precipitation brings to the continent surface 108,000 km³ of water with low mineral contents. This represents a dual energy storage: water is elevated against the gravitational force storing potential energy. Average continent height is 823 m, although altitude precipitation is uneven. A second energy source is chemical water potential that will lower as more substances are dissolved along the lithosphere course.

If precipitation is snow, and no melting occurs in the area, water effects are limited to rock abrasion and transport, with limited chemical alteration, generating a peculiar morphology on exposed rocks and deposits. When ice finally melts fluvial processes reshape glacial geomorphology. When glaciers reach the sea they sometimes form an ice field that breaks up into icebergs. Icebergs represent an ice transport mechanism that carries out rock fragments to some distances over the sea, following the currents that drive ice floes away from their origin. The bottom of Labrador Channel is littered with drifting materials and boulders carried out by ice foes.

The action of water on rock surface depends on the mineral composition. When a single or few mineral species dominate (as in limestone, dolomite, gypsum, halite and carnalite or quartzite), the rock mass dissolves and is carried out in solution by water. Joints, fractures, strata discontinuities, easy water circulation inside rock openings, pipes, vertical chimneys, open a network of ducts and cavities inside the mass. Limestone formations are outstanding for the development of caves, ducts, inner lakes and the formation of deposits inside then. In a similar way, circulating water opens ducts and cavities inside the ice mass in glaciers.

When parent rock is composed of several minerals, alteration rate depends on temperature and water availability. Easily altered minerals yield ions, salts, or lattice arrangements where some ions or elements are substituted for others. The rock and the circulating solution interact with one another for as long as water moves through it in a large series of processes involving exchange, solution, substitution, oxidation, hydration, carbonation, precipitation, eventually erosion or deposition.

Under different climates (temperature, water precipitation), rock alteration proceeds differently, with a different array of sensitive *vs.* resistant minerals under each condition. Thus the water solution will carry away a variant balance of elements and a new assemblage of unaltered resistates will stay longer on the rock surface.

The partition of rainfall water between infiltration and runoff depends on the surface balance: where rainfall excess over infiltration feeds runoff flow. The sheet of surface running water will exhibit a low residence time, moving downslope with little time to develop a chemical interaction with surface minerals. On the other hand, as water concentrates on a few lines of flow, velocity will rise and turbulence appears, permitting the turbulent transport of soil particles. As runoff channels concentrate higher debits, water velocity increases and channel slope, width and depth adjust to water flow in a process that will extend to the whole watershed.

Runoff water initially moves resistate minerals as sands and rock fragments, as gravels. Other materials from rock alteration are also available to transport, such as clay particles, salts and ions. Organic compounds and particles, including minute organisms, will also be carried downslope by water turning potential to kinetic energy.

Rock materials under adequate conditions move downward by mass wasting processes such as solifluxion, creeping, slumping and others. The sheet of loose materials is regolith. Its rate of displacement depends on the slope angle, easing as it decreases. The increase of water content reduces viscosity of regolith, favouring creeping. In solifluxion it is through the volume changes of water in thawing-melting cycles that makes regolith progress. It becomes even more important for slumping where a regolith sheet, in the form of a large slate, slips on a surface, often lubricated by the presence of water. In a mudflow it is the regolith saturated in water that flows downslope forming a semi-liquid current of high density and destructive effects that will stop its progression when reaching a flat area, where no more potential energy conversion is possible.

Slopes are the key morphological elements of landscapes and it is in the geological processes building them that surface water is present at every level. Alteration and erosion prevail at the top of the slope on subhorizontal to moderately steep segments. Transport dominates the intermediate steeper segments. Sedimentation will occur at the lower part of the slope that tends, again, to be subhorizontal and where a channel usually drains the excess water feeding the drainage network.

Infiltrated waters will develop soil structure through a sequential exchange of materials with soil horizons from accumulation of organic matter and elution in the upper (A) horizon, accumulation in the underlying (B) horizon and fresh rock surfaces where alteration was starting at the deeper (C) horizon.

Water shapes the morphology and dominates mineral alteration and transport according to climate. In each case a sequence of different habitats is generated, to which organisms adjust. In surviving, communities often regulate the habitat.

The above description underlines the continuity of processes involving continental waters, both in time and space. Water performs as a ubiquitous media renewing itself in volume and composition and connecting biosphere systems. Drainage channels, rivers and lakes behave as a connection network for living organisms. But in addition, water acts as a barrier limiting energy exchange, mineral cycling, gas diffusion and organism displacement. One such a case is ice, since under low temperatures runoff and infiltration come to a stand still. Snow or ice covers are ecosystem shields preventing animal displacement or exploitation of resources (such as foraging) across them.

The lower density of ice leaves it floating on water, severely restricting exchanges of the water body, or the sea, with atmosphere. Stratification is an important outcome of water density that depends on suspended sediments load, salt contents and temperature. For pure water and at moderate pressures maximum densities are attained at 4°C. An energy input is needed to raise to the surface the lower and denser layers close to 4°C, above the upper and lighter ones. In consequence, stratification is a stable arrangement of the water column for as long as densities remain unchanged and no energy input upturns the column.

Surface heating for incoming radiation reduces water density in summer, reinforcing stratification. Cooling by irradiation, evaporation or exchange with a cool air mass, brings the column to a point of instability when surface densities exceed the value of deeper levels. Cool waters sink bringing to the surface warmer ones until the column reaches a new equilibrium or the whole water mixes and the same temperature extends from surface to bottom. If cooling proceeds, water masses approaching 4°C attain the highest densities (other variables being equal) and collect at the bottom of the lake basin. Lower temperatures induce less dense water bodies that now pile above the densest 4°C waters until the surface is finally frozen.

Stratification of the column whether caused by temperature or solid content differences confines water environments to sub-horizontal layers. Of these it is only the superficial one that receives light and exchanges gases with the atmosphere. Lower layers, deprived of light, are inadequate for photosynthetic organisms, and $O_2$ production ceases, but decomposition proceeds, often resulting in anoxic conditions and a higher nutrient content. Stratification confinement induces chemical and biological differentiation of water environments leaving an upper illuminated epilimnion where primary productivity persists. The upper layer behaves as the collector of metabolic energy for the whole column. It is the input of organic matter synthesized at the euphotic section that maintains all trophic levels in the dark section of the column.

Minute scale stratification operates in ponds. The seas and the oceans also exhibit the same pattern of stratification where living forms gather at two levels: close to the surface where light supplies the energy and on the bottom surface where the shower of organic particles provides energy to the biological community.

For as long as water was available life thrives. But it is also water that separates or connects, that makes terrestrial media favourable or adverse for each species.

## 4   COMMUNITY LEVELS

The ecological effects of water are not mediated by a single species, however important it was for biomass or for cover. It is the community as the ecological assemblage of plants, animals, and micro organisms, interacting above and below the soil surface, the water column and the bottom surface that regulates matter, energy and, to some degree, structure. The outcome of regulation is most varied, as shown with a few examples.

In temperate arid areas with scarce and irregular precipitations, such as the Negev Desert in East Mediterranean, soil surface is largely covered by a crust of lichens and terrestrial algae that absorbs some water when rainfall occurs but prevents infiltration favouring runoff that concentrates in scattered points where perennials grow, thus producing patchy vegetation. Should the lichen crust be disturbed, runoff would not concentrate and perennials fade away.

In mid latitudes, the regions dominated by high atmospheric pressure often show deserts (Belnap & Lange, 2001). They exhibit scanty perennial vegetation, crusts of lichens or algae and buried seed banks of annual species. These ephemeral plants quickly germinate after a rainfall and produce seeds in just a few weeks of growth for as long as soil moisture lasts.

In cool and moist areas mineralization of organic matter proceeds slowly, building up strong organic horizons. Primary productivity of vegetation remains low, but formation of peat horizons frequently occur on flat areas, forming mires with water logged soils. Peat levels rise above the soil level, forming large heaps of ombrogenic peat fed only by rainwater and some dust brought in by winds. In boreal regions bogs, mires and peat deposits are widespread, often displacing forest and grassland vegetation through water logging of soils.

Under wet tropical climates the two major factors enhancing productivity are favourable to vegetation: temperatures are high and water supply is plentiful during the whole year or a large part of it. The combination provides at community level, a plentiful primary productivity and also larger productivities of consumers and decomposers. The accumulation of biomass per unit surface in the form of timber, leaves and stems, epiphytes, algae, moss and lichens, soil fungi, moulds, invertebrates or micro organisms is also high. The deposits of necromass as shed-off organs, seeds and fruits, twigs and branches are impressive and pile up on the soil surface.

Per unit surface of ground, the fluxes of energy, water and minerals are high although they exhibit a dynamic equilibrium through the operation of multiple regulatory loops based on large biological diversities. There are many species present at each level; even the large trees usually belong to several species that rise to tens of meters, with trees of all ages that can reach the figure of 1000 per hectare. In a single tree may coexist several thousand invertebrate species (Erwin, 1983).

This rich and diverse system tropical forest is sensitive to intervention and disturbances because the uncoupling of regulatory processes easily unbalances it. Logging activities destroy the structure, leaving many epiphytes and animals with a scarcity of ecological niches. The drop in organic matter input in the soil does not impair the strong mineralization, which results in a depletion of nutrients and a change in the surface soil horizon. Heavy rainfall regimes lead to soil leaching. On steep slopes this may favour erosion and the onset of a vegetation dominated by species of large leaves and fast to very fast growth and the presence of vines and climbers, often making huge shields, hanging screens that profoundly alter structure and functioning of the community. For partially unknown reasons, many animal species are particularly sensitive to disturbance and community fragmentation suffices from their disappearance.

In temperate or warm climates where a dry season occurs, fires play a significant ecological role. The Mediterranean type climate is such an example, occurring in the Mediterranean basin and other mid latitude areas such as California, central Chile, West Australia and West South Africa. In all these cases wildfires are naturally present, biomass and necromass accumulation and nutrient deposits are lost in a recurrent way through fire. Loss of the vegetation structure and the deposits of cinders on the soil surface ease runoff transport of nutrients when rainfall reassumes after the dry season.

Wildfire occurrence is not limited to Mediterranean type climates: a continental climate with a dry hot summer is prone to fires. Subtropical and tropical climates with a dry season also produce natural wildfires. In a sense, it is water through rainfall patterns that controls wildfire occurrence. The Mediterranean shrub vegetation of *chemise, garriga, frygana, jaral,* apparently has gone one step further, integrating the fire event and the ensuing erosion process of autumn rainfall into a mechanism of survival that prevents other plant communities from invading its area (Herrera, 1992).

## 4.1 *Wetlands: the productive waters*

Shallow aquatic ecosystems on continents or coastal areas exhibit communities with a high number of species, some in large populations. The reason, apparently, has to do with their high primary productivity that maintains a wide base for trophic chains. Accumulation of organic matter modifies nutrient balances, water regimes and the soil level of the wetland. Fermentation of organic debris suffices to raise the temperature of their deposits, helping some large reptiles such as alligator or caiman species to hatch their eggs. Wetland biomass, dominated by photosynthetic organisms, includes phytoplankton in the water column and vascular species rooting in the bottom and emerging to the surface. Many life forms may coexist from trees such as mangroves, swamp cypress, tamarisks, poplars, to minute sedge and rush species. Wetland plants often exhibiting leaves inside the water column such as *Utricularia* or some *Potamogeton* species or spreading over the surface such as in *Ranunculus.* Floating leaflets of *Marsilea* or round leaves of *Nuphar* and *Ninphaea* may almost cover water surface, growing to an immense size of *ca.* 2 m across in *Victoria regia* leaves. Free floating plants such as *Eichornia crassipes, Trapa natans* or *Pistia stratiodes* form a thick surface cover. Smaller species also form almost continuous blankets such as species of gen. *Salvinia, Lemma, Ricciocarpus. Wolffia arhiza,* a minute water lentil, with *leaves* 0.5–1 mm across, is the smallest vascular plant. Some of the aquatic species have become aggressive invaders spreading to new areas: *Elodea canadensis* in European channels; the water lilies *Eichornia spp.* in tropical wetlands, and the contemporary invader in South Spain: *Azolla caroliniana,* the mosquito fern.

Animal groups benefit from the ample primary productivity in wetlands. Zooplankton and macroinvertebrates that sustain vertebrate trophic chains of fish, amphibians, some reptiles such as water turtles and crocodiles and many snake species live in tropical wetlands. Mammals are present in small numbers with several rodents, bears, ungulates, manatees, a few large cats, and others. The hippopotamus (*Hyppopotamus amphibius*) is the largest wetland mammal. Wetlands offer a substantial habitat for many bird species exploiting primary productivity (algal mats, fruits and seeds, tubers, leaves) or the net productivity of invertebrates, fishes and amphibians. Large concentrations of birds forming breeding colonies in trees, shrubs, banks, or amidst the emergent vegetation are a key character of many wetlands. Huge bird gatherings numbering into the hundreds of thousands, possibly millions, generate an important effect on vegetation and on the circulation of nutrients of the system.

Wetlands offer a breeding ground for many birds, fish and other living creatures that will later spread to broader areas, possibly migrate to different regions. Other species will not breed but come to the area to exploit the temporary bonanza of resources. Or endure a development phase in their lifecycle such as the exchange between fluvial and marine environments or the larval phase of many insect species. The wetlands keep unusual habitats where endemic species survive and diverse communities have been maintained over long periods in spite of climatic changes, thus preserving old ecological legacies.

Considering above arguments, the efficient removal of nutrients and sediments, and the nursery effect on commercial species, coastal wetlands have been rated on an economic basis, among the most important ecosystems to mankind in the comparative study of world ecosystems carried out by Constanza et al. (1997).

## 5 MAN'S INFLUENCE

The advent of cultural man started a cascade of ecological phenomena that impaired the primeval functioning of ecosystems. The industrial revolution that flourished mid 17th century in Central Europe expanded environmental impacts beyond regional boundaries. It was during the second half of the 20th century when global impacts on atmosphere composition, ozonosphere, ocean fisheries, forest cover and biodiversity were first acknowledged. The early years of the 21st century reinforced the scientific evidence for a global climate change that also induced sea level variations. World water resources entered an era of incertitude.

Water availability severely limits biosphere productivity in warm and dry climates. The historical dependence of agriculture and husbandry on precipitation was addressed by a rich array of water technologies. This led to powerful water cultures emerging as actors of History in the 4th Millennium BC. The early Empire development has been related to large scale water management for irrigation (Wittfogel, 1957). In the vicinity of the Mediterranean basin, Mesopotamia, the Nile Valley, the Fertile Crescent and the Arabian Peninsula agricultural empires flourished prior to the 1st Millennium BC. This led to important cultural and social developments, large population increases and depletion of local resources (fauna, forests, mines) at an early phase.

Technology limitations prevented significant river impoundments but river diversion by channels was feasible, greatly increasing crop production. Wetland drainage through channel building turned swamps into agricultural lands. The combination of both strategies (such as in the Tigris-Euphrates marshes in Mesopotamia, the Nile-Fayum in Egypt) represents an early achievement for water technology. Another important step was the construction of subterranean channels (quanat, foggara, mayra, viaje, galería), preventing evaporative losses and permitting the exploitation of distant aquifers.

The combination of wells, vertical and horizontal water-wheels, channels, quanats and drainage networks spread over the Mediterranean in successive waves, traversing the basin from East to West. Significant advancements on water technologies were introduced under the Roman Empire (approximately 3rd century to 4th century BC): long channels, aqueducts, tunnels, lead pipes, water decanters, fountains, urban drainage systems (cloacae), sewers, field drains (cuniculi). From the 7th century, Islam expansion in the Mediterranean introduced new irrigation technologies from Mesopotamia and the Middle East including large water wheels kept running by river flow, chain pumps, water distributors for field irrigation and gave new impulse for quanat construction.

For centuries, available water technologies changed little but there was a steady improvement in channel design, creating a fresh drainage network in Central Europe and England, and finally developing internal transport systems. Dam building also improved, permitting some water regulation for agriculture. The Dutch polder development coupled to water abstraction by windmills permitted marsh reclamation for agriculture. European technology spread to America after 15th century.

Water wheel mills were common under the Roman Empire for grain grinding, a tiresome domestic task. They regained importance in Europe with the economic development occurring *ca.* 1000. The small mills (1–2 kW power output per wheel) were used for grain grinding. But step by step other energy demands were met by hydraulic wheels: metal processing, paper production, wool cleaning and preparation, spinning, and other tasks. Tide mills were added to river mills during medieval times. Water offered an adequate source of mechanical energy for industrial purposes, fuelling the early steps of the Industrial Revolution. Turbines were added early in the 19th century.

The advent of the steam engine in the last quarter of the 18th century completely altered the picture: the Watt-Boulton engines were successful in every industrial branch and were soon adapted to transport. The outcome had profound effects on industrial production, which became cheaper and of a higher quality. New scales of production, new industrial processes and a different type of human labour also emerged. From the environmental side, a few points should be underlined.

The application of steam engines to pumps allowed the drainage of wetlands on an unprecedented scale. It also permitted the extraction of water from rivers, providing water for industrial, urban and agricultural uses. Mining was severely limited by water infiltration into deep galleries. Pumping the excess water paved the way for deeper wells and again the steam engine permitted the easy lift of ore, gear and man power. Intense mining resulted in widespread washing of ore and the building up of slag heaps. Coal regions lost their rivers to contamination from coal dust. Coal and mineral sulphur mines released acid effluents from galleries and tips, also degrading riverine ecosystems to a state beyond recognition: a flow of diluted sulphuric acid rich in metal cations. The Rio Tinto River in Southwest Spain has suffered this type of contamination for centuries and has turned to an almost abiotic river. The longstanding contamination has suppressed original flora and fauna only surviving primitive chemosynthetic life forms that thrive in the acidic waters and a heath (*Erica andevalensis*) growing on the toxic slates left by Roman Empire miners.

The spread of railway networks was made possible by the steam engine. In turn it was this easy land transport, combined with steamers on the high seas that favoured the distribution of manufactured products, raw materials and the exploitation of minerals, forests, fisheries, along with development of range and agriculture at rates never before attained. For Europe, the second phase of the Industrial Revolution was a period of enormous industrial development, urban growth, social change, and migratory movements. Other continents underwent a period of invasion from European powers with large settlement of Europeans in mid-latitudes. The vast human movement created new cities from scratch that quickly became large then important trade and industrial centres. Buenos Aires and New York were old historical settlements in America. But Chicago or Montevideo, small villages at the beginning of 19th century grew to the million mark by 1900, surpassing Buenos Aires and approaching New York in population during the following century. The Great Plains, to the West of the Great Lakes were entirely transformed from forest and prairie to corn and cattle in less than a century: the Mississippi basin was turned into a different entity.

From old historical cities large modern cities evolved, fed by migration, trade and industry. Their vast populations were supplied with water from reservoirs or pumped from rivers.

Domestic waste water, industrial effluents and street drainage were collected and discharged into neighbouring rivers. Certain micro-organisms survive easily in urban sewage, some tolerating the process of river self-purification, so as to reach the next water supply downriver, affecting another population. Cholera and typhus outbreaks were commonplace. Parasitic worms remained in the population through repeated contagion.

Aquatic flora and fauna endured an impact related to the ratio of wastewater discharge to river flow. Large settlements such as Paris and London turned their rivers into stinking sewers almost void of fish. Industrial regions such as the Ruhr destroyed the biota of their rivers. Industrial ports contaminated coastal waters and estuaries. Water supplies became impossible and many water resources such as fish or sea food, vanished. At present, the large urban and industrial impact of sewage on continental and littoral waters has been controlled ... in developed countries. Polluting activities survive (or have been transferred) to less demanding countries where politicians accept a relatively high degree of disturbance, usually because they lack funds to tackle the problem. Water, in many forms, has entered our societies as a valuable asset.

Water has been the berth of Life, it paved the way for the evolution of life forms and the building up of communities and ecosystems. For an enormous period of time, natural regulation commanded the biosphere; it was only during Holocene, 10,000 years ago, that man began to gain control. The last 100 years have witnessed overriding changes imposed by human activities on the systems of Planet Earth.

Chapter 17 of *Silent Spring*, the magnificent book of Rachel Carson (1962), begins with a vigorous presentation of our responsibilities: "We stand now where two roads diverge. But unlike the roads in Robert Frost's familiar poem, they are not equally fair. The road we have long been travelling is deceptively easy, a smooth superhighway on which we progress with great speed, but at its end lays disaster. The other fork of the road – the one *less travelled by* – offers our last. Our only chance to reach a destination that assures the preservation of our Earth. The choice, after all, is ours to make".

## 6   WATER SIGNALS

Rachel Carson advice has been followed and new water policies have been implemented all over the world. However, the combined demands of agriculture, new urban settlements, an expanding industry and a growing population widen the gap between preservation and disturbance of natural waters.

Several processes affecting water resources are far from equilibrium and may serve as *signals* of water mismanagement, pointing out to fresh technical undertakings and coming political issues. To mention but a few, see Table 1.

Human impact on the biosphere waters was responsible for the expansion of many biotopes and ecosystems and the creation of bridges among them. Table 2 identifies a few examples.

Table 1 of *water signals* focuses on negative impacts. To balance the view, the social benefits of water exploitation must be incorporated, waving a new list of water signals.

Living standards, health, labour conditions, life expectancy, all improved with the era of economic development started by Industrial Revolution. The appreciation for springs, cascades, rivers, lakes and the maritime littoral has spread to most of the contemporary world population. Water resorts and water sports have became commonplace in Europe where littoral cities witnessed a steady growth during the second half of the 20th century. However, many human communities are lagging behind others in economic and social development and human welfare.

Table 1.   Signals of water mismanagement due to several processes affecting water resources.

| Water signals | Comments |
| --- | --- |
| Eutrophication of continental waters | Diffuse contamination from agriculture is growing as chemical fertilizers are more widely used.<br>Leachates from uncontrolled rubbish dumps find their way into waters.<br>Waste water treatment plants rarely controls P in effluents sufficiently to prevent water enrichment. 1–2 mg/L concentrations in effluents or an 80% reduction in wastewater have been set as an European standard for some countries. |
| Eutrophication of littoral waters | As population shifts to the sea shore, point sources of untreated waste waters released into the sea keep growing.<br>Tourist settlements with fluctuating occupation rarely possess waste water treatment plants adequate to the variable load.<br>In poor regions the water works design usually includes sewage collection and its deposition in rivers or coastal waters with no treatment. |
| New organic molecules are added to urban waste water in significant amounts such as hormones, beta-blockers, antibiotics, or tensioactives. | Waste water treatment cannot prevent the growing concentration of biologically active molecules in effluents.<br>Hormones at concentrations found in wastewater have measurable effects on aquatic organisms such as toads, inducing significant development alterations. |
| Growing water demands from agriculture, industry and urban developments. Water requirements may be close to 10,000 km$^3$/yr, roughly 1/10 of continental precipitation.<br>The increase of rent in population and the development of urban water supplies greatly increase water demands in an early development phase. | New demands tend to be met by more abstraction from aquifers and increased river or lake pumping. River impoundment for watershed regulation starts far reaching changes in river ecology.<br>In large rivers with important sediment loads, the loss of sedimentation unbalances coastal areas close to the estuary, also lowering biological productivity in them.<br>Overexploitation of aquifers has several consequences other than consumption of water resources. Desiccation of connected springs, upwellings and wetlands. Loss or reduction of habitats and populations, generally reducing biodiversity.<br>In littoral areas, aquifer salination occurs due to marine water intrusion. |
| Repeated introduction of alien fish species in continental waters for sport or commercial fisheries. | In some cases, such as the Nile perch (*Lates niloticus*) in Victoria Lake, led to a biodiversity collapse. In many other cases, native fish |

*(Continued)*

Table 1.  (*Continued*)

| Water signals | Comments |
| --- | --- |
| | populations retreat when aggressive newcomers compete for resources or prey on them. |
| Intense trade of aquatic organisms with little control. Domestic aquaria house many menacing invaders. | Other than introductions, there is a problem with illegal trade of protected species and the destruction of populations of tropical native species of fishes and amphibians. |
| The use of rivers and lakes as cooling circuits for power plants creates long-standing heated plumes or raised river temperatures. Nuclear plants raise water concentration of tritium and other isotopes, some of them causing health impacts. | A rise in temperature favours the introduction of tropical or subtropical organisms into cool or temperate regions.<br>The addition of chemicals to cooling circuits to prevent animal growth damages biota on the marine littoral and in lakes and rivers. |
| Untreated sludge from chemical plants may introduce to continental waters some elements, ions or molecules rarely confronted by ecosystems, causing harmful ecological effects.<br>Desalination plants release a concentrated brine creating a plume of hyper saline waters close to the seashore. | Ecological concentration mechanisms can raise harmless concentrations of chemicals in the environment to significant or toxic levels along trophic chains. Hg concentration in Minamata Bay (Japan) from water to fish and sea food in human diet was an early warning of the process.<br>Large discharges of waste water, even when contamination was caused by organic matter or naturally abundant compounds (such as nitrates, nitrites, ammonia, phosphates), easily exceeds regulatory ability of aquatic systems, causing alterations and the eventual collapse of the ecosystem.<br>Mineral oil can be dispersed, oxidised, degraded, and assimilated by natural systems, in moderate quantities. But point concentrations may be very high in oil fields, chemical plants, or after accidental releases from tankers and pipelines. It is oil concentration what makes ecological disasters, fouling soils, rivers and coastal areas. Hyper saline waters are harmful to most marine species. |
| World biotopes are sensitive to interference in the water cycle: overexploitation of aquifers feeding wetlands, reduction of river flow and sediment transport, fragmentation of rivers with dams and impoundments, drainage of wetlands, eutrophication and contamination of continental waters. And above all, the climatic change. | Biotopes, populations, aquatic ecosystems, natural landscapes and cultural landscapes response to interventions are most varied from small readjustments to collapse and substitution by a different system.<br>The release of species from other biogeographical regions often aggravates the situation of local endemic taxa. |
| Human cultures have created a rich legacy on water issues. Technical changes in the exploitation and loss of traditional resources (such as fish, flax, watermills), severs cultures at a world scale. | New balances among cultural survival, health, comfort and efficient use of resources must be developed.<br>The paradigm of *sustainability* needs profound reshaping of traditional cultures and, the more so, of contemporary western technical culture. Sustainability of water resources merits an independent assessment. |

Table 2. Ecological changes induced by water resource exploitation.

| Process | Comments |
| --- | --- |
| Growing water demands | Hot deserts, range and dry lands show limited production and biomass due to water scarcity. Irrigation favours crop and husbandry productivity as well as the expansion of associated wild species of forbs and soil invertebrates. When large irrigation schemes are applied to hot countries there is a drop in peak temperatures in the area. Sometimes irrigation helps introduce more advanced agriculture practices and protects soils from desertification. |
| River impoundment | Dams interrupt the *river continuum* and insert lakes (reservoirs) in it. Still stratified waters with reduced suspended solid load and deep columns (as compared to rivers) offer new habitats for aquatic species. Birds, insects, reptiles and amphibian faunas of river and reservoirs differ. Fish communities of lentic and lotic systems are markedly different. |
| The release of species from other biogeographical areas favours introductions | The shifting of aquatic species across terrestrial or oceanic barriers, insurmountable to them as individuals, opened up a new era to biodiversity. Fish species introduced to adequate media may expand adding to the local diversity. Introductions are deleterious to some native wildlife, but often favour other species until a new equilibrium is reached. In the future, world continental waters will probably lose most local wildlife species but will be enriched by fast growing species coming from every continent. Some introductions bring important benefits to local inhabitants: edible fish for commercial fishing and sport. Fast growing fish and sea food for aquaculture. Filtering fish to abate water eutrophication in lakes. Mosquito fish *Gambusia* to fight malaria. Others introductions are detrimental to humans such as parasites (*Bilharzia*) or aquatic pests. |
| Urban expansion offers new habitats to many wild species | Modern cities have created new aquatic environments that have been colonised by wild species. Pipes carrying irrigation or river waters may exhibit a thick cover with an association of filtering animals such as *Hydroidea* (*Cordilophora caspia*) and molluscs (*Mytilopsis leucophaeta, Corbicula fluminea*). Sewers offer a network of channels and pipes with a steady flow of water rich in organic matter and with limited temperature fluctuations. Decomposers thrive in this warm, moist and well fed environment, such as microbes, protozoa, fungi and some worms. And larger animals such as insects and rats. Tropical cockroaches have adopted sewers as an expansion of their habitat all over the world. Small alligators or crocodiles, aquatic snakes, otters, some fish as mullet (*Liza*) species, survive in large sewers. Sources of disturbance are the occasional water flashes caused by an intense precipitation, and the release of chemicals to sewers. |
| Air conditioning and new urban environments | Air filters collect particles and condensation provides water in devices with large surfaces, offering an aerated water medium enriched in organic matter. This environment suits well *Legionellacea*, a large family of bacteria often causing epidemic outbreaks of respiratory diseases. This happens when microbes are dispersed to the air in droplets from the air conditioning systems. Water pockets in roofs, collectors, pipes, offer a temporary water environment that can be exploited by bacteria and algae. Basements, wells, gardens, fountains, river banks, piers and harbours, boats, offer an expanded array of wet urban habitats to biological forms. |

*(Continued)*

Table 2.    (*Continued*)

| Process | Comments |
|---|---|
| Navigation | Boat hulls behave like wandering rocks for marine sessile organisms. On some occasions, the distribution of marine species from shallow waters over deep waters was made possible through this mechanism. *Caulerpa taxifolia* has been expanding on the Mediterranean from harbour to harbour likely fixed on sport sailing boats. *Enteromorpha intestinalis*, a marine green algae, has reached Switzerland and colonised the Rhine River with freshwater forms fixed to commercial boats operating in this water way. |
| | Mississippi River is behaving as a continental loop for freshwater species from the Great Lakes to the Caribbean Sea. |
| | When ships in ballast emptied tanks in harbours the waters often contain live specimens from distant areas that can be introduced. There is growing concern about the aggressive *Decapoda* (crustacean) species introduced in fluvial ports and marine areas through this mechanism. |
| | Used tyres and metal scrap cargo in ships is known to offer water pockets for reproduction of mosquito species and their introduction to new areas. |

We are forcing the water cycle to meet human needs often overlooking the basic principles governing the life of organisms, the assembly of their communities, the organization of ecosystems and the multiple roles of water in the biosphere. The onset of climatic change dramatically signals how far human interference pervaded the water cycle of biosphere; and the risk of severing the *unity of natural waters*.

Sustainability policies must take full advantage of self regulatory processes in natural systems, bringing them to social uses by means of new technologies founded on hydrology, limnology and ecology. As Ian McHarg advised, long ago (1969): *Design with Nature!*

# REFERENCES

Belnap, J. & Lange, O.L. (2001). *Biological soil crusts: Structure, function and management.* Springer–Verlag, Berlin. 503 pp.

Carson, R. (1962). *Silent Spring.* Houghton Mifflin. Boston, USA.

Cloudsley-Thompson, J.L. (1991). Ecophysiology of Desert Arthropods and Reptiles. Springer–Verlag, Berlin. 216 pp.

Constanza, R.; d'Arge, R.; de Groot, R.; Farber, S.; Grasso, M.; Hannon, B.; Limburg, K.; Naeem, S.; O'Neill, R.V.; Paruelo, J.; Raskin, R.G.; Sutton, P. & van den Belt, M. (1997). The value of the world's ecosystem services and natural capital. *Nature*, 387: 253–260 (May 15).

Erwin, T. (1983). Beetles and other insects of tropical forest canopies at Manaus sampled by insecticidal fogging. In: S.L. Sutton; T.C. Whitmore & A.C. Chadwick (eds.). *Tropical rain forest: ecology and management.* Blackwell, London, UK: 59–75.

Herrera, C.M. (1992). Historical effects and sorting processes as explanation for contemporary ecological patterns: character syndromes in Mediterranean woody plants. *The American Naturalist*, 140: 421–446.

McHarg, I. (1969). *Design with Nature.* Doubleday and Company. New York, USA.

Mojzsis, S.J.; Arrhenius, G.; McKeegan, K.D.; Harrison, T.M.; Nutman, A.P. & Friend, C.R.L. (1996). Evidence for life on Earth before 3800 million years ago. *Nature*, 384: 55–59 (November 7).

Wittfogel, K. (1957). *Oriental despotism: a comparative study of total power.* Yale University Press. New Haven, Connecticut, USA.

CHAPTER 15

# Water and nature: a critical link for solving the water management crisis

G. Bergkamp

*IUCN – The World Conservation Union, Gland, Switzerland*

ABSTRACT: Environmental aspects of water management have long been neglected. During the last two decades, however, progress has been made in specific cases. More recently, a range of practical tools and mechanism have come to be used to incorporate the values of ecosystem services in water resources planning, decision-making and management. These include environmental flow management, economic valuation of ecosystem goods and services and payments for maintenance of environmental services. Nature, by using these new approaches, is becoming part of solving the water management crisis.

Keywords: *Water and nature, environmental flows, ecosystem valuation, ecosystem services, protection of wetlands*

## 1 INTRODUCTION

Early whistle-blowers of the degradation of the world's water resources, like Aldo Leopold and Rachel Carson, are emblematic for the way the modern world has dealt with its relationship with water and nature. Since the publication of Carson's book (1962), thousands of dams have been constructed worldwide. Today, 70% of rivers are fragmented by large scale infrastructure, affecting migration of fish and other species (Bergkamp *et al.*, 2000). During the last 100 years more than 50% of the world's wetlands have been drained and turned into agricultural production areas. Up until the 1980s most developed and developing societies alike choose to ignore the early warning signs related to a mismanagement of the world's water resources. They preferred to develop their water resources in ways that would maximise short-term economic growth rather than safeguard their resource base for future generations.

During the last two and a half decades, societies have started to respond, however, to signals of degradation. Especially in the developed world, more and more information became available on the seriousness of the situation. This triggered policy-level responses. For example, it was in the late 1980s and early 1990s that with the *Brundtland report*, *Caring for the Earth* and *Agenda 21*, environmental issues and sustainable development were put on the policy agendas at the national level. At that time, the water *society* at large was still mainly focused on a *Green Revolution* (1960–1970) and the *Water Supply Decade* (1981–1990). Only in the early 1990s, with the *Dublin Principles* and *Chapter 18 of Agenda 21*, a start was made catalysing change towards a more sustainable management of water resources.

Since the early 1980s, significant changes have occurred in the environmental aspects of water management. With increased knowledge about the effects of the development of water resources and industries came a growing demand for reducing those impacts. Some significant improvements in water quality and in the protection of wetlands were made. For example, in OECD countries, pollution of rivers and lakes with phosphate has been reduced by 40%. Since the early 1970s, more than hundreds of thousands of hectares of wetlands have been protected under the *Ramsar Convention*. These changes are steps in the right direction.

Much of these improvements were built upon an improved understanding of causes of pollution and degradation, and political will. The latter was often driven by motives that were not necessary environmental in the first place. For example, the deteriorated quality of bathing water in many places in Western Europe caused great public concern, which caused politicians to act. Water quality standards and monitoring were put in place, while new water treatment systems were built and existing ones upgraded.

## 2    ENVIRONMENTAL FLOWS: A NEGOTIATED APPROACH TO RIVER FLOW MANAGEMENT

Water resource managers are now increasingly coming to terms with the need to take a holistic view of the river ecosystem. This includes taking care of aquatic ecosystems and the resources they provide for long-term economic viability and allocating appropriate amounts of water to them.

Ecosystems provide goods and services to humankind which are relevant to water supplies. Upper watershed forests regulate water flow, ensure recharge of aquifers and reduction of silt loads in rivers. Wetlands can purify water and filter out chemical substances harmful to humans or crops. Rivers and lakes also provide sources of protein through fish and other animals. Often these services are of direct benefit to local communities living from the resource (IUCN, 2000).

The concept often used for this allocation is *environmental flows*. It is defined as the water regime provided within a river, wetland or coastal zone to maintain ecosystems and their benefits, where there are competing water uses and where flows are regulated. Environmental flows must be seen within the context of applying *Integrated Water Resources Management* (IWRM) in catchments and river basins. Environmental flows will only ensure a healthy river, however, if they are part of a broader package of measures, such as soil protection, pollution prevention, and protection and restoration of habitats (Dyson *et al.*, 2003).

Taking steps to manage for environmental flows brings into focus the struggle over access to and ownership of water and water rights. In systems where water is already over-allocated, the challenge of environmental flows may include reallocating or conserving water from existing private users and returning it to the river. Before starting to work on environmental flows, one therefore needs to realise that a wide range of stakeholders will have to be involved.

The degree of *good health* at which the river will be sustained is a societal judgement that will vary from country to country and region to region. What the appropriate environmental flow is for a particular river will thus depend on the values for which the river system is to be managed. Those values will determine the decisions about how to balance environmental, economic and social aspirations and the uses of the river's waters.

To set an environmental flow, one needs to identify clear stream objectives as well as water abstraction and use scenarios. Objectives need to have measurable indicators that can form the basis for water allocations. Useful objectives can be, for example, *maintaining the brown trout at 1995 levels, preserving at least 75% of downstream mangrove forests*, or *maintaining river nitrate levels below a particular standard.* Where a system is seriously over-committed and values do not allow a sufficient reallocation of resources to restore the *entire system*, certain river stretches or wetland sites may be targeted for protection and specific water allocations. For rivers with high biodiversity values, for example, an environmental flow might be provided to preserve the natural state of the river system.

An environmental flow programme will not necessarily have ecological gains as the only or even the primary outcome. It will need to strike a balance between water allocations to satisfy the ecological water requirement and other water use needs like those of hydropower generation, irrigation, drinking water or recreation. Developing an environmental flow programme therefore means articulating the core values on which to base decisions, determining what outcomes are sought and defining what trade-offs those will entail. In this way, environmental flow setting and implementation provide an opportunity for basin and watershed stakeholders to start discussing the existing allocation patterns that might not be sustainable.

## 3   COUNTING ECOSYSTEMS AS PART OF WATER SERVICE INFRASTRUCTURE

Ecosystems provide a wide range of economic goods and services for human production and consumption, such as fish, timber, fuel wood, medicines and pasture. As pointed out above, water allocations need to be made to maintain those services. On the supply side, ecosystems such as forests and wetlands generate economic services which maintain the quality and quantity of water available for human use or mitigate flooding and drought. Healthy ecosystems are thus important to secure water supplies, while water supplies to maintain ecosystem health are critical.

A recognition of the relationship between ecosystems status and water service infrastructure has long been missing from water resources planning and infrastructure investment programmes. Economic arguments are, however, an important determinant of how water is allocated and used and how priorities for water investments are set. To date little appreciation exists for the fact that ecosystems are economic users of water and form equally a critical part of the water supply chain. Though they are essential for water service delivery, they often remain a sharply under-funded part of investment in the water sector (Emerton & Bos, in prep.).

To improve on this practice, it is critical to determine the benefits from ecosystems in monetary terms. Economic valuation of ecosystem benefits can be a powerful tool to incorporate those values into economic planning and decision-making. It provides a basic and comparable indicator of economic values which can be used to judge ecosystem use alongside other economic sectors and activities.

A wide range of techniques now exist to value different components of the total economic value of ecosystems. The total economic value includes the direct values, e.g. of products such as fish and timber, and the indirect values, e.g. of services such as flood control and nutrient retention. They also include option values, the premium placed on maintaining resources for future use. Finally they include the intrinsic values of resources and ecosystems irrespective of their use, such as cultural and spiritual significance.

Available techniques for determining ecosystem values include, for example, determining the market-price of ecosystems goods and services, and valuing the effect of change on production output of an ecosystem. A range of experiences is now available around the world. For example, in Guatemala the effect of rainforest degradation on stream flow and agricultural profits was found to be worth US$ 15,000–53,000 in two river basins (Brown *et al.*, 1996). In the Zambezi river basin, estimating the total value of its major wetland areas indicated that they have a marginal value of US$ 145 million/yr, or an average of US$ 48/ha (Seyam *et al.*, 2001).

A major challenge arises once values have been determined. How to ensure those values are used in decision making? How to make use of those values when decision need to be made to developing a hydropower scheme and preserving river values for downstream users? Several techniques exist to translate these data into information, indicators and criteria that can be used to weigh different development options. They include, for example, cost-benefit analysis and multi-criteria analysis. In Northern Cameroon these techniques were used to determine the cost and benefits of restoration of the 3000 km² Waza Logone floodplain. The study showed that the cost of *not* having an annual floodplain inundation is more than US$ 2 million/yr due to loss of fish, fodder and grazing grounds. Restoring the floods could be done at an investment cost of US$ 10 million/yr, which would result in benefits of US$ 2.5 million/yr to the region's economy. The investment would pay itself back in five years (IUCN, 2001).

Valuing the benefits and the costs of ecosystems goods and services can thus provide a meaningful contribution to solving the water management crisis. By determining these values and incorporating them into planning and decision-making, more optimal water investment programmes can be designed and implemented.

## 4   PAYMENT FOR ECOSYSTEM SERVICES

Increasingly, potential beneficiaries of ecosystem services make payments to ecosystem managers. These payments reward land owners and resource users for managing land and water resources in a way most favourable to the protection of the ecosystem services of interest. The majority of compensation mechanisms in place have been established by the public sector, which has been buying services such as maintenance of hydrological flow from private land users. Many of the existing schemes relate ecosystems services directly to the provision of water supplies to cities, hydropower stations or irrigated farmland.

A well known example is the New York City Water Supply that started to invest in watershed protection as a cost-effective means to comply with water quality regulations. The City receives its water supply from two watersheds located 125 miles (200 km) north of the City. Over a ten year period, the city invested US$ 1500 million to protect the watershed natural purification processes. Measures included US$ 472 million to control point-source pollution through the upgrade and rehabilitation of city-owned sewage treatment plants, US$ 300 million for land acquisition, and US$ 40 million for *Best Management Practices* in agriculture and forestry. The annual costs of operating the watershed protection program have been reduced to US$ 20–30 million, well below the operational costs of a filtration facility of US$ 6000–8000 million plus an estimated US$ 200–300 million in annual operating expenses (Perrot Maître & Davis, 2001).

In many parts of the world, hydropower companies have decided to make voluntary contributions to watershed protection. This is the case for example in Quito, Ecuador, where since 2000, the water utility and the hydroelectric utility have decided to transfer 1% of their revenues from water and electricity sales into a *Trust Fund* that will be used for watershed protection. It is expected that over US$ 2 million/yr will be generated.

In Costa Rica, five hydroelectric companies are investing in watershed protection. They annually allocate US$ 10–42/ha or over US$ 600,000/yr to a government fund which is used to compensate forest owners for forest protection activities. Companies do so for two reasons: (1) they expect, through watershed protection, to ensure a regular supply of water throughout the year; and (2) they can increase revenues from electricity production and reduce dredging costs associated with reservoir sedimentation.

Bottling companies are also heavily dependent on a secure water supply of good quality. Protecting the resources is a fundamental element of their success. Over the last years, a number of companies have bought land around their springs to reduce the risks of water contamination by outside activities. For example, *Nestlé* (*Vittel S.A.*) invested US$ 9 million to purchase land around the *Vittel* springs and provided a US$ 24.5 million of subsidies to land users to adopt *Best Management Practices*. In doing so, the company is investing in protecting the water purification services provided by forest and pasture lands. Research at the *French National Agronomic Institute* (INRA) demonstrated that the programme is cost-efficient as long as one hectare of well-managed pasture is able to filter and produce 3000 $m^3$/yr of drinkable water.

## 5 CONCLUSIONS

The above examples show practical approaches that make ecosystems services part of the solution, rather than being perceived as part of the problem. If we are to finance the sustainable management of our water resources, we will need to convincingly demonstrate that investing in their protection and wise use makes good economic and financial sense. Compensating those land and water managers that protect the resource and maintain a critical element of the water supply chain is likely to become more common practice in the future.

## REFERENCES

Bergkamp, G.; McCartney, M.; Dugan, P.; McNeely, J. & Acreman, M. (2000). *Dams, Ecosystem Functions and environmental restoration.* World Commission on Dams Thematic Reviews – Environmental Issues II.1. World Commission on Dams, Cape Town, South Africa: 1–186.

Brown, M.; de la Roca, I.; Vallejo, A.; Ford, G.; Casey, J.; Aguilar, B. & Haacker, R. (1996). *A Valuation Analysis of the Role of Cloud Forests in Watershed Protection: Sierra de las Minas Biosphere Reserve, Guatemala and Cusuco National Park, Honduras.* RARE Center for Tropical Conservation, Fundación Defensores de la Naturaleza and Fundación Ecológica.

Carson, R. (1962). *Silent spring.* Houghton Mifflin, New York, USA.

Dyson, M.; Bergkamp, G. & Scanlon, J. (eds.) (2003). *Flow – The essentials of environmental flows.* IUCN, Gland, Switzerland: 1–118.

Emerton, L. & Bos, E. (in prep.). *Value – Counting ecosystems as an economic part of water infrastructure.* IUCN, Gland, Switzerland.

IUCN (2000). *Vision for Water and Nature. A world strategy for Conservation and sustainable management of water resources in the 21st century.* World Water Vision – Environment component. World Water Council/IUCN. Gland, Switzerland: 1–52.

IUCN (2001). *Economic Value of Reinundation of the Waza Logone Floodplain, Cameroon.* Projet de Conservation et de Développement de la Région de Waza-Logone, Maroua, Cameroon.

Perrot-Maître, D. & Davis, P. (2001). *Case studies: Developing Markets for Water Services from Forests.* Forest Trends, Washington, D.C., USA. Web Page: (http://www.forests-trends.org).

Seyam, I.M.; Hoekstra, A.Y.; Ngabirano, G.S. & Savenije, H.H.G. (2001). *The Value of Freshwater Wetlands in the Zambezi Basin.* Paper presented at Conference on Globalization and Water Resources Management: the Changing Value of Water. AWRA/IWLRI–University of Dundee.

VIII

New Technologies to Cope with
Water Scarcity

CHAPTER 16

# Water recycling – A relevant solution?

T. Asano
*Department of Civil and Environmental Engineering, University of California, Davis, USA*

ABSTRACT: Water reclamation, recycling, and reuse have become important elements in integrated water resources management. The role of water recycling and reuse is reviewed in this chapter with respect to the evolution of the practice, applications, technology, health and regulatory requirements, and planning. The future of water recycling and reuse are also discussed in the context of a relevant solution to water shortages and water pollution.

Keywords: *Planning, public health, recycling, water resources management, water reclamation, wastewater treatment, water reuse*

## 1 INTRODUCTION

For water supplies to be sustainable, the rate at which water is withdrawn from water sources needs to be in balance with the rate of renewal or replenishment of these water sources. In addition to a balance of water quantity, water quality must also be sustainable or recoverable. Precipitation serves to replenish water sources by augmenting surface water supplies and recharging groundwater. Shifts in land use patterns (e.g. urbanization, agriculture, dams and reservoirs) alter the rate, extent, and spatial distribution of freshwater withdrawal, consumption, and replenishment. Water that is withdrawn for societal needs is also a potential source of water replenishment that should be considered in the sustainability equation (Levine & Asano, 2004).

The sustainability of water resources is of particular importance in light of projected increases in global population. It has been reported that the current world population of 6200 million is increasing at a rate of about 1.2% per year (United Nations, 2003) with the highest rates of population growth occurring in urban areas where supplies of freshwater tend to be limited. On a global scale, about 3800 km$^3$/yr of water is withdrawn to meet societal needs. Increasing urbanization has resulted in an uneven distribution of population and water, thus imposing unprecedented pressures on limited water supplies. These pressures are exacerbated during periods of drought. When viewed on a global scale, the total available volume of renewable freshwater is several times more than is needed to sustain the current world population. However, only about 31% of the renewable water is directly accessible to population centers due to geographical constraints and seasonal variations (Postel, 2000).

Historically, after water has been used for societal needs, it has been labeled as *waste* water and treated to the extent perceived necessary for discharge into receiving water or on land. During most of the 20th century, the emphasis of wastewater treatment was on pollution abatement, protection of public health, and prevention of environmental degradation through removal of biodegradable material, nutrients, and pathogens. However, over the last few decades, the potential for recovering water from [waste]water has been recognized. In fact, in many parts of the world, it is no longer practical or possible for water to be used only once.

Reclaimed water is a valuable water resource for non-potable and subpotable water uses. By using treated wastewater effluent to augment existing water supplies, water resources can be conserved more economically and effectively. Water recycling and reuse, in conjunction with conservation, efficient management and use of existing water resources results in a sound and integrated approach to water resources management. In many industrialized nations there are growing problems associated with developing adequate water supplies in an environmentally sound manner. Water shortages, especially during periods of drought, have made it necessary for many municipalities to mandate the reduction of water consumption. In addition, the costs of municipal and industrial wastewater disposal for water quality protection and pollution abatement are increasing. In developing countries, particularly those in arid parts of the world, reliable low cost, low technology methods are needed for acquiring new water supplies and protecting existing water sources from pollution.

In the planning and implementation of water reclamation and reuse, the intended water reuse applications dictate the extent of wastewater treatment required, the quality of the finished water, and the method of distribution and application. As technology continues to advance and the reliability of water reuse systems is demonstrated, it is likely that water recycling and reuse will continue to expand in the future and become a more relevant solution in the context of integrated water resources management.

The purpose of this chapter is to provide an introduction to the major elements of water reclamation, recycling, and reuse with an emphasis on significant developments that have paved the way for present and future applications. The major topics covered are: (1) evolution of water reuse; (2) water recycling and reuse terminology; (3) the role of water recycling in the water cycle; (4) water reclamation and reuse benefits; (5) water reuse applications; (6) overview of wastewater treatment technology; (7) health and regulatory requirements; (8) water reuse planning; and (9) unresolved problems and research needs. The goal of this chapter is to stimulate the readers a new thinking in water resources management and to illustrate the relevant role played by water reclamation, recycling, and reuse in water resources planning and management.

## 2   EVOLUTION OF WATER REUSE

Indications for utilization of wastewater for agricultural irrigation extend back approximately 5000 years in Crete, Greece (Angelakis & Spyridakis, 1995). In more recent history, during the 19th century, the introduction of large-scale wastewater carriage systems for discharge into surface waters led to indirect use of sewage and other effluents, downstream, for inadvertent potable water supplies. This unplanned reuse, coupled with the lack of adequate water and wastewater treatment, resulted in catastrophic epidemics of waterborne diseases such as Asiatic cholera and typhoid during the 1840s and 1850s. However, when the water supply link with these diseases became clear, engineering solutions were implemented that included the development of alternative water sources using reservoirs and aqueduct systems, the relocating of water intakes upstream and wastewater discharges downstream as in the case of London, and the progressive introduction of water filtration during the 1850s and 1860s (Young, 1985; Barty-King, 1992).

## 2.1   *Water reuse in the USA and the European Union*

The development of programs for planned reuse of wastewater within the USA began in the early part of the 20th century. The State of California pioneered efforts to promote water reclamation and reuse. The city of Bakersfield has used reclaimed water since 1912 to irrigate corn, barley, alfalfa, cotton, and pasture. The first reuse regulations were promulgated in 1918 by the State of California. Some of the earliest water reuse systems were developed to provide water for irrigation with projects implemented in both Arizona and California in the late 1920s. In the 1940s chlorinated wastewater effluent was used for steel processing, and in the 1960s urban water reuse systems were developed in Colorado and Florida.

During the last quarter of the 20th century, the benefits of promoting water reuse as a means of supplementing water resources have been recognized by most state legislatures in the USA as well as by the European Union. For example, in 1970 the California State Water Code stated that "it is the intention of the Legislature that the State undertake all possible steps to encourage development of water reclamation facilities so that reclaimed water be available to help meet the growing water requirements of the State" (California Water Code, 1988 Amendments). In the same context, the European Communities Commission Directive (91/271/EEC) declared that "treated wastewater shall be reused whenever appropriate. Disposal routes shall minimize the adverse effects on the environment" (EEC, 1991).

In recent years, increased interest in water reuse in many parts of the world is occurring in response to growing pressures for high quality, dependable water supplies by agriculture, industry and the public; a situation that is exacerbated by rapid population growth in urban areas, drought and other climatic irregularities. Today, technically proven wastewater treatment and water purification processes exist to produce water of almost any quality desired.

## 2.2   *Milestone events*

Milestone events that have been significant for the evolution of water reclamation, recycling, and reuse are itemized on the timeline in Figure 1. Microbiological advances in the late 19th century precipitated the *Great Sanitary Awakening* (Fair & Geyer, 1954) and the advent of disinfection processes. The development of the activated sludge process in 1904 was a significant step towards advancement of wastewater treatment and pollution control and the development of biological treatment systems. In 1918, the California State Board of Public Health adopted its first regulations addressing the use of sewage for irrigation.

Technological advances in physical, chemical, and biological processing of water and wastewater during the early part of the 20*th* century led to the *Era of Wastewater Reclamation, Recycling, and Reuse*. Since the 1960s, intensive research efforts, fueled by regulatory pressures and water shortages, have provided valuable insight into health risks and treatment system design concepts for water reuse. In 1965, the Israeli Ministry of Health issued regulations to allow the reuse of secondary effluents for crop irrigation with the exclusion of vegetable crops that are eaten uncooked. In 1968 extensive research on direct potable reuse was conducted in Windhoek, Namibia. In the USA, a milestone event was the passage of the Federal Water Pollution Control Act in 1972 (later renamed: the *Clean Water Act*) "to restore and maintain the chemical, physical, and biological integrity of the Nation's waters" with the ultimate goal of zero discharge of pollutants into navigable, fishable, and/or swimmable waters. During the 1970s and 1980s several comprehensive research and demonstration projects were conducted to evaluate the potential for non-potable and potable reuse with a major emphasis

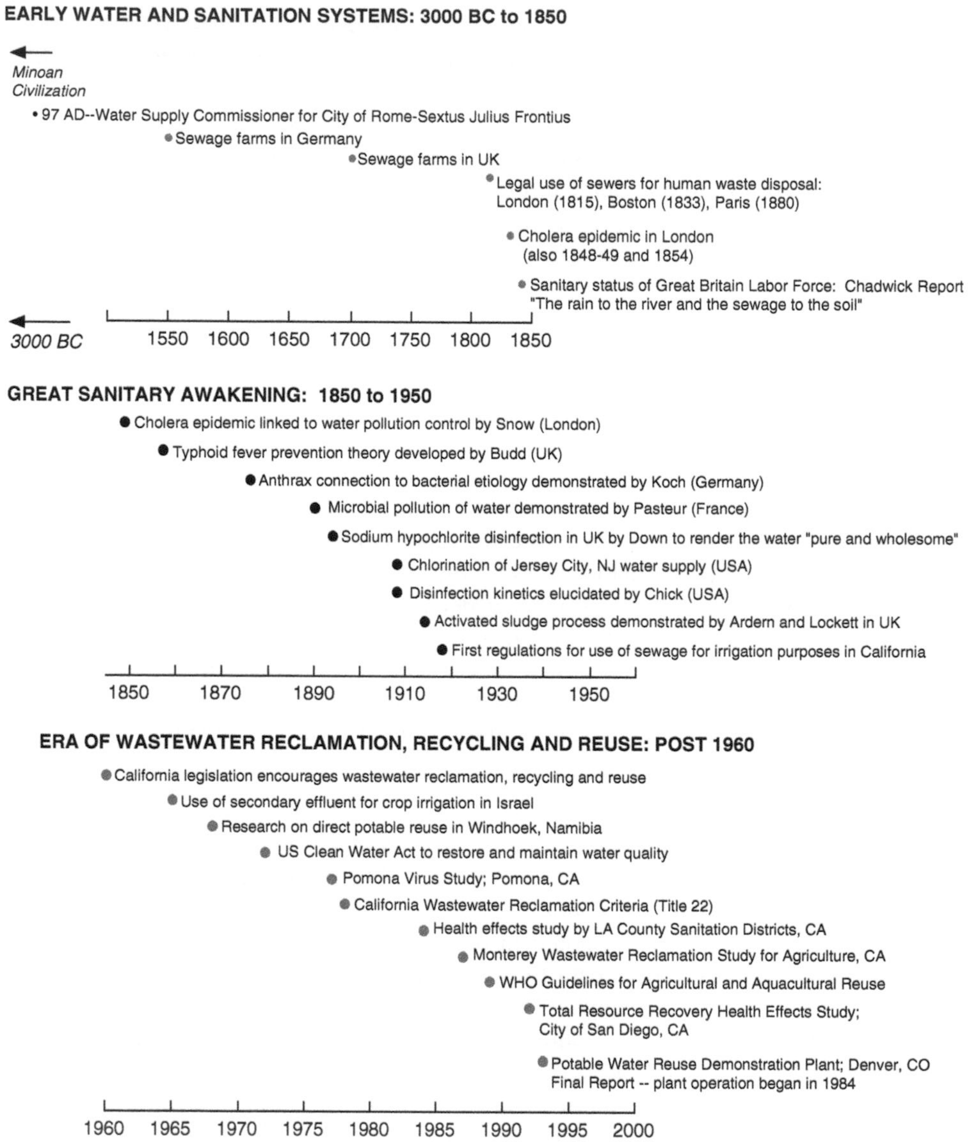

**EARLY WATER AND SANITATION SYSTEMS: 3000 BC to 1850**

←

*Minoan*
*Civilization*

• 97 AD--Water Supply Commissioner for City of Rome-Sextus Julius Frontius

● Sewage farms in Germany

●Sewage farms in UK

● Legal use of sewers for human waste disposal:
London (1815), Boston (1833), Paris (1880)

● Cholera epidemic in London
(also 1848-49 and 1854)

● Sanitary status of Great Britain Labor Force:  Chadwick Report
"The rain to the river and the sewage to the soil"

←
*3000 BC*    1550    1600    1650    1700    1750    1800    1850

**GREAT SANITARY AWAKENING:  1850 to 1950**

● Cholera epidemic linked to water pollution control by Snow (London)

● Typhoid fever prevention theory developed by Budd (UK)

● Anthrax connection to bacterial etiology demonstrated by Koch (Germany)

● Microbial pollution of water demonstrated by Pasteur (France)

● Sodium hypochlorite disinfection in UK by Down to render the water "pure and wholesome"

● Chlorination of Jersey City, NJ water supply (USA)

● Disinfection kinetics elucidated by Chick (USA)

● Activated sludge process demonstrated by Ardern and Lockett in UK

● First regulations for use of sewage for irrigation purposes in California

1850    1870    1890    1910    1930    1950

**ERA OF WASTEWATER RECLAMATION, RECYCLING AND REUSE: POST 1960**

● California legislation encourages wastewater reclamation, recycling and reuse

● Use of secondary effluent for crop irrigation in Israel

● Research on direct potable reuse in Windhoek, Namibia

● US Clean Water Act to restore and maintain water quality

● Pomona Virus Study; Pomona, CA

● California Wastewater Reclamation Criteria (Title 22)

● Health effects study by LA County Sanitation Districts, CA

● Monterey Wastewater Reclamation Study for Agriculture, CA

● WHO Guidelines for Agricultural and Aquacultural Reuse

● Total Resource Recovery Health Effects Study;
City of San Diego, CA

● Potable Water Reuse Demonstration Plant; Denver, CO
Final Report -- plant operation began in 1984

1960    1965    1970    1975    1980    1985    1990    1995    2000

Figure 1.   Milestone events in the evolution of wastewater treatment, reclamation, recycling, and reuse (Adapted from Asano & Levine, 1996, 1998).

on quantifying health risks and defining treatment and technological requirements. These research efforts have resulted in increased implementation of water reuse projects in various regions and the evolution of new reuse alternatives.

The continued testing and implementation of treatment systems and new applications have helped to overcome many technical barriers to water reuse projects. Improvements in treatment process reliability, risk assessment, and public confidence in reuse systems in conjunction with increasing water demands and pollution control requirements have promoted the integration of water reuse into water resources management strategies throughout the world.

## 3 WATER RECYCLING AND REUSE TERMINOLOGY

The terminology used in water reclamation, recycling, and reuse has evolved from sanitary and environmental engineering practice. The water potentially available for reuse includes discharge from municipalities (sewage, municipal wastewater), industries (industrial wastes), agricultural return flows, and storm water. Of these, return flows from agricultural irrigation and storm water are usually collected and reused without further treatment.

*Water reclamation* involves the treatment or processing of wastewater to make it reusable, and *wastewater reuse* or *water reuse* is the beneficial use of the treated water. In recent years, term *water* recycling and reuse, instead of *wastewater*, is preferred from the public acceptance point of view. Reclamation and reuse of water frequently require water conveyance facilities for delivering the reclaimed water and may require intermittent storage of the reclaimed water prior to reuse.

*Reclaimed water* can be used directly (*direct reuse*) without passing through a natural body of water for applications such as agricultural and landscape irrigation. *Indirect reuse* includes mixing, dilution, and dispersion of reclaimed water by discharge into an impoundment, receiving water, or aquifer prior to reuse. Indirect reuse, through discharge of a treated effluent to receiving water for assimilation and withdrawal downstream, while important, does not normally constitute *planned reuse*. For example, the diversion of water from a river downstream of a discharge of treated water constitutes an incidental or *unplanned reuse*. The topics covered in this chapter are related to deliberate or *planned reuse* as defined above.

### 3.1 *Unplanned indirect water reuse*

Unplanned indirect water reuse, through effluent disposal to streams and groundwater basins, has been a long accepted practice throughout the world. Many communities situated at the end of major waterways: New Orleans, USA (the Mississippi River); London, England (the River Thames); the cities and towns in the Rhine River Valley, Germany; and Osaka, Japan (the Yodo River); ingest water that has already been used many times through repeated river withdrawal and discharge. For example, it has been estimated that more than 80% of the water in the Santa Ana River in Southern California originates from wastewater discharged by upstream municipal wastewater treatment plants. Similarly, river beds or percolation ponds may recharge underlying groundwater aquifers with waste-containing water, which in turn is withdrawn by subsequent communities for domestic water supply.

It is also increasingly important to differentiate between *potable* and *non-potable* reuse applications. *Potable water reuse* refers to the use of highly treated reclaimed water to augment drinking water supplies. Direct potable reuse is the incorporation of reclaimed water into a potable water supply system, without relinquishing control over the resource. Indirect potable reuse includes instead an intermediate step in which reclaimed water is mixed with surface or groundwater sources prior to drinking water treatment. *Non-potable water reuse* includes all water reuse applications other than potable water reuse and it constitutes a large majority of water recycling and reuse in the world.

## 4  THE ROLE OF WATER RECYCLING IN THE WATER CYCLE

The inclusion of planned water reclamation, recycling, and reuse in water resource systems reflects the increasing scarcity of water sources to meet societal demands, technological advancement, public acceptance, and improved understanding of public health risks. As the link between wastewater, reclaimed water, and water recycling and reuse has become better defined, increasingly smaller recycle loops are possible.

Traditionally, the hydrologic cycle has been used to represent the continuous transport of water in the environment. The water cycle consists of fresh and salt water surface resources, subsurface groundwater, water associated with various land use functions, and atmospheric water vapor. Many sub-cycles to the hydrologic cycle exist including the engineered transport of water such as canals and aqueducts. Water reclamation, recycling, and reuse, represent significant components of the hydrologic cycle in urban, industrial, and agricultural areas.

A conceptual overview of the cycling of water from surface and groundwater resources to water treatment facilities, irrigation, municipal, and industrial applications, and to wastewater treatment and water reclamation/reuse facilities is shown in Figure 2. Water reuse may involve a completely controlled *pipe-to-pipe* system with an intermittent storage step, or it may include blending of reclaimed water with non-reclaimed water either directly in an engineered system or indirectly through surface water supplies or groundwater recharge. The major pathways of water reuse include groundwater recharge, irrigation, industrial use, and surface water replenishment.

Surface water replenishment and groundwater recharge also occur through natural drainage and through infiltration of irrigation and storm water runoff. The potential use of reclaimed water for potable water treatment is also shown albeit rare and small quantity. The quantity of water transferred via each pathway depends on the watershed characteristics, climatic and geohydrological factors, the degree of water utilization for various purposes, and the degree of direct or indirect water reuse.

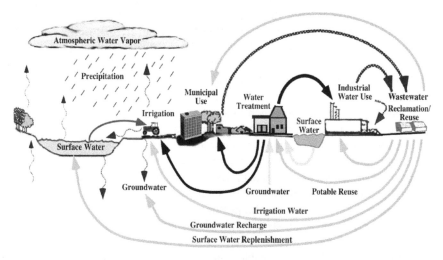

Figure 2.    The role of engineered wastewater treatment, reclamation, and reuse facilities in the cycling of water as a sub-set of the hydrologic cycle (Adapted from Asano & Levine, 1998).

The water used or reused for agricultural and landscape irrigation includes agricultural, residential, commercial, and municipal applications. Industrial reuse is a general category encompassing water use for a diversity of industries that include power plants, pulp and paper industries, and other industries with high rates of water utilization. In some cases closed-loop recycle systems have been developed that treat water from a single process stream and recycle the water back to the same process with some additional make-up water. In many industrialized countries, zero liquid discharge regulations mandated water recycling and reuse within industrial facilities. In other cases, reclaimed municipal wastewater is used for industrial purposes such as in cooling towers. Closed-loop systems are also under evaluation for reclamation and reuse of water during long-duration space missions by the National Aeronautics and Space Administration (NASA).

### 4.1 *The water quality changes*

The water quality changes during municipal use of water in a time sequence is conceptually shown in Figure 3.

Through the process of water treatment, a drinking water is produced which has an elevated water quality meeting applicable standards for drinking water. The municipal and industrial uses degrade water quality, and the quality changes necessary to upgrade the wastewater then become a matter of concern of wastewater treatment. In the actual case, the treatment is carried out to the point required by regulatory agencies for protection of other beneficial uses. The dashed line in Figure 3 represents an increase in treated water quality as necessitated by water reuse. Ultimately as the quality of treated water approaches that of unpolluted natural water, the concept of water reclamation and reuse is generated. Further advanced wastewater reclamation technologies, such as carbon adsorption, advanced oxidation, and reverse osmosis, will generate much higher quality water than conventional drinking water, and it is termed *repurified* water. Today, technically proven wastewater reclamation or purification processes exist to provide water of almost any quality desired.

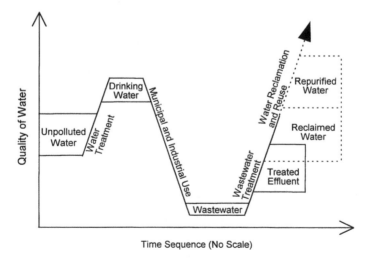

Figure 3.   Water quality changes during municipal uses of water in a time sequence.

## 5    WATER RECLAMATION AND REUSE BENEFITS

Water reclamation and reuse make the best use of existing resources by: (1) conserving high-quality water supplies by substituting reclaimed water for applications that do not require that quality; (2) augmenting potable water sources and providing an alternative source of supply to assist in meeting both present and future water needs; (3) protecting aquatic ecosystems by decreasing the diversion of freshwater, reducing the quantity of nutrients and other toxic contaminants entering waterways, and reducing the need for water control structures; and (4) complying with environmental regulations and reuse liability by better managing water consumption and wastewater discharges to meet regulatory limitations. Benefits of water reclamation and reuse and factors driving its future are summarized in Table 1.

Table 1.    Benefits of water reclamation and reuse and factors driving its future[a].

---

*Potential benefits of water reclamation and reuse*
- Water recycling conserves water supplies:
  Water recycling increases the total available water supply. High-quality water supplies can be conserved by substituting reclaimed water where appropriate.
- Water recycling is environmentally responsible:
  Water recycling can preserve the health of waterways, wetlands, flora and fauna. It can reduce the level of nutrients and other pollutants entering water ways and sensitive marine environments by reducing effluent and stormwater discharge.
- Water recycling makes economic sense:
  Reclaimed water is at the door step of the urban development where water supply reliability is most crucial and water is priced the highest.
- Water recycling can save resources:
  Recycled water originating from treated effluent contains nutrients; if this water is used to irrigate agricultural land, less fertilizer needs to be applied to the crops. By reducing pollution and nutrient flows into waterways, tourism and fishing industries are also helped.

*Factors driving its further use*
- Increasing pressure on existing water resources due to population growth and increased agricultural demand.
- Growing recognition among water and wastewater managers of the economic and environmental benefits of using recycled water.
- Recognition that recycled water can be a reliable source of water supply even in drought years.
- Increasing awareness of the environmental impacts associated with overuse of water supplies.
- Greater recognition of the environmental and economic costs of water storage facilities such as dams and reservoirs.
- Preference to recycling over effluent disposal, coupled with tighter controls on the quality of any effluent discharged to the environment.
- Community enthusiasm for the concept of water recycling.
- The growing numbers of successful water recycling projects in the world.
- The introduction of new water charging arrangements that better reflect the full cost of delivering water to the consumers, and the widespread use of these charging arrangements.
- Increased costs associated with upgrading wastewater treatment facilities to meet higher quality water quality standards.

---

[a] Compiled from various sources including Asano (1998), and *Queensland Water Recycling Strategy* (Queensland Government, 2001).

## 6   WATER REUSE APPLICATIONS

To provide a framework for evaluating water reuse it is important to correlate major water use patterns with potential reuse applications. On the basis of water quantity, agriculture and landscape irrigation accounts for 54% of total freshwater withdrawals in the USA. The second major user of water is industry, primarily for cooling and process needs. However, industrial uses vary greatly and additional wastewater treatment beyond secondary treatment is usually required. Thus, the effective integration of water reuse into water resource management is based on the *quantity* of water required for a specific application and the associated water *quality* requirements. The major factors that influence water reuse implementation are: (1) water quality requirements for intended water reuse applications; (2) existing or proposed wastewater treatment facilities; (3) requirements for degree of treatment process reliability; (4) potential health risks mitigation; and (5) public perception and acceptance.

Significant regional and seasonal variations in water use patterns exist. For example, in urban areas, industrial, commercial and non-potable urban water requirements account for the major water demand. In agricultural and arid zones, irrigation is dominant. Irrigation requirements tend to vary seasonally whereas industrial water needs are more consistent. The feasibility of water reuse for a given watershed depends on the practical extent that reclaimed water could augment existing water supplies through substitution of water in commercial, industrial, and agricultural applications.

The applications for municipal water reuse parallel major water use applications. An overview of the seven major categories of water reuse is given in Table 2. These categories are arranged in descending order of current and projected volume of reclaimed water with applicable water quality requirements.

Much of the attention in recent years on the reclaimed water applications, however, has been for its use in the urban environment, such as irrigation of parks and playgrounds and its potential for groundwater recharge. It is interesting to note that Japan's water recycling has been decisively directed toward non-potable urban applications, such as toilet flushing, urban environmental water, and industrial reuse. Comparison of water reuse in California and Japan is shown in Figure 4. Only 16% of total water reuse in Japan (*versus* 68% in California) is for agricultural and landscape irrigation.

Table 2.   Categories of water reuse from treated municipal wastewater.

| Category of wastewater reuse | Treatment goals | Example applications |
| --- | --- | --- |
| *Urban use* unrestricted | Secondary, filtration, disinfection $BOD_5 \leq 10$ mg/L Fecal coliform: ND/100 mL Turbidity $\leq 2$ NTU $Cl_2$ residual: 1 mg/L pH: 6 to 9 | Landscape irrigation: parks, playgrounds, school yards, fire protection, construction, ornamental fountains, aesthetic impoundments. In-building uses: toilet flushing, air conditioning. |

*(Continued)*

Table 2.    (*Continued*)

| Category of wastewater reuse | Treatment goals | Example applications |
|---|---|---|
| restricted access irrigation | Secondary and disinfection<br>$BOD_5 \leq 30$ mg/L<br>TSS $\leq 30$ mg/L<br>Fecal coliform $\leq 200/100$ mL<br>$Cl_2$ residual: 1 mg/L<br>pH: 6 to 9 | Irrigation of areas where public access is infrequent and controlled: freeway medians, golf courses, cemeteries, residential, greenbelts. |
| *Agricultural irrigation*<br>food crops | Secondary, filtration, disinfection<br>$BOD_5 \leq 10$ mg/L<br>Turbidity $\leq 2$ NTU<br>Fecal coliform: ND/100 mL<br>$Cl_2$ residual: 1 mg/L<br>pH: 6 to 9 | Crops grown for human consumption and consumed uncooked. |
| non-food crops and food crops consumed after processing | Secondary, disinfection<br>$BOD_5 \leq 30$ mg/L<br>TSS $\leq 30$ mg/L<br>Fecal coliform $\leq 200/100$ mL<br>$Cl_2$ residual: 1 mg/L<br>pH: 6 to 9 | Fodder, fiber, seed crops, pastures, commercial nurseries, sod farms, commercial aquaculture. |
| *Recreational use*<br>unrestricted | Secondary, filtration, disinfection<br>$BOD_5 \leq 10$ mg/L<br>Turbidity $\leq 2$ NTU<br>Fecal coliform: ND/100 mL<br>$Cl_2$ residual: 1 mg/L<br>pH: 6 to 9 | No limitations on body-contact: lakes and ponds used for swimming, snowmaking. |
| restricted | Secondary, disinfection<br>$BOD_5 \leq 30$ mg/L<br>TSS $\leq 30$ mg/L<br>Fecal coliform $\leq 200/100$ mL<br>$Cl_2$ residual: 1 mg/L<br>pH: 6 to 9 | Fishing, boating, and other non-contact recreational activities. |
| Environmental reuse | Site specific treatment levels<br>pH<br>Dissolved oxygen<br>Coliform, nutrients | Use of reclaimed wastewater to create artificial wetlands, enhance natural wetlands and sustain stream flows. |
| Groundwater recharge | Site specific | Groundwater replenishment. Salt water intrusion control. Subsidence control. |
| Industrial reuse | Secondary and disinfection<br>$BOD_5 \leq 30$ mg/L<br>TSS $\leq 30$ mg/L<br>Fecal coliform $\leq 200/100$ mL | Cooling-system make-up water, process waters, boiler feed water, construction activities and washdown waters. |
| Potable reuse | Safe drinking water requirements | Blending in municipal water supply reservoir. Pipe to pipe supply. |

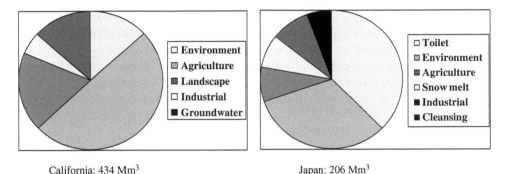

California: 434 Mm$^3$                Japan: 206 Mm$^3$

Figure 4.  Comparison of water reuse in different categories in California and Japan.


## 7   OVERVIEW OF WASTEWATER TREATMENT TECHNOLOGY

The effective treatment of wastewater to meet water quality objectives for water reuse applications and to protect public health is a critical element of water reuse systems. Conventional municipal wastewater treatment consists of a combination of physical, chemical, and biological processes and operations to remove solids, organic matter, pathogens, and sometimes nutrients from wastewater. General terms used to describe different degrees of treatment, in order of increasing treatment level, are preliminary, primary, secondary, and advanced treatment. A disinfection step for control of pathogenic organisms is often the final treatment step prior to distribution or storage. Water reclamation and reuse treatment systems have developed from technologies used for conventional wastewater treatment and drinking water treatment.

The goal in designing a water reclamation and reuse system is to develop an integrated cost-effective treatment scheme that is capable of meeting water quality objectives reliably.

### 7.1   *Water quality objectives*

The contaminants in reclaimed water that are of public health significance include biological and chemical constituents. Water quality control measures reference parameters that can be classified as organic matter, pathogenic organisms, nutrients, toxic contaminants, or dissolved minerals. Conventional wastewater treatment systems are typically designed to meet water quality objectives based on biochemical oxygen demand ($BOD_5$), total suspended solids (TSS), total or fecal coliform, and turbidity.

Where reclaimed water is used for applications that have potential human exposure routes, the major acute health risks are associated with exposure to biological pathogens including bacterial pathogens, helminths, protozoa, and enteric viruses. From a public health and process control perspective in developed countries, the most critical group of pathogenic organisms is enteric viruses due to the possibility of infection from exposure to low doses and the lack of routine, cost-effective methods for detection and quantification of viruses. In addition, treatment systems that can remove enteric viruses effectively will most likely be effective for control of other pathogenic organisms. Thus, it is essential to produce virtually virus-free effluent for water reuse applications that have the potential for significant human exposure or contact; e.g. spray irrigation of food crops eaten uncooked, parks and playgrounds, and unrestricted recreational impoundments where swimming may take place.

The degree of treatment required in individual water treatment and wastewater reclamation facilities varies according to the specific reuse application and associated water quality requirements. The simplest treatment systems involve solid/liquid separation processes and disinfection whereas more complex treatment systems involve combinations of physical, chemical, and biological processes employing multiple barrier treatment approaches for contaminant removal. An overview of the major technologies that are appropriate for water reclamation and reuse systems is shown in Figure 5. A summary of the major unit operations and processes used in water reclamation is given in Table 3.

## 8    HEALTH AND REGULATORY REQUIREMENTS

In every water reclamation, recycling and reuse operation, there is some risk of human exposure to pathogenic organisms and trace amount of harmful chemicals. Potential health impacts are, however, related to the degree of human exposure to the reclaimed water and the adequacy, effectiveness, and reliability of the treatment system. To protect public health without unnecessarily discouraging water reclamation and reuse, regulatory approaches stipulate water quality standards in conjunction with requirements for treatment, sampling, and monitoring. To minimize health risks and esthetic problems, tight controls are imposed on the delivery and use of reclaimed water after it leaves the treatment facility. Since major issues surrounding water reclamation and reuse are often related to health, considerable amounts of research have been conducted in the general area of public health. The major microbiological health hazards associated with water consumption and contact with reclaimed water originate from fecal contamination.

In the USA, comprehensive Federal standards for water reclamation and reuse do not exist to date, and water reclamation criteria are developed by individual States, often in conjunction with regulations on land treatment and disposal of wastewater. Some of the major differences among the approaches taken by individual States are associated with the degree of specificity provided in the rules. Also, discrepancies exist from place to place in terms of monitoring and treatment requirements (Crook & Okun, 1987).

Reclaimed water quality criteria for protecting health in developing countries must be established in relation to the limited resources available for public works and other health delivery systems that may yield greater health benefits for the funds spent. Confined sewage collection systems and wastewater treatment are often non-existent in these countries and reclaimed water often provides an essential water resource and fertilizer source. Thus, for most developing countries, the greatest concern for the use of wastewater for irrigation are caused by the enteric helminths such as hookworm, *Ascaris*, *Trichuris*, and under certain circumstances, the beef tapeworm. These pathogens can damage the health of both the general public consuming the crops irrigated with untreated wastewater and sewage farm workers and their families.

The degree of treatment required and the extent of monitoring necessary depend on the specific application. In general, irrigation systems are categorized according to the potential degree of human exposure. A higher degree of treatment is required for irrigation of crops that are consumed uncooked, or use of reclaimed water for irrigation of locations that are likely to have frequent human contact.

To illustrate alternative regulatory practices governing the use of reclaimed wastewater for irrigation, the major microbiological quality guidelines by the World Health Organization (WHO, 1989) and the State of California's (2001) current *Water Recycling Criteria* are

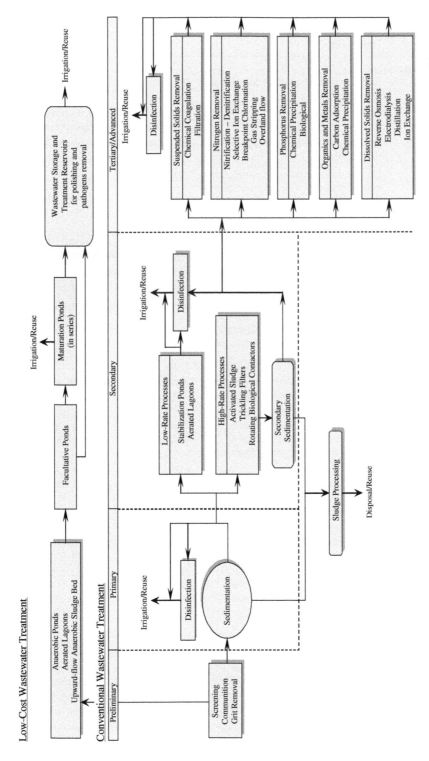

Figure 5.   Generalized wastewater treatment operations and processes, and water reclamation and reuse schemes (Pettygrove & Asano, 1985; Asano, 1998).

Table 3.    Overview of unit operations and processes used in water reclamation.

| Process | Description | Application |
|---|---|---|
| **Solid/liquid separation** | | |
| Coagulation | Addition of chemicals to destabilize suspended matter. | Promote particle destabilization to improve flocculation and solids removal. |
| Flocculation | Particle aggregation. | Particle agglomeration upstream of liquid/solid separation processes. |
| Filtration | Particle removal by porous medium. | Removal of particles larger than about $3\,\mu M$. |
| Sedimentation | Gravity sedimentation of particulate matter, chemical floc, and precipitates from suspension by gravity settling. | Solids removal. |
| **Biological treatment** | | |
| Aerobic biological treatment | Biological metabolism of waste solids by bacteria in an aeration basin. | Removal of organic matter from solution by synthesis into microbial cells. |
| Oxidation pond | Ponds with 2 to 3 feet of water depth for mixing and sunlight penetration. | Reduction of suspended solids, BOD, fecal bacteria and ammonia. |
| *Disinfection* | The inactivation of pathogenic organisms using oxidizing chemicals, ultraviolet light, caustic chemicals, heat, or physical separation processes. | Protection of public health. |
| **Advanced treatment** | | |
| Activated carbon | Process by which contaminants are physically adsorbed onto the carbon surface. | Removal of hydrophobic organic compounds. |
| Air stripping | Wastewater is distributed over a packing through which forced air is drawn to extract ammonia from the water droplets. | Used to remove ammonia nitrogen and some volatile organics. |
| Ion exchange | Exchange of ions between an exchange resin and water using a flow through reactor. | Softening and removal of selected ionic contaminants; effective for removal of cations such as calcium, magnesium, iron and anions such as nitrate. |
| Lime treatment | The use of lime to precipitate cations and metals from solution. | Used to stabilize lime-treated water, to reduce its scale forming potential. |
| Reverse osmosis | Pressure membrane to separate ions from solution based on reversing osmotic pressure differentials. | Removal of dissolved salts from solution. |

compared in Table 4. The WHO guidelines emphasize that a series of stabilization ponds is necessary to meet microbial water quality requirements. In contrast, the California criteria stipulate conventional biological wastewater treatment followed by tertiary treatment including filtration and chlorine disinfection to produce effluent that is virtually pathogen-free. Microbiological monitoring requirements also vary: the WHO guidelines also require monitoring of intestinal nematodes, whereas the *California Water Recycling Criteria* rely on treatment systems and monitoring of the total coliform density for assessment of microbiological quality.

Table 4. Comparison of microbiological quality guidelines and criteria for irrigation by the World Health Organization (WHO, 1989) and the State of California's (2001) current *Water Recycling Criteria*.

| Category | Reuse conditions | Intestinal nematodes[a] | Fecal or total[b] coliforms | Wastewater treatment requirements |
|---|---|---|---|---|
| WHO | Irrigation of crops likely to be eaten uncooked, sports fields, public parks. | <1/L | <1000/100 mL | A series of stabilization ponds or equivalent treatment. |
| WHO | Landscape irrigation where there is public access, such as hotels. | <1/L | <200/100 mL | Secondary treatment followed by disinfection. |
| California | Spray and surface irrigation of food crops, high exposure landscape irrigation such as parks. | No standard recommended | <2.2/100 mL[b] | Secondary treatment followed by filtration and disinfection. |
| WHO | Irrigation of cereal crops, industrial crops, fodder crops, pasture, and trees. | <1/L | No standard recommended | Stabilization ponds with 8–10 days retention or equivalent removal. |
| California | Irrigation of pasture for milking animals, landscape impoundment. | No standard recommended | <23/100 mL[b] | Secondary treatment followed by disinfection. |

[a] Intestinal nematodes (*Ascaris* and *Trichuris* species and hookworms) are expressed as the arithmetic mean number of eggs per liter during the irrigation period.

[b] *California Water Recycling Criteria* is expressed as the median number of total coliforms per 100 mL, as determined from the bacteriological results of the last 7 days for which analyses have been completed.

## 9 WATER REUSE PLANNING

Over the past decade, the impetus for water reuse has, in general, resulted from four motivating factors: (1) availability of high quality effluents; (2) increasing cost of freshwater development; (3) desirability of establishing comprehensive water resources planning, including water conservation and water reuse; and (4) avoidance of more stringent and expensive water pollution control requirements such as needs for advanced wastewater treatment facilities (Asano & Mills, 1990; Asano, 1991).

A common misconception in planning for water reuse is that reclaimed water represents a low-cost new water supply. This assumption is only true when water reclamation facilities are conveniently located near large industrial or agricultural users, and when additional wastewater treatment is not required. The conveyance and distribution systems for reclaimed water represent the principal cost of most water reuse projects. Recent experience in California indicates that approximately US\$ 2.50 in average capital costs are required for making each 1 $m^3$/yr of reclaimed municipal wastewater available for water reuse. Assuming a facilities life of 20 years and a 9% interest rate, the amortized cost of this reclaimed water is in the neighborhood of US\$ 0.24/$m^3$, excluding operation and maintenance (O & M) costs.

The optimum water reuse project is best achieved by integrating both wastewater treatment and water supply needs into planning. Thus, the facilities planning for water reuse should consist of: (1) wastewater treatment and disposal needs assessment; (2) water supply and demand assessment; (3) detailed reclaimed water market analysis; (4) engineering and economic analyses of alternatives; and (5) implementation plan with financial analysis. These planning steps are described in more detail in this section and important factors to consider in planning and implementation are highlighted.

## 9.1   Planning basis

The typical framework for analysis is first to establish clearly defined objectives and identify whether a project is intended as primarily single purpose or as multiple purpose, that is, to serve two or more basic functions. Generally water reclamation projects serve the functions of either water pollution control or water supply. Because most public works agencies, or subdivisions of agencies, are established as single purpose entities, planning for water reclamation projects is usually initiated with a single purpose in mind. For example, a city wastewater department is confronted with the need to meet more stringent effluent discharge requirements and will investigate water reclamation and reuse as one of the pollution control options. On the other hand, a water department may be faced with a falling groundwater table and look upon water reuse as a means of satisfying some of water demands by supplementing existing water supply.

In recent years, however, there are at least two simultaneous trends in the USA and other countries which should be forcing us to view water reuse as fundamentally serving multiple purposes: (1) standards for the discharge of wastewater are becoming increasingly more stringent; and (2) freshwater resources are becoming increasingly stressed to meet growing and competing water demands.

Many projects intended originally as single purpose inevitably have spillover benefits. If these would be recognized at the outset of project planning, the scope could be expanded. By recognizing their multiple benefits and beneficiaries, there are additional options available in terms of sharing project responsibility and costs and achieving the optimum balance of benefits (i.e. realizing maximum net benefits). The point of emphasizing the multiple purpose concepts is that the traditional perspective of a single purpose agency and funding program is often becoming outmoded and a disservice to meeting increasingly complex needs of society with an environmentally conscious public.

The project study area is another critical planning issue. The typical approach is to equate the study area with the project sponsor's jurisdictional boundaries. However, this approach can have serious pitfalls. The project study area should include all of the area that can potentially benefit from reuse of effluent from a particular wastewater treatment plant. Because water supply is typically dependent on water resources outside of the project study area, it is essential to look beyond the local area to obtain an understanding of the water resources situation. For example, over-drafted groundwater basins may be having their most serious impacts on communities great distances beyond the local area. Thus, implementing water reuse in the project area could result in a water supply savings in another area.

Planning for water reuse typically evolves through three stages: (1) conceptual level planning; (2) preliminary feasibility investigation; and (3) facilities planning. At the conceptual level planning, a potential project is sketched out, rough costs estimated, and a potential reclaimed water market identified. Based on the preliminary feasibility investigation, if water reuse appears to be viable and desirable, then detailed planning can be pursued, refined facilities alternatives developed, and a final facilities plan proposed.

## 9.2 *Reclaimed water market assessment*

A key task in planning a water reclamation project is to find potential customers that are capable and willing to use reclaimed water. The approach to take in marketing the reclaimed water depends on two factors: (1) project purpose – is the intent solely to treat and dispose of the wastewater or also to obtain optimum water supply benefit? And (2) user option – will the use of reclaimed water be voluntary or mandatory?

If the primary purpose is to treat and dispose of wastewater on land, then planners usually seek land application sites where water can be applied at high rates, usually in excess of optimum crop uptake rates, at least cost. Unless the system is designed with backup wastewater disposal methods, the users will have to make a long-term commitment to accept the treated effluent and may not have full control over the quantities of water delivered. If users cannot be found to accept treated effluent on a voluntary contractual basis, the wastewater agency will itself have to purchase wastewater application sites and apply the reclaimed water or lease the land to a private farmer.

Projects designed with the primary purpose of water supply can usually be operated more flexibly if alternate disposal, such as stream discharge, is available to dispose of effluent that cannot be reused. The reclaimed water can be marketed on a voluntary basis. However, if water supply is critical, the managing agency may elect to impose the use of reclaimed water in place of freshwater where it is environmentally and humanly safe to do so.

Whether a user is capable of using reclaimed water depends on the quality of effluent available and its suitability for the type of use involved. Willingness to use reclaimed water depends on whether its use is voluntary and, if it is, on how well reclaimed water competes with freshwater with respect to cost, quality, and convenience. It is essential to have a thorough knowledge of the water supply situation, especially if reclaimed water is to be marketed on a voluntary basis.

### 9.2.1 *Market assessment*

The water reuse market assessment consists of two parts: determination of background information, and a survey of potential reclaimed water users and their needs. Important water supply information includes a complete background on all of the wholesale and retail water agencies in the planning area, their boundaries, quality of water served, prices charged, and willingness to allow reclaimed water use in their jurisdictional areas. Because the introduction of reclaimed water could reduce freshwater revenues, at least in the short run, there might be resistance to implement wastewater reuse by some agencies. There should be the willingness to consider the freshwater revenue impacts in the analysis and appropriate revenue and cost sharing to obtain the full cooperation of all affected agencies.

Without much investigation it is possible to list most of the potential reclaimed water use categories in the study area, e.g. landscape irrigation, industrial cooling, and irrigation of food crops. On the basis of the use categories, health and water pollution control regulatory authorities should be consulted to obtain their respective requirements. These would include wastewater treatment requirements, on-site facilities modifications (e.g. backflow prevention devices), and use area controls (e.g. no irrigation in areas of direct human contact). Technical experts, such as farm advisors, can be consulted to determine acceptable water quality for various use categories.

### 9.2.2   Identification of users

It is then possible to begin identifying and contacting individual potential users of reclaimed water. Access to records of water retailers can be especially helpful. Several years of actual water use records are helpful to ensure that planning is not misled by data from unusually wet or dry precipitation years. It is important to obtain actual prices paid for water or, if a user has its own supply, its fixed and variable costs. Potential users should be contacted and the reuse sites visited to determine potential site problems or on-site water system modifications needed to accommodate use of reclaimed water. These factors have cost implications which must be assessed in the planning stage. The concern, needs, and financial expectations of users must be identified. Group presentations with potential users may be useful to disseminate information and make technical experts accessible to respond to questions.

### 9.3   Monetary analyses

While technical, environmental, and social factors are considered in project planning, monetary factors tend to be overriding in the key decisions of whether and how to implement a wastewater reuse project. Monetary analyses fall into two categories: economic analysis and financial analysis, the distinction between which is critical. The economic analysis focuses on the value of the resources invested in a project to construct and operate it, measured in monetary terms and computed in the present value. On the other hand, the financial analysis is based on the market value of goods and services at the time of sale, incorporating any particular subsidies or monetary transfers which may exist. Whereas economic analysis evaluates wastewater reuse projects in the context of impacts on society, financial analysis focuses, instead, on the local ability to raise money from project revenues, government grants, loans, and bonds to pay for the project.

The basic result of the economic analysis is to answer the question, *should* a reuse project be constructed? Equally important, however, is the question, *can* a reuse project be constructed? Both orientations, therefore, are necessary. However, only water reuse projects which are viable in the economic context should be given further consideration for a financial analysis (Mills & Asano, 1998).

### 9.3.1   Economic analysis

The role of an economic analysis is to provide a basis for justifying a wastewater reuse project in monetary terms. A project is considered justified if its total benefits exceed its total costs. If several alternatives can meet the same objective, then the alternative providing the maximum net benefit is the economically justifiable project. While the benefit-to-cost ratio is a common measure of economic justification, it is not the best measure to determine the optimum project size.

An important aspect of the economic analysis is that it takes into consideration all costs and benefits associated with the alternatives under consideration, placing all alternatives on equal footing for comparison. Also, this analysis is completely independent of financing considerations. To identify all costs and benefits it is essential to look beyond the boundaries of the agency doing the planning. For example, an agency may be seeking a new source of water supply from outside of its boundaries. To perform an economic comparison it would be necessary to identify the construction, operation, and maintenance costs of this supply.

### 9.3.2 *Cost and price of water*

Another important aspect of an economic analysis is that it considers only the future flow of resources invested in or derived from a project. Past resources investments are considered sunk costs that are irrelevant to future investment decisions. Thus, debt service on past investments is not included in an economic analysis. A common error in this respect is to confuse water price with water cost. Water price is the purchase price paid to a water wholesaler or retailer to purchase water and usually reflects a melding of current and past expenditures for a combination of projects, as well as water system administration costs, which are generally fixed costs. The costs of relevance to an economic analysis would be only costs for future construction, operations, and maintenance. If a water reuse project were to be compared to a particular new water supply development, the relevant costs would be the future stream of costs: (1) to construct new freshwater facilities; and (2) to operate and maintain all of the facilities needed to treat and deliver the new increment of water supply developed. This stream of costs may bear no resemblance to the present and future price, at the wholesale or retail level, charged for water.

In contrast, water prices embody debt service on existing facilities, and future projections are an average price to recover costs for both existing facilities and future additions. Typically, water price will be much lower than the marginal cost of developing a new water supply, because the cost of each new source of supply is increasingly expensive due to inflation and the greater difficulty in developing new supplies.

### 9.3.3 *Financial analysis*

The financial analysis addresses the question whether a wastewater reuse project is financially feasible. The project sponsor will need a source of capital and sources of revenue to pay for debt service and operational costs for both the proposed reuse project and any existing facilities. Fixed costs for existing facilities, while irrelevant in the economic analysis, must be considered in a financial analysis if they are a continuing financial obligation.

The water reuse project sponsor is not the only important party to consider in a financial analysis. Of particular importance is the participation of the user of the reclaimed water. The user will be expecting a net cost of reclaimed water that is no more than it would have paid for freshwater. For example, a reclaimed water customer may have to invest in piping modifications or a dual water supply system to accommodate the reclaimed water. On the other hand, a farmer may be able to save on fertilizer costs by taking advantage of nitrogen and phosphorus contained in reclaimed water. A prospective user will expect the difference in price between freshwater and reclaimed water to reflect any added costs or savings.

Because the sale of reclaimed water may reduce revenue from freshwater sale, there may be a need to evaluate the effect on the freshwater retailer and freshwater prices. It may be necessary to allocate some of the reclaimed water revenue to compensate for the freshwater revenue loss. On the other hand, if reclaimed water offsets the purchase or development of more expensive freshwater, then it may be appropriate for freshwater revenue to be used to subsidize the wastewater reuse project.

It is not uncommon that potential users may have different sources of water or have different rate schedules. It is important, therefore, not to assume that there is an average price that all users are paying for freshwater. Failure to take into account the financial situation of each user could result in the loss of key reclaimed water customers. The initial market assessment should have included this financial data. In conclusion, there should be flexibility in tailoring a financial scheme to fit each situation best.

## 9.4   Other planning factors

A number of factors besides the monetary aspects have to be evaluated during the planning for a water reuse project, such as environmental impacts. However, factors of particular significance in project development are related to engineering and public health. Engineering involves more than water distribution system design. A water reuse project is a relatively small-scale water supply project with considerations of matching supply and demand, appropriate levels of wastewater treatment, reclaimed water storage, and supplemental or backup freshwater supply.

### 9.4.1   Water demand characteristics

In freshwater systems, water demands are first projected and then water supplies are developed to meet the demand. The reverse procedure is often applied in water reuse system planning. The wastewater supply rate is accepted as a given and the reclaimed water demand is added to the system until the economic optimum is met. For example, landscape irrigation demand in California is seasonal. However, wastewater production is nearly constant year-round. Reclaimed water supply may be sufficient to meet annual demands, but only if seasonal storage is available. Seasonal storage, however, is costly and, in urban settings, even impossible to site.

Another option is to include fewer users in the system such that the peak demands can be met entirely by the reclaimed water supply without seasonal storage. This could, however, result in the waste of as much as 40% of the available reclaimed water. What will probably be the optimum situation is to add users in excess of supply and meet the peak demands with supplemental freshwater. There may still be some supply that cannot be used or economically stored during low water demand periods, but this lost supply can be reduced substantially because of the availability of a supplemental supply in the peak season. Some projects have incorporated an added benefit by utilizing a poor quality water supply unsuitable for potable use, such as an abandoned groundwater basin, to supplement reclaimed water.

### 9.4.2   Supplemental water supply

Supplemental freshwater can be blended with reclaimed water in the distribution system or on the user's site. Because of public health concerns about potential cross connection or backflow of reclaimed water into potable water supply systems, it may be necessary to provide an air gap between the supplemental supply and the reclaimed system.

Even if supplemental water is not needed to meet demands, there may still be a need to provide an emergency backup water supply during periods of treatment plant upset or equipment failure. Because a backup water supply would be utilized in place of, rather than simultaneously with, reclaimed water, there are more options for introducing it into the system. With appropriate backflow prevention the reclaimed water distribution system can be connected to the potable system during the emergency period. It should be noted that with the availability of a backup water supply, there is less need for equipment redundancy in the reclaimed water system to ensure 100% reliability.

### 9.4.3 *Water quality*

If a significant market could be added by upgrading reclaimed water quality, project alternatives should be developed for various treatment levels. The levels of wastewater treatment and water quality for landscape and agricultural irrigation uses are normally governed by health-related regulations, though crop sensitivity to effluent constituents such as salts and boron should be investigated (Ayers & Westcot, 1985; Pettygrove & Asano, 1985).

Industrial users will have more stringent physical and chemical water quality requirements that will affect levels for wastewater treatment. With some exceptions, it is impracticable to serve more than one quality of water. Thus, the level of wastewater treatment provided may be higher than many of the users actually require. If there is a reclaimed water market for two levels of water quality, it should be considered whether the distribution system can be separated so that the higher and more expensive treatment can be sized to serve only those users needing such higher water quality.

While the emphasis on the wastewater disposal or water supply aspects will vary depending on whether a project is single or multiple purposes, the nature of water reuse is such that both aspects must at least be considered. Even if it is determined that a water reuse project is not feasible at the conclusion of the study, it is still advisable to publish the information and data collected and the analysis performed to arrive at this conclusion. Water reuse is good public policies in appropriate situations and the public interest in them will continue to recur as long as water supply needs are perceived to be critical, such as in drought years. Documentation of even unsuccessful reuse planning is helpful in responding to public inquiry and in orienting future planning efforts.

## 10 WATER REUSE: A RELEVANT SOLUTION

Inadequate water supply and water quality deterioration represent serious contemporary concerns for many municipalities, industries, agriculture, and the environment in various parts of the world. Several factors have contributed to these problems such as continued population growth in urban areas, contamination of surface water and groundwater, uneven distribution of water resources, and frequent droughts caused by the extreme global weather patterns. For more than a quarter century, a recurring thesis in environmental and water resources engineering has been that improved wastewater treatment provides a treated effluent of such quality that it should not be wasted but put to beneficial use. This conviction in responsible engineering, coupled with the vexing problem of increasing water shortage and environmental pollution, provides a realistic framework for considering reclaimed water as a water resource in many parts of the world. Water reuse has been dubbed as the *greatest challenge of the 21st century* as water supplies remain practically the same and water demands increase because of increasing population and per capita consumption. Water reuse accomplishes two fundamental functions: (1) the treated effluent is used as a water resource for beneficial purposes; and (2) the effluent is kept out of streams, lakes, and beaches; thus, reducing pollution of surface water and groundwater.

### 10.1 *The main concerns in water reuse*

The main concerns in water reuse are: (1) reliable treatment of wastewater to meet strict water quality requirements for the intended reuse; (2) protection of public health; and (3) gaining public acceptance. Water reuse also requires close examinations of infrastructure and facilities planning, and water utility management involving effective integration of water and reclaimed water supply functions. Whether water reclamation and reuse will be appropriate in a specific locale depends upon careful economic considerations, potential uses for the reclaimed water, and stringency of waste discharge requirements. Public policy wherein the desire to conserve and reuse rather than develop additional water resources with considerable environmental expenditures may be an important consideration. Through integrated water resources planning, the use of reclaimed water may provide sufficient flexibility to allow a water agency to respond to short-term needs as well as increase long-term water supply reliability.

## 10.2    *Groundwater recharge and direct potable water reuse*

Groundwater recharge with reclaimed water and direct potable water reuse share many of the public health concerns encountered in drinking water withdrawn from polluted rivers and reservoirs. Three classes of constituents are of special concern where reclaimed water is used in such applications: (1) enteric viruses and other emerging pathogens; (2) organic constituents including industrial chemicals, home care products, and medicines; and (3) heavy metals. The ramification of many of these constituents in trace quantity are not well understood with respect to long-term health effects, and, as a result, regulatory agencies are proceeding with extreme caution in permitting water reuse applications that affect potable water supplies. In all the cases in the USA where potable water reuse has been contemplated, alternative sources of water have been developed in the ensuing years and the need to adopt direct potable water reuse has been avoided. As the proportional quantities of treated wastewater discharged into the nation's waters increase, much of the research which addresses groundwater recharge and potable water reuse is becoming of equal relevance to *unplanned indirect potable reuse* such as municipal water intakes located downstream from wastewater discharges or from increasingly polluted rivers and reservoirs.

## 10.3    *Locally controllable water resource*

Reclaimed water is a locally controllable water resource that exists right at the doorstep of the urban environment, where water is needed the most and priced the highest. Closing the water cycle not only is technically feasible in industries and municipalities but also makes economic sense. While potable reuse is still a distant possibility and may never be implemented in the USA, groundwater recharge with advanced wastewater treatment technologies is a viable option backed by the decades of experiences in Arizona, California, New York, and Texas, as well as in Australia, Israel, Germany, the Netherlands, and the UK. It is important to recognize that public acceptance of water reuse projects is vital to the future of water reclamation, recycling, and reuse; the consequences of poor public perception could jeopardize future projects involving the use of reclaimed water. Indeed, water recycling and reuse provide a relevant solution to water resources development and management but with more challenges in costs, health protection, and public acceptance.

## REFERENCES

Angelakis, A.N. & Spyridakis, S.V. (1995). The Status of Water Resources in Minoan Times: A Preliminary Study. In: A.N. Angelakis, A. Issar & O.K. Davis (eds.), *Diachronic Climatic Impacts on Water Resources in Mediterranean Region*. Springer-Verlag, Heidelberg, Germany.

Asano, T. (1991). Planning and Implementation of Water Reuse Projects. *Water Science and Technology,* 24(9): 1–10.

Asano, T. (ed.) (1998). *Wastewater Reclamation and Reuse.* Water Quality Management Library. Vol. 10, CRC Press. Boca Raton, Florida, USA.

Asano, T. & Levine, A.D. (1996). Wastewater Reclamation, Recycling, and Reuse: Past, Present and Future. *Water Science and Technology,* 33(10–11): 1–14.

Asano, T. & Levine, A.D. (1998). Wastewater Reclamation, Recycling, and Reuse: An Introduction, In Asano, T. (ed.), *Wastewater Reclamation and Reuse.* Water Quality Management Library, CRC Press. Boca Raton, Florida, USA.

Asano, T. & Mills, R.A. (1990). Planning and Analysis for Water Reuse Projects. *Journal American Water Works Association,* January: 38–47.

Ayers, R.S. & Westcot, D.W. (1985). *Water Quality for Agriculture.* FAO Irrigation and Drainage Paper 29, Rev. 1. Food and Agriculture Organization of the United Nations, Rome, Italy.

Barty-King, H. (1992). *Water the Book, an Illustrated History of Water Supply and Wastewater in the United Kingdom.* Quiller Press Limited, London, UK.

Crook, J. & Okun, D.A. (1987). The Place of Non-potable Reuse in Water Management. *Jour. Water Pollution Control Fed.,* 59(5): 236.

EEC (European Communities Commission) (1991). Council Directive regarding the treatment of urban wastewater (91/271/EEC). *Official Journal of the European Communities,* No. L 135, 30 May: 40–50.

Fair, G.M. & Geyer, J.C. (1954). *Water Supply and Wastewater Disposal.* John Wiley and Sons, New York, USA.

Levine, A.D. & Asano, T. (2004). Recovering Sustainable Water from Wastewater. *Environmental Science & Technology,* 38(11): 201A–208A.

Mills, R.A. & Asano, T. (1998). Planning and Analysis of Water Reuse Projects. In Asano, T. (ed.), *Wastewater Reclamation and Reuse.* Water Quality Management Library. Vol. 10, CRC Press, Boca Raton, Florida, USA.

Pettygrove, G.S. & Asano, T. (eds.) (1985). *Irrigation with Reclaimed Municipal Wastewater: A Guidance Manual.* Lewis Publishers, Inc., Chelsea, Michigan, USA.

Postel, S.L. (2000). Entering an era of water scarcity: the challenges ahead. *Ecological Applications,* 10(4): 941–948.

Queensland Government (2001). *Queensland Water Recycling Strategy: An Initiative of the Queensland Government.* The State of Queensland, Environmental Protection Agency, Queensland, Australia.

State of California (2001). *Water Recycling Criteria, an excerpt from the California Code of Regulations, Title 22, Division 4.* Environmental Health, Department of Health Services, Sacramento, California, USA.

United Nations (2003). *World Population Prospects: the 2002 Revision – Highlights.* United Nations Population Division, Department of Economic and Social Affairs, United Nations, New York, USA.

WHO (World Health Organization) (1989). *Health Guidelines for the Use of Wastewater in Agriculture and Aquaculture, Report of a WHO Scientific Group.* Geneva, Switzerland.

Young, D.D. (1985). Reuse via rivers for water supply, In: *Reuse of Sewage Effluent,* Proceedings of the International Symposium organized by the Institution of Civil Engineers. London, 30–31 October 1984. Thomas Telford, London, UK.

CHAPTER 17

# Urban and industrial watersheds and ecological sanitation: two sustainable strategies for on-site urban water management

D. Del Porto
*Ecological Engineering Group, Concord, Massachusetts, USA*

ABSTRACT:  Many of the water crises in urban areas are based not so much on the availability of water but rather on our failure to manage how water is used. This is due to the failure to recognize stormwater as a water source and using water to transport unwanted wastes. Wastewater recycling is an important part of an integrated water strategy – but not the final answer. That integrated strategy includes stormwater collection, source separation, wastewater and graywater recycling for potable and non-potable use, and advanced conservation.

Keywords:  *Integrated water planning strategy, stormwater, ecological sanitation, wastewater, watershed, recycling, reuse, conservation*

## 1  INTRODUCTION

"Problems cannot be solved at the same level of awareness that created them".

(Albert Einstein)

### 1.1  *The urban dilemma*

As the world's population increases, migrates to coastal areas and builds new cities, urban planners are confronted with insufficient supplies of fresh water to meet the increasing demands. As a consequence, supply augmentation, water demand management practices, and water recycling are expected to supplement the gap between available supply and new demand. The problem is not limited to arid zones. It is a function of population explosions and the inability to manage demand. "Even in water-rich countries such as Canada, lack of quality control for drinking water has led to death and illness in several communities, and some provinces suffer chronic shortages of water for agriculture; and almost everywhere capital costs for infrastructures to supply and remove water are growing" (Brooks, 2003).

Much of the water crises in urban areas are based not so much on the availability of water but rather on our failure to manage how it is used. Both the failure of urban planners to recognize stormwater as a resource and the continued use of water to transport unwanted wastes are at the root of the problem.

### 1.2  *Problems with the current approach*

Addressing urban water supplies is quickly becoming a high-priority challenge. The United Nations Population Fund estimates that by 2025, 61% of the world's population of 8000 million will live in cities (United Nations, 2004), creating population densities throughout the world heretofore unseen.

The problem is exacerbated as urban development follows the western model of big pipes bringing fresh water from long distances and then treating influent wastewater and discharging into receiving waters. Or, it flows to sophisticated expensive recycling treatment plants and then distributed through more big pipes to where the water was wasted in the first place.

Continuing this model, urban planners will have to seek:

– New supplies from other aquifers and surface waters (yet these resources are already overtaxed in many areas and may rob water from outlying areas and agriculture).
– Desalinization (very energy intensive).
– Wastewater recycling (energy and infrastructure intensive).

It is clear that wastewater recycling will be a necessary strategy in the years ahead. However, it is only one component of an ecological and integrated strategy that makes best use of local sites as well as local sources of water and local effluents that can be reused.

Continuing to combine wastes, including excreta, toxics, and heavy metals, in wastewater flows, then turning to advanced ultra-filtration at the end of the pipe will not be feasible for many cities, especially those in developing countries.

To identify water recycling as the best solution to the world's water supply challenges is to use the same mentality that created today's many water problems.

The practice of centrally collecting combined effluents from a wide variety of sources containing many constituents, and then treating it with end-of-pipe solutions has evolved in an effort to avoid the complexity of using many smaller, more local and effluent-specific strategies. Yet nature's model shows us that complexity is the best way to manage resources.

For this reason, urban wastewater recycling can only be viewed as an important part of a much larger picture of integrated water planning. Such a strategy, which decentralizes water collection, use and treatment, nevertheless still benefits from central management. Also, in contrast to the *big pipe* approach, this strategy requires us to heed the tenet *start at the source*. By first seeking more local opportunities – both on the site of the water use and wastewater discharge and according to what exactly is discharged – we may find ecological and low-entropy possibilities for reducing water demand and treating effluents in more ecologically effective ways. We will call this integrated source- and site-based water planning.

### 1.2.1  *Untapped water sources*

One aspect of integrated source – and site-based water planning is identifying all of the water sources available to a particular site of water use.

Consider the following example: on average, New York City receives about 937 Mm$^3$/yr (246,500 million gal/yr) of stormwater, or 2.57 Mm$^3$ (675 million gal) per averaged day (New York City has 836.1 km$^2$ of land area and receives 122.6 mm/yr of total precipitation). While some is lost through evaporation, most is piped away into the adjacent rivers and out to sea.

New York City uses about 4.16 Mm$^3$/d (1100 million gal/d). According to the New York City Water Department, the city saved 1.5 Mm$^3$ (400 million gal) of water by replacing 1.3 million old water-wasting toilets, installing locks on fire hydrants in neighborhoods throughout the five boroughs, and implementing a leak detection program to inspect underground water mains for leaks.

Therefore, if the stormwater had been collected, it would have satisfied 61.4% of the current demand. Using the current price: US$ 1.39 per m³ (3.94 per 100 ft³) (New York City Water Department) for water service, the value of the stormwater is US$ 3.56 million per day or US$ 1300 million per year.

Stormwater could satisfy 100% of the demand if: (1) demand is reduced by an additional 22% by continued conservation; and (2) recycled wastewater supplied 20% of the demand.

Yet the city draws its water from various watersheds in distant mountains, requiring miles of pipes and straining the communities along the watershed. New York City might instead look to capturing this stormwater, using some of the thousands of millions of dollars spent for collection and distribution pipe to fund cisterns, rooftop collection, and simple rainwater treatment to augment water supply.

### 1.3   *Recycled wastewater*

While recycling wastewater is an important step in the conservation of fresh water, it is technologically complex, expensive, and unacceptable to many communities, such as San Diego, California, according to the USA-based WateReuse Association (Web Page). In addition, there is a growing concern for the impacts of recycled water on valuable aquifers due to contamination by constituents that are currently not monitored by groundwater discharge permits. According to the California Department of Health Services (Web Page), these aquifer anti-degradation concerns are reported as action levels. Action levels are health-based advisory levels for chemicals in drinking water that lack maximum contaminant levels (MCLs).

### 1.4   *The QWERTY syndrome*

We humans have a curious tendency to continue old practices into the future long after the need for such practices has passed. Take the example of typing keyboards. The upper left row of letters begins with the letters *QWERTY*. This arrangement was developed in 1872 to slow down typists; otherwise the keys would jam on the early mechanical typewriters. We no longer have mechanical typewriters, yet we continue to use the old keyboard configurations to this day. It may be that this *QWERTY syndrome* is the root cause of some of today's most fundamental water management problems.

There are two *QWERTY*-related water issues that will be the focus of this chapter:

– *Urban and industrial stormwater management attitudes*: we must reverse the attitude that stormwater in an urban and industrial setting is not a resource, but rather a problem due to flooding and so we must quickly pipe it away.
– *Historical linkage of water and sanitation*: one of the main impediments to sustainable water management is the historic linkage of water and sanitation. We must uncouple water from sanitation as much as possible or forever be trapped in flushing away unwanted residuals, valuable nutrients (found in urine), and pathogens with large volumes of water. An example is New York City's expensive flush. As of the 2000 census, the population of New York City was 8.1 million. Assuming that the average person flushes a toilet 5.1 times per day (Vickers, 2001) and that the average water use per flush is 11.36 L (3 gal), then New York

City flushes about 0.47 Mm$^3$ (123.75 million gal) worth US$ 0.65 million of drinking water into the sewers every day.

### 1.4.1    *The fundamental problem*

Essentially, we are using a valuable resource – drinking water treated and delivered at significant expense – to dilute and dispose of another potentially valuable resource, human excreta. To this we add industrial and household chemicals and stormwater drainage. Then we pay a very high cost to transport this combined effluent to a facility which attempts to separate all of those constituents, clean them to a degree mandated by national law, and discharge the remaining water back into the environment – usually rivers, oceans, and the ground. In most cases, the same nutrients and toxic chemicals that went into the wastewater mix are still present in what leaves the treatment plant. Growing realization of the effects of this is prompting regulators to mandate further treatment – and that is making this *combine-dilute-treat-and-dispose* approach very expensive.

However, even with ever-tightening requirements for advanced wastewater treatment, many of the world's waterways do not meet the goals of swimmable and fishable quality set by the USA 1972 *Clean Water Act* nor the United Nation's *Agenda 21* (Brooks, 2003). And, while much progress has been made over the last 32 years, there are many who ask if there might not be a better way.

Every year we are learning more about the longer-term effects of partial treatment and disposal, our present approach to cleaning wastewater. In USA, the 1972 *Clean Water Act* only required the reduction of suspended solids, biological oxygen demand and fecal coliform bacteria to protect lakes and rivers. But now, responsible regulators worldwide are mandating the removal of nutrients, toxic chemicals, parasites, viruses, radioactive wastes and other constituents. At the same time, planners are asking: in a world where drinking water is increasingly expensive and scarce, can we use this valuable resource for flushing toilets?

It is clear that better ways are needed; the good news is that many are here and more are emerging. However, the answer is not merely a matter of more clean-up at the end of the sewage pipe. A larger, more strategic solution is called for based on a broad approach:

– Advanced water conservation.
– Collect and store stormwater.
– On-site wastewater treatment.
– Recycling.

## 2    BETTER WAYS EMERGING: FROM DISPOSAL TO UTILIZATION

More regulations requiring better treatment, and increasing costs for water and wastewater treatment are prompting a reframing of the wastewater issue. In the current system, we create *wastes* that we want *disposed of*. A better strategy is to put these outputs to use, just as they are in nature's model. In balanced ecosystems there is no waste: the outputs of one organism are the inputs of another.

The solution to both of these problems is the development of a new ecological paradigm for integrated water management. To accomplish this integration we must investigate the new initiatives that address these issues. Key initiatives can be summarized as urban and industrial watersheds and ecological sanitation.

## 2.1   *Urban and industrial watersheds*

The ecological engineering approach for water management presented in this report will not only demonstrate environmental sustainability for the 21st century, but will also be testaments to these visions. The principles of industrial ecological and sustainable development do not preclude economic development or manufacturing efficiency. Rather, recycling, reuse and conservation result in reduced dependence on scarce resources, higher efficiency in manufacturing processes, and long-term cost savings.

Terrestrial life efficiently self-organizes around water and the mineral-rich landforms through which water flows. The ecosystem that best describes life and all related activities within the water and landform context is a *watershed*. A natural watershed synthesizes inputs of rainwater, solar energy and minerals from within its physical, chemical, and biotic communities to produce an array of nutrients, raw materials and products that sustains a certain quality of life for all its inhabitants. Concerning society's welfare, almost all essential landscape functions (i.e. fertile soil regeneration, climate stabilization, etc.) for are connected to water ecology. In the same way, the industrial watershed is defined by its rainwater, wastewater and process water flows and its interaction with the surrounding watershed.

### 2.1.1   *Ford Motor Company Rouge Facility as an Industrial Watershed*
### *(Del Porto & The Ecological Engineering Group, 2001)*
According to *Environment Canada* (Web Page), "an automobile coming off the assembly line, for example, will have used at least 80,000–120,000 L (21,000–32,000 gal) of water to produce its ton of steel and 40,000 L (10,600 gal) more for the actual fabrication process. Many thousands more liters of water are involved in the manufacture of its plastic, glass, fabric components".

The Ford Rouge Plant in Dearborn, Michigan – with its 243 ha (600 acres) of natural and constructed landforms, inputs of water, energy, minerals, capital, information, and nutrients from its 30,000 employees, and outputs of products that sustain the Dearborn community – is in fact a constructed ecosystem that we will call an *Industrial Watershed*.

### 2.1.2   *A water* eco-nomic *example*
With increasing prices and competition, water may arise as an important cost factor within the production process. Therefore, new water management strategies, not only in conservation but also in terms of reuse, are needed. Following figures show what the gain of water from the industrial watershed could mean to the production of automobiles.

### 2.1.3   *Source: stormwater*
About 2 Mm$^3$/yr (530 million gal/yr) of water falls on the 243 ha (600 acres) Rouge Plant site and discharges to the Rouge River. Average precipitation is 825.5 mm/yr (32.5 inches/yr) based on 30-year average (NRCS/US Department of Agriculture weather data). Increasingly this water is coming under state and federal jurisdiction as stormwater discharge controls are implemented to protect the Rouge River. The federal government, in order to limit pollutants in stormwater discharges from industrial facilities, implemented the National Pollutant Discharge

Elimination System (NPDES), Phase I Storm Water Program, which includes an industrial stormwater permitting component. The requirement for a NPDES industrial stormwater permit has added an additional financial component to the Ford water management costs. While the Michigan counterpart is voluntary, one can predict mandatory compliance will be coming in the near future.

### 2.1.4    Application: process water

General Motors de Mexico (Ramos Arizpe Complex) recently received the Stockholm Industry Water Award for demonstrating sustainable water and wastewater treatment and recycling techniques. Using a variety of physical, chemical and biological wastewater treatment processes, the facility was able to recover and reuse 70% of its industrial wastewater. The facility also increased its production seven-fold, while reducing the average amount of well water needed to produce a vehicle from $32\,m^3$ (8452 gal) to $2.2\,m^3$ (581 gal).

The success of General Motors may serve as an incentive and model for other industrial operations that seek to minimize environmental impact while maximizing production and quality of life.

If Ford, using General Motors water usage, would collect 90% of the stormwater that would provide $1.8\,Mm^3$ (477 million gal) which could be used to make 820,998 vehicles ($477 \times 10^6/581$) without taking any from the Dearborn Water Company. Because the same water can be treated (in respect to its constituents) and reused with about 90% recovery efficiency, another 738,898 automobiles could be made ($429.3 \times 10^6/581$) after treatment. The remaining treated water could be sold back to Ford Steel or the Dearborn Water Department for a profit or used to grow some of the raw materials. Therefore, 1,559,896 vehicles per year could be made with only one recycle pass and still have recycled water to use for other profitable purposes!

### 2.1.5    Source: waste water

In 2001, the price of water service from the Dearborn Water Department via the Detroit Water and Sewer Department was comprised of the cost to treat and deliver potable water from Lake Huron and the Detroit River, as well as the cost of collecting and treating it as wastewater once it has been used (see Detroit News Online, Web Page). According to Dearborn, Michigan's 1999 annual report, the city spent US$ 25.3 million on the city's water distribution and sewage collection systems. Of the total bill, wastewater management is more than 66% of the total. Upgrading the aged infrastructure will cost US$ 5300 million, which in the short-term will increase the sewage portion of the total bill by 13.7%.

The 2001 unit cost for water of US$ 0.87 per $m^3$ (US$ 3.30 per 1000 gal) does not reflect the cost of infrastructure upgrade. Therefore, the unit cost does not reflect all the new marginal costs and will certainly escalate dramatically in the future.

## 2.2    Water-wise design through ecological engineering

The management of the industrial watershed to sustain the highest quality of life for the lowest cost of living involves adopting an engineering model that is founded on the ecological paradigm. The transformation of an industrial site into a highly efficient and therefore profitable industrial watershed will require the ecological engineering of all aspects of the facility.

Ecological engineering is predicated on the knowledge that the self-organizing order found in stable ecosystems is so universal that it can be applied as an engineering discipline to solve the pressing problems of global pollution, goods and services production, and efficient resource-utilization, while providing a high quality of life.

When possible, we must recycle, reuse or utilize effluents. Using them strategically, such as for landscape irrigation, flushing toilets, and evaporative cooling, helps save both water supply and wastewater treatment costs – and prevents effluents from becoming pollution.

Just as *reduce, reuse and recycle* has become the *credo* of responsible solid waste management, this five-pronged strategy will become more obvious and important to the world's population than presently can be imagined.

Increasing costs of water and wastewater treatment infrastructure are driving the interest for integrated water management, by viewing urban, domestic, commercial, industrial and agricultural landscapes and hardscapes as watersheds that collect water, use water, and dispose of stormwater and wastewater within larger regional watersheds.

### 2.3 *Ecological sanitation*

According to the *Deutsche Gesellschaft für Technische Zusammenarbeit* (GTZ, Web Page), "Conventional forms of centralized sanitation are coming under increasing criticism. Especially because of the enormous investment involved, the huge operating and maintenance costs, high water consumption and other drawbacks, they are not suitable as a blanket solution for developing countries, particularly in arid climate zones. Even conventional individual disposal systems, such as latrines and cesspits, make poor alternatives – especially in view of increasing population densities and the substantial groundwater pollution they cause. Moreover, all conventional types of wastewater and sewage disposal systems usually deprive agriculture, and consequently food production, of the valuable nutrients contained in human excrement. A more holistic approach towards sustainable sanitation is offered by the concepts referred to as *ecological sanitation*. The key objective of this approach is not to promote a certain technology, but rather a new philosophy of dealing with what has been regarded as wastewater in the past. The systems of this approach are based on the systematic implementation of a material-flow-oriented recycling process as a holistic alternative to conventional solutions. Ideally, ecological sanitation systems enable the complete recovery of all nutrients from faeces, urine and greywater to the benefit of agriculture, and the minimization of water pollution, while at the same time ensuring that water is used economically and is reused to the greatest possible extent, particularly for irrigation purposes".

### 2.4 *On-site and waterless sanitation: a component of ecological sanitation*

On-site options include composting toilets and constructed ecosystems for graywater that use up the effluent on site. In a Canadian office building and in a Swedish apartment building, waterless composting system process toilet wastes, while graywater is filtered and used to irrigate the landscapes around the buildings. In homes with on-site *waterless toilet systems* (WTS) in Toronto, Canada and in Massachusetts, graywater is utilized by water-loving plants in planter beds and greenhouses, which are integrated into the homes.

The United States Environmental Protection Agency has found that decentralized waste treatment systems are a viable alternative to centralized systems when analyzing their effectiveness and economics.

WTS (also known as dry, composting and biological toilets and non-liquid saturated systems) are gaining popularity because, among wastewater treatment technologies, they are one of the most direct ways to avoid pollution and conserve water and resources. Most users who install WTS do so simply because they need to have a toilet system where a conventional water-borne system is unavailable or cannot be installed due to environmental constraints.

Long used by developing countries, parks, subsistence homeowners, and vacation cottage owners around the world, WTS are now making their way into mainstream year-round homes in North America, for many reasons:

- Micro-flush (one pint or less) toilets are increasingly used with WTS, making these systems more socially acceptable.
- Graywater (wash water) systems are increasingly permitted by public health officials.
- Service contracts are available for maintaining WTSs.
- Water shortages threaten at least one-third of the world. Some estimates place it at one-half.
- Many states are tightening on-site wastewater system standards, so that many of the USA's millions of septic systems are now considered inadequate, and therefore in noncompliance. As a result, many property owners are seeking ways to supplement their septic systems, so they can avoid installing new ones. Diverting excrement and flush water from the flow removes more than 90% of the pollution, leaving only graywater to manage.
- Population densities are increasing in cities and coastal areas, intensifying the challenge of managing human waste.
- Owners are converting vacation homes into year-round residences. These homes are often in remote and environmentally sensitive natural areas, such as seacoasts, lakes and mountains, with limited capacity for wastewater disposal.
- Individuals and institutions are increasingly interested in sustainable technologies, as the public's awareness of sustainability issues grows.

### 2.4.1   Sewer-less society

According to the *United States Environmental Protection Agency* (USEPA), and the United States Census Bureau, on-site systems are increasingly chosen over central sewer systems by property owners and municipalities because they cost less than a central sewer system (USEPA, 1997).

Public health specialists at development agencies worldwide are promoting effective and ecological on-site waste treatment systems that save water and help prevent the spread of fecal-oral disease.

At the same time, the acceptance of WTS as a technology has grown tremendously. They are far more efficient, refined and proven. Every year, more states change laws and regulations to permit them. Even researchers at Harvard University have decided that this is the technology of the future, and have developed a high-tech prototype *smart* WTS with solid-state sensors and microchips that control the process.

Thanks to these developments, WTS – long considered appropriate only for remote applications – may soon be widely viewed as a conventional wastewater treatment technology with obvious advantages for the present and the future (Del Porto & Steinfeld, 1999).

### 2.5   Innovative financing and management

The United States Environmental Protection Agency and regulators worldwide are recommending the formation of on-site management districts in response to poorly maintained or inadequate conventional on-site systems. These would involve a central organization that manages a district's on-site systems, so no matter what system a property owner used, an agency would be accountable for its performance.

The formation of these districts would allow on-site systems to receive the federal funds for design, construction and maintenance that were once provided only for central wastewater treatment plants.

Recently New York City instituted a financial incentive for individual property owners to recycle wastewater and stormwater (New York City Department of Environmental Protection, Web Page). The program, called the Comprehensive Water Reuse Program (CWRP), provides a 25% reduction in water rates and a reduced sewer bill to property owners who install wastewater-recycling equipment (for onsite non-potable reuse) and rainwater collection equipment as well as take other water-saving measures.

## 2.6 *Comparing two water management scenarios (Steinfeld, 2004)*

### 2.6.1 *A hypothetical big pipe water recycling scenario*
An urban neighborhood in New York City comprises industrial, residential, and commercial buildings. The water supply transferred from a far distant watershed basin, is now completely recycled water from one of the city's treatment plants. Miles of pipes and pumping stations carry away the wastewater to a treatment facility converted to a full recycle facility three (and far more) miles away (1 mile = 1609 m). Stormwater is piped to rivers and out to sea. To supply recycled wastewater back to this block, pipes and pumping stations have to be dug and installed, with great disruption to the city, to pipe water via the existing supply lines or new ones. Cost of the ultra-filtration and distribution is considerable.

Compare and contract the former to the following:

### 2.6.2 *Hypothetical integrated source – and on site – water scenario*
The same urban neighborhood is offered incentives to conserve and recycle water. Several buildings are retrofitted with low-flush dual-flush urine-diverting toilets flushed with treated graywater collected from sinks and showers. The flushed urine – which accounts for most of the nitrogen in wastewater – is used to fertilize and irrigate surrounding landscaping, green strips, parks, and green corridors via a hidden subsurface piping system. Excess nutrient-rich urine is piped to greenhouses on industrial roof tops and brownfields to grow food or oils used as a substitute for imported petroleum. Stormwater is collected from roofs, streets and parking lots and treated and disinfected to be used for process water for flushing toilets and urinals, evaporative air conditioning, boiler feed stocks, etc.

Graywater and combined wastewater is treated on site with a polyculture of treatment systems, including planted systems such as *Solar Aquatics/Living Machines* and constructed ecosystems (a category that includes planted evapotranspiration systems and constructed wetlands) that double as landscaping, recreation area, gardens, bird habitat, orchards, and public art. These systems create an incentive for planners to create green spaces, such as terraces and public gardens. The beauty of these ecological designs increase land values and foot traffic in the neighborhoods.

Toilets and kitchens effluents drain to treatment tanks for these appliances only. Septage is taken to a septage composting facility. Recipients of this composted septage (sludge) are relieved to know that no industrial chemicals, toxics or heavy metals are in this material, unlike today's composted sewer sludge. Or, dry or micro-flush toilets drain to aerobic composters which are periodically removed to a central processing facility.

Industrial flows with toxics and heavy metals are treated separately by systems designed for their specific constituents (because the solution to pollution is not dilution). Cooling water and condensate is recovered and recycled.

To minimize combined sewer overflows, the rooftops are retrofitted with green eco-roofs to absorb stormwater and slow-release it for evaporation. Some of this water can be drained to cisterns. This also reduces heat islands in the city, insulates buildings, and extends the life of roofing. Raingardens, pervious paving, and other *lower-impact development* (LID) techniques are used to infiltrate stormwater, reducing flows to the wastewater treatment plant (see Low Impact Development Center, Web Page). Brownfields are used to infiltrate effluents and stormwater. Unused buildings may house cisterns and treatment works. Marginal buildings may be dismantled to make way for collection, treatment, and planted treatment systems.

Some uses are seasonal. Rainwater may be used during wet months and treated graywater used during dry months.

A city management team periodically checks on the systems. This staff formerly worked in the central plant or is funded with avoided central treatment costs.

The result: much-reduced flow to the wastewater treatment plant and reduced demand for water. Also, this scenario embeds incentives for building users to reduce toxics and perhaps water usage. This also makes for a more secure water system, as water supply and treatment is no longer entirely centralized, so that disease outbreak, terrorist action, or treatment plant failure do not affect users as much.

A decentralized approach may also offer a solution to one of the hurdles to implementing full water recycling:

– Security against terrorist attacks on central systems.
– Liability is dispersed, so that any disease outbreak can be more isolated.
– Users of these water sources, now more aware of and invested in their sources, are far more confident in using recycled water and more likely to safeguard it.

It is clear that combining excreta, toxics, heavy metals, and many other constituents in a water-carriage wastewater infrastructure and then attempting remove it all with advanced filtration may be unfeasible for many developing countries and perhaps the first world too. The advantages of an integrated site- and source-based approach are too great to ignore.

### 2.6.3   *Costs for an integrated urban water management plan*
A typical Los Angeles single-family household pays only US$ 63 per month for combined water, stormwater and wastewater services. It is estimated to cost an additional US$ 45 per month to implement an integrated water management plan (City of Los Angeles, Web Page).

## 3   CONCLUSION

Many of the water crises in urban areas are based not so much on the availability of water but rather on our failure to manage how water is used. Specifically, the failure to recognize stormwater as a resource and the practice of using water to transport unwanted wastes.

Wastewater recycling will be a necessary strategy in the years ahead. However, it is only one component that should be part of an integrated strategy that makes best use of local sites, as well as local sources of water and local effluents that can be used.

Continuing to combine wastes – including excreta, toxics, and heavy metals – in wastewater flows, then using advanced ultra-filtration at the end of the pipe to treat them will not be feasible for many cities, especially those in developing countries.

To identify wastewater recycling as the best solution to the world's water supply challenges is to use the same mentality that created today's many water problems.

The common urban approach of centrally collecting combined effluents from a wide variety of sources, then treating this soup with end-of-pipe solutions, has evolved in an effort to avoid complexity. Yet nature's model shows us that complexity is the best way to manage resources.

## REFERENCES AND WEB PAGES

Brooks, D.B. (2003). *Another path not taken: a methodological exploration of water soft paths for Canada and elsewhere*. Report to Environment Canada. Friends of the Earth Canada, Ottawa, Canada.

California Department of Health Services (Web Page). *Drinking Water Action Levels*. Drinking Water Program, Division of Drinking Water and Environmental Management. [http://www.dhs.ca.gov/ps/ddwem/chemicals/AL/actionlevels.htm]

City of Los Angeles (Web Page). Integrated Resources Program. *Information sheet of the Integrated Resources Plan (IRP)*. Bureau of Sanitation Department of Public Works Program. [http://www.ci.la.ca.us/SAN/irp/documents/factsheet.pdf]

Del Porto, D. & Steinfeld, C. (1999). *The Composting Toilet System Book*. The Center for Ecological Pollution Prevention, Concord, Massachusetts, USA: 235 pp.

Del Porto, D. & The Ecological Engineering Group (2001). *Consulting report for Ford Motor Company*. Concord, Massachusetts, USA.

Detroit News Online (Web Page). Detroit Water Department in midst of $4.3 billion construction effort. *2 May 2001*. [http://detnews.com/2001/detroit/0105/04/s06-219086.htm]

Environment Canada (Web Page). *Industrial use*. The management of water, Freshwater Website. [http://www.ec.gc.ca/water/en/manage/use/e_manuf.htm]

GTZ (*Deutsche Gesellschaft für Technische Zusammenarbeit*) (Web Page). *Ecosan: ecological sanitation*. [http://www2.gtz.de/ecosan/english/subject.htm]

Low Impact Development Center (Web Page). Beltsville, Maryland, USA. [http://www.lid-stormwater.net]

New York City Department of Environmental Protection (Web Page). City Introduces Innovative New Comprehensive Water Re-Use Program. *DEP News*, Press Release, *2 April 2004*. [http://www.nyc.gov/html/dep/html/press/04-16pr.html]

Steinfeld, C. (2004). *Scenarios for Urban Integrated Water Planning*.

United Nations (2004). *UN Report says world urban population of 3 billion today expected to reach 5 billion by 2030*. Press Release, POP/899. 24 March. Department of Economic and Social Affairs (DESA), United Nations Population Division, New York, USA. [http://www.un.org/esa/population/publications/wup2003/pop899_English.doc]

USEPA (United States Environmental Protection Agency) (1997). *Response to Congress on Use of Decentralized Wastewater Treatment Systems*. USEPA Office of Wastewater Management, Washington, D.C., USA. Available at: [http://www.epa.gov/owmitnet/mab/smcomm/scpub.htm]

Vickers, A. (2001). *Handbook of Water Use and Conservation*. Waterplow Press, Amherst, Massachusetts, USA: 464 pp.

WateReuse Association (Web Page). WateReuse Association testifies before Subcommittee on Water and Power. *15 April 2003*. [http://www.watereuse.org/Pages/pr041503.htm]

CHAPTER 18

# The potential for desalination technologies in meeting the water crisis

J. Uche, A. Valero & L. Serra
*CIRCE Foundation, University of Zaragoza, Spain*

ABSTRACT:  This chapter presents the potential paper of desalination as a new and non-finite source of fresh water that plays an important role in solving fresh water scarcity in some oil-rich regions and the Middle East, and in many arid and semiarid countries. The most widespread desalination technologies, the weight of desalination in the different regions of the world and the future trends, the economic costs of desalted water and the uses derived from those costs, and the environmental charges of desalination and their corrective measures to be undertaken have been reviewed carefully in the paper. Finally, the future improvements expected for desalination technology, and their economic and environmental consequences are also briefly commented in the chapter.

Keywords:  *Desalination, water resources, seawater, brackish water, brine*

## 1  INTRODUCTION

Water scarcity will soon be a serious problem, especially considering the rapidly world population growth and raising water consumption per capita. So, it is becoming an increasing problem in many regions of the planet. For any society, $1000\,m^3/yr$ per person is considered the *standard* benchmark level below which chronic water scarcity is considered to impede development and harm human health. However, we are using now about $4000\,km^3/yr$ (World Water Council, 2000). The picture of the present situation is really dark, as figures show that 30% of the world population is suffering water stress, 3000 million people lack an appropriate water sanitation and 12% of the world population uses about 85% of the water supplied (Freshwater Action Network, 2002), while the world's renewable fresh water resources than can be technically used are about $10,000\,km^3/yr$ (Shiklomanov, 1998). All these data clearly indicate that fresh water demand will increase in the next years.

Desalination, which is of continuous increasing importance, is a possible and very interesting alternative for increasing fresh water resources, and it is already a means of augmenting them in many parts of the world, especially in the Middle East but also for the European Union and particularly in the Mediterranean countries and island territories. By the beginning of the year 2002, a worldwide total of 15,233 desalinating units with a total capacity of $32.4\,Mm^3/d$ of fresh water have been installed or were under construction (Wangnick, 2002a). That represents more than 30-fold increase in global capacity over three decades. Yet desalinated water is only about 0.2% of the fresh water used worldwide.

As Leon Awerbuch (2002) comments, "Desalination is the only realistic hope to create new water resources in the midst of the doom and gloom of water crisis and water pollution. The

19th century was the century of *gold*. The 20th century was driven by *oil*. In the 21st century, *water* will be the most important resource. Desalination will deliver the promise not only to create new water but also to produce fresh water at dramatically reduced cost".

Nevertheless, desalination is an energy intensive process. Producing the total amount of desalted water in all of the facilities in the world requires a huge amount of energy, which is approximately equivalent to the 0.3% of primary energy in terms of fossil fuels consumed all over the world (Uche *et al.*, 2003). To underline how important energy is in desalination, if all the water consumed in the world came from desalination plants, the required oil would surpass the current yearly oil consumption. The world's overall energy demand, 80% proceeding from fossil fuels, is continuously increasing and it will be maintained in future. Environmental problems associated to that energy consumption should also take into account in the near future. Thus, decreasing the production of whatever good or commodities, e.g. fresh water, also reduces the energy consumption. The present situation demands for imaginative solutions, for instance process integration or the use of renewable energy sources to produce desalted water, in order to prevent from more profound economic and social effects. Unfortunately, the development of renewable-driven desalination is still severely impeded (if not stopped) by the pressure from contemporary economic factors and political inertia. If our technology continues along the present unsustainable path, not only it is essential to have an orderly transition in the energy used for desalination (from fossil fuels to renewable resources) but the whole industry needs to gear itself towards enhanced efficiency, waste minimization and less environmental impact.

This lecture reviews the state of the art of desalination technology, including the most widespread technologies, the countries in which desalination is really an important portion of their water resources, the economic trend of desalted water produced and that will be produced in the next future, and obviously the environmental loads of producing desalted water by different methods. Finally, the lecture will introduce the improvements that allow desalination to be introduced in areas where nowadays it is not considered.

## 2    DESALINATION TECHNOLOGIES

All major desalination technologies have demonstrated their ability and reliability to produce fresh water in an economical manner. Each technology has found their supporters and users, stimulating competition between them and provoking the continuous improvement of all of those technologies.

### 2.1    *Multistage flash (MSF) distillation* (Figure 1)

Multistage flash distillation is the most widely form to produce water from seawater, especially common wherever the temperature, salt content, biological activity or pollution level of seawater is high, as in the Middle East. This process has been in large-scale commercial use for over 30 years, coupled to power stations. In general, MSF plants are more common because they are simple and robust, although their specific consumption may be higher than other technologies. Other advantage of MSF plants are their unit size, considered of large scale (more than 50,000 m$^3$/d of capacity per unit).

The Multistage Flash process is described as follows: seawater pumped through heat exchanger tubes installed in the various evaporator stages, is heated to a certain temperature.

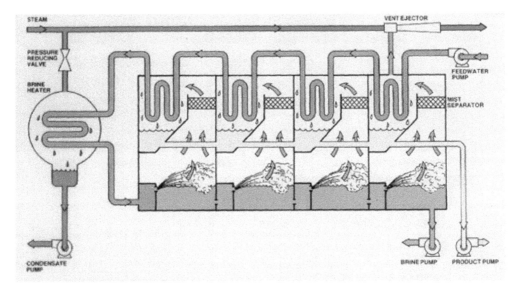

Figure 1.   Sketch of a typical MSF unit.

Final heating is performed by steam (coming from a power station) in a brine heater. The hot seawater then goes into flash chambers where the pressure is maintained below the equilibrium pressure corresponding to the temperature at which the brine enters. Part of the brine flashes into vapor and after passing a demister, it condenses outside the tubes while heating the seawater flowing through the tubes. The multistage flash distillation unit contains cells assembled in series, at a different pressure. The water produced in each stage is collected in a trough mounted below the tube bundle which collects the fresh water end product. These widely used units perform recycle brine in order to reduce the quantity of the make-up seawater needed to produce fresh water. The concentrated seawater is also removed from the last stage by a pump or by gravity.

## 2.2   *Multi-effect distillation (MED)* (Figure 2)

Contrary to MSF, in Multi-effect distillation evaporation takes place on surfaces, by exchanging the latent heat through the heat transfer surface between condensing vapor on one side and evaporating brine on the other.

   The MED plant also has several stages, each with a heat exchanger tube bundle. Seawater is sprayed onto the tubes and the condensing heating steam inside the tubes evaporates part of the seawater on the outside. The steam produced is used as heating steam in the next stage, where it condenses inside the tubes. The condensate is the water product. Obviously, the boiling temperatures and pressures in the different evaporators cannot be the same. The first stage is heated by external steam from a heat recovery steam or a back-pressure steam turbine, but in most cases, MED plants are equipped with thermal vapor compressors for better efficiency. The steam produced in the last stage is condensed on the outside of exchanger tubes in a separate condenser, which is cooled by incoming seawater. Part of the heated seawater is then used as feedwater. Product water and concentrated seawater are then pumped out from the last stage of the evaporator.

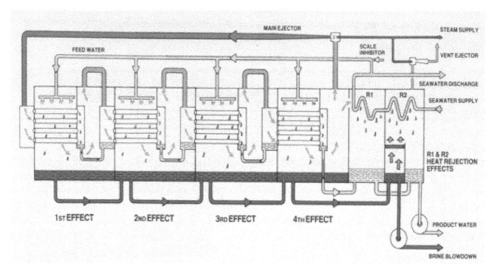

Figure 2.   Scheme of a MED unit with thermocompressor (TVC).

The major advantage of MED with respect to MSF is the ability to produce more water per steam consumed (Performance Ratio). MED plants could reach a value of 15 while MSF almost 10, and includes lower specific power consumption ($<1.5\,kWh/m^3$) than MSF ($>3\,kWh/m^3$). Furthermore, their efficiency does not depend on the steam temperature coming from the turbines as MSF plants, so MED plants are the most promising distillation technologies for desalination. However, the unit size of MED plants is up to the third with respect to MSF units at present.

## 2.3   *Vapor compressor (VC) distillation* (Figure 3)

Vapor compression distillation is similar to multi-effect distillation, but the main difference is that the vapor produced by the evaporation of the brine inside is not condensed in a separate condenser. In this case, that vapor enters in a centrifugal, single-stage type designed for high-volumetric flows, and this high-energy compressed steam is discharged into the evaporator onto the outside of the enhanced surface tubes, where it condenses and provide its latent heat energy to the boiler seawater inside the tubes.

Note that the process is very efficient thermodynamically, because most of the shaft work required by the compressor is used to avoid the *boiling point elevation* of seawater. In comparison with thermal desalination plants, no cooling water is required resulting in smaller intake and pumping systems and lower energy requirements (up to $9\,kWh/m^3$ of product). Unfortunately, the maximum size of VC plants is only about $3000\,m^3/d$ but they could grow in capacity and number of effects in order to improve its efficiency and reduce its cost.

## 2.4   *Reverse osmosis (RO)* (Figure 4)

Seawater reverse osmosis plants allow to demineralize water in a reliable manner. In RO desalination, seawater (or brackish water in case of inland territories that suffer from saline aquifers) is pretreated to avoid membrane fouling. It then passes through filter cartridges (a safety

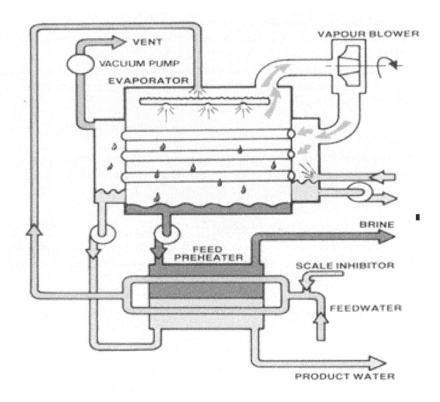

Figure 3.   Typical one-stage vapor compressor (VC) distillator.

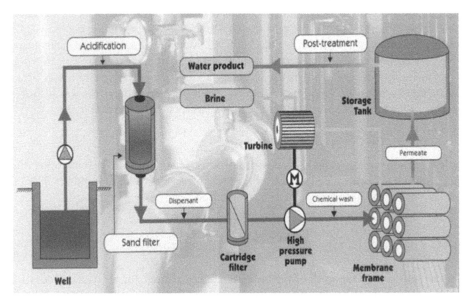

Figure 4.   Diagram for a seawater reverse osmosis (RO) unit.

device) and is sent by a high-pressure pump through the membrane modules (permeators). Because of the high-pressure, pure water permeates through the membranes and the seawater is concentrated. The water product flows directly from the permeators into a storage tank, and the concentrated seawater (at high pressure) is sent via an energy recovery system back into the sea.

Seawater RO process uses only electrical energy, and the higher consumption occurs in the high-pressure pump. Nowadays, energy recovery systems, as those:

- Pelton or Francis turbines, which takes profit of the pressurized brine as water stored in reservoirs.
- Inverse pumps (or turbochargers), that allow to impulse a great part of the energy required for the high-pressure pump, therefore they are mounted in a unique shaft.
- Pressure exchangers, translating the brine pressure into the feed water, by means of ceramic rotors or a set of valves and closed cylinders.

Could recover more than the 95% of the energy stored in the reject brine, so the total energy consumption for seawater could be from 3 to 5 kWh/m$^3$, depending on the feed water salinity and temperature. The modularity of RO systems also provokes that those systems have been rapidly installed in all over the world.

Seawater RO is a membrane technology. Other membrane processes as nanofiltration (NF), ultrafiltration (UF) and microfiltration (MF) are rapidly growing to be used in brackish desalination or waste water reuse, with very low energy consumptions (less than 1 kWh/m$^3$).

### 2.5   *Electrodyalisis (ED)* (Figure 5)

This process is used to demineralize brackish water by making different ions migrate through selective membranes in electric field made by the direct difference of voltage potential between two electrodes connected at the boundaries of the membranes. Whenever salt water is flowing in a cell, the cations are attracted by the anode and the anions by the cathode. If not constrained, these ions discharge on the electrodes of opposite sign. In return, if a set of selective and permeable membranes is placed between the electrodes, salt concentration decreases in some compartments of the cell where salt water becomes even more concentrated. This process is suitable for desalinating brackish waters with an average salt content between 1 to 3 g/L (for other salinities the process is not profitable) with a very low power consumption (less than 1 kWh/m$^3$) and a salt rejection of more than 80%. For seawaters, the process is also feasible but with very high energy costs, so it is generally discarded for those proposes.

### 2.6   *A comparison between desalination processes*

Each desalination process is highly recommended in one area, but under some other circumstances should also be convenient to replace the previous process with some others. The Table 1 resumes the main characteristics of the referred processes.

## 3   DESALINATION IN THE WORLD AND FUTURE TRENDS

The information about the state of the art of the world market for all types of desalination plants can be found in the *IDA Worldwide Desalting Plants Inventory*, compiled by Klaus Wangnick (2002a). The last inventory report (no. 17) shows that by the beginning of the year

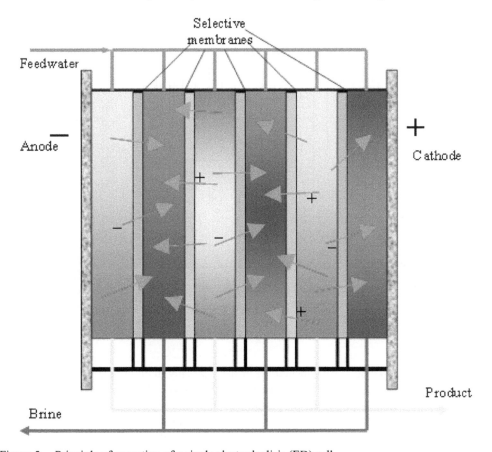

Figure 5.   Principle of operation of a single electrodyalisis (ED) cell.

Table 1.   A comparison between the most widespread desalination processes.

|  | MSF | MED | VC | RO | ED |
|---|---|---|---|---|---|
| Energy required | thermal | thermal | mechanical | mechanical | electrical |
| Operation temp. (°C) | 110 | 70 | 70 | 45 | 45 |
| Power cons. (kWh/m$^3$) | 3–5 | 1–2 | 8–12 | 3–6 | 0.8–1.5 |
| Raw water quality (ppm) | >50,000 | >50,000 | >50,000 | <50,000 | <3000 |
| Product quality (ppm) | <50 | <50 | <50 | <500 | <500 |
| Unit capacity (m$^3$/d) | 50,000 | 20,000 | 5000 | 200,000 | 10,000 |
| Plant reliability | high | medium | low | high | high |
| Increase capacity | difficult | difficult | difficult | easy | easy |
| Surface required | high | medium | low | low | low |

2002, a worldwide total of 15,233 desalinating units with a total capacity of 32.4 Mm$^3$/d – or 7127 MIGD (million imperial gallons per day) – have been installed or are under construction. Seawater plants (SWDP) produced 19.1 Mm$^3$/d of that capacity and 13.3 Mm$^3$/d are other types of water. The total installed capacity by the end of the year 2001 was 31.4 Mm$^3$/d (Wangnick,

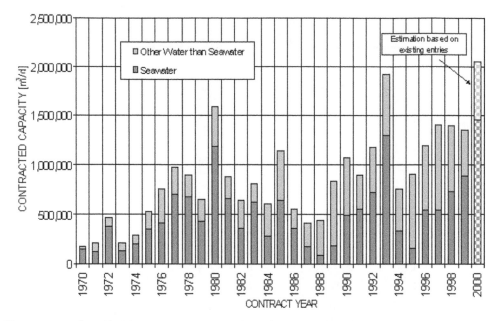

Figure 6.    Total worldwide cumulative contracted desalination capacity (all waters) (Wangnick, 2001).

2000) in over 14,060 units if we consider the projects already in bidding and planning stages. The evolution of the annually contracted capacity can be shown in Figure 6.

According to that inventory, the proportions of the various processes have been changed along the time. Now, 60% were membrane plants and the rest were thermal plants, but ten years ago that proportion was 50% to 50%. If only seawater desalting plants are considered, the figures were 60% for thermal plants and 40% for membrane plants, but the ratio was 80% for thermal plants and 20% for membrane plants ten years ago. This is due to the introduction of RO membranes valid for the Gulf seawaters. In terms of desalination processes, 42.87% of the total installed or contracted capacity is based on the MSF process, reflecting a continuing decline. On the other hand, RO process increased its share to almost 39%. For plants of more than 4000 m$^3$/d the MSF technology raises that proportion to 57.5% and RO decreases to 27.6%. Figure 7 shows the proportion of different processes in continents.

### 3.1    Middle East

In the Middle East, Saudi Arabia, the United Arab Emirates and Kuwait have a dominating role in constructing seawater desalination plants. Particularly, the *Gulf Cooperation Council* States (GCC) are the biggest users of desalination technology with more than 50% of the total capacity in the world. Thermal processes are preferred here whereas reverse osmosis dominates elsewhere in the world. That use of thermal plants is based on the existing good experience in the region with such processes, and the difficulty of the Gulf seawaters (high temperature and salinity and high amount of biological matter) for using the RO process, as well as the low cost of energy.

The Kingdom of Saudi Arabia, supported by several seawater multistage flash plants (MSF) desalination units combined with power generation in dual purpose power and desalination

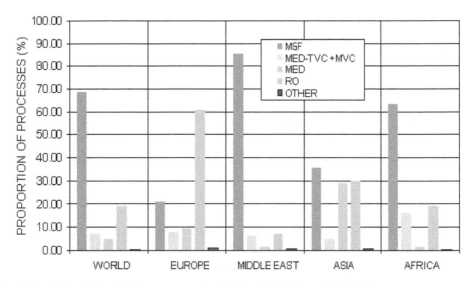

Figure 7. Quote of desalination processes in each continent (Wangnick, 2001).

plants, had in 2001 the 17.3% of the total installed capacity in the world. The largest plant in the world is Al Jubail Phase II complex, with a production of 910,000 m$^3$/d (200 MIGD in 40 MSF units) of distilled water and 1300 MW of electric power operating since 1982.

The third producer of desalted seawater in the world are the United Arab Emirates (UAE), with a rank of 16.3% of the total installed capacity. This figure indicates that the desalted seawater per capita is almost 2 m$^3$/d in that country. The Al-Taweelah complex contains the largest MSF units constructed up to now, with a capacity of 57,600 m$^3$/d (12.6 MIGD).

Kuwait is one of the GCC countries that only incorporates MSF distillers to provide fresh water. The number of MSF units installed in 2003 was 88, with a total capacity of 1.43 Mm$^3$/d (315.6 MIGD) (Al-Fraij *et al.*, 2003) representing almost the 5% of the total worldwide installed capacity at that year.

Some other Middle East countries as Qatar, Bahrain (with a total capacity of 335,000 m$^3$/d; Khalaf, 2003) and Oman complete the set of countries that depend fundamentally of desalted seawater resources. The majority of installations are composed by dual-purpose power and desalination plants that allow the combined production of water and energy in the same plant. In those cases, some vapor coming from steam turbines or heat recovery steam generators (in case of gas turbines or combined cycles) is used to feed the MSF or MED units coupled with the power plants. Therefore, they are considered as huge cogeneration plants providing power and water to the end consumers. Table 2 resumes the contribution of desalination for the most important GCC countries.

The future trend of desalination industry is marked with the privatization of the water and energy suppliers, taking into account the continuous growing demands of water and energy and the constraints that suffer the public financial resources. One of the most interesting innovations which are being included in that area is the concept of hybrid plants (power station + MSF/MED + RO/VCD), which combines the advantage of distillation and membrane processes. This combination is unique in those countries because the power demand decreases up to 40% in winter, and that power surplus could be consumed in RO or VCD units

Table 2.   Water resources in the GCC countries (Al-Weshah, 2002).

| | Conventional resources (Mm$^3$) | | Non-conventional resources (Mm$^3$) | | |
|---|---|---|---|---|---|
| Country | Surface water | Groundwater | Desalination | Agricultural drainage reuse | Total |
| UAE | 185 | 130 | 405 | 108 | 828 |
| Bahrain | 0.2 | 100 | 75 | 17.5 | 192.7 |
| Saudi Arabia | 2230 | 3850 | 795 | 131 | 7006 |
| Oman | 918 | 550 | 47.3 | 21.5 | 1536.8 |
| Qatar | 1.4 | 50 | 131 | 33 | 215.4 |
| Kuwait | 0.1 | 160 | 338 | 30 | 578.1 |
| Yemen | 3500 | 1400 | 9 | 52 | 4961 |
| Total | 6834.7 | 6240 | 1850.3 | 392 | 15,318 |

and that additional desalted water can be stored or consumed in the summer period. The use of softening membranes like nanofiltration (NF) or ultrafiltration (UF) is another technique to improve the productivity of existing MSF units avoiding the scale formation at higher operation temperatures (*Top Brine Temperatures*) than conventional plants.

### 3.2   *USA, Latin America and Caribbean*

IDA Inventory Report no. 16 (Wangnick, 2000) reports that USA has the second rank in the world capacity, with 16.7% of the overall score and more than 3000 plants. However, seawater desalination is not representative in that country whereas desalination of brackish aquifers or degraded aquifers by seawater intrusion (ED, RO), as well as membrane softening techniques (membrane softening MS, NF or UF) which remove bacteria, viruses and ions are also spread in the most arid and southern coastal states, that will account for more than 45% of the nation's total population growth between now and 2025.

The future of desalination in the USA is very limited for evaporation techniques due to its lack of competence with respect to membrane technologies (Birkett, 2002). Moreover, the future of the ED/EDR is similar to the previous processes taking into account the economy of the RO/NF/UF and MF and its advantage in stopping pathogens that ED techniques are not capable to do. RO techniques will be widely used in big plants (more than 10,000 m$^3$/d) to reduce salinity of brackish waters or reuse waste water directly. Seawater RO plants are planned only for big plants in Texas, California and Florida. Finally, the potential of the MF/UF/NF is huge providing high-quality water coming from degraded surface or well waters, considering very big plants of more than 100,000 m$^3$/d.

Regarding the region of Latin America and Caribbean (Andrews & Verbeek, 2002), only in the Caribbean desalination is really important to supply water driven by the growing tourism and decreasing costs. The total installed capacity in the Caribbean is 724,000 m$^3$/d by the year 2000. The most important facilities are settled at the Netherlands Antilles with a capacity of 290,000 m$^3$/d (it means almost 1350 L/d per person), Virgin Islands (118,000 m$^3$/d supposing 1350 L/d per person) and Trinidad, with more than 110,000 m$^3$/d. In consonance with the Canary Islands, the Caribbean has been the cradle to prove the technological developments of desalination along the time: the first seawater RO market, the first low-temperature MED units, the implementation of the BOOT (Build, Own, Operate and Transferred) contracts.

The future in this area is the continuous growth of desalination, the majority of them based on BOOT contracts till reaching self-sufficient islands, favored by the decreasing trends of seawater desalinated costs. A proof is the increase of the total capacity in 57% during the period 1996–2000. That expansion provided municipal drinking water (most of them ground waters affected by saline intrusion) as well as industrial process water in some cases (Trinidad island).

On the other hand, desalination in Latin America does not represent an equal weight that in the Caribbean (530,000 m$^3$/d), although the capacity increase in the recent years is also significant. The two main contributions come from Chile (131,000 m$^3$/d, mainly for industrial purposes) and Mexico (a capacity of 285,000 m$^3$/d for industrial uses and tourism resorts). The future of desalination in that area is focused on the use of membrane systems for removing contamination (one of the main problems of Latin America is providing a safe supply of water), more than reduce the salt content. Seawater RO plants will growth but not significantly for supplying industries and covering the tourist demands in the new complex appearing in that semi-continent.

### 3.3   *East and Southeast Asia*

The majority of desalination plants in China have industrial or power generation uses (Goto, 2002) consuming brackish, river, pure or waste waters in RO processes, and only two municipalities and two tourism facilities exist. The total capacity at the end of 1999 was 375,000 m$^3$/d, but only a continuous growth is foreseen in the industrial development of the country.

Japan leads the installed capacity in the region with more than 1 Mm$^3$/d and more than 1200 units. Only two big plants (40,000 m$^3$/d and 50,000 m$^3$/d) are used for domestic use, but several small plants are sometimes used in remote areas such as small islands. On the other hand, electronics industry consumes ultra pure water and softening techniques are applied from pure or river waters in a great scale.

Korea had 263 desalination units in 1999 and a capacity of 448,000 m$^3$/d without domestic use and RO processes consuming brackish or river waters. The industry of Taiwan required, in 1999, 104 desalination plants with a total capacity of 164,000 m$^3$/d, but 4 new desalination plants are projected for industrial parks with a total capacity of 138,000 m$^3$/d (Hsu, 2001).

In the Southeast, only Indonesia has a great capacity (179,000 m$^3$/d at the end of 1999) which is used for industrial and power applications with only one municipality.

Summarizing, the future of desalination industry in that area is focused on the industry requiring ultra pure waters, although demand for desalination will occur in some areas of highly dense population and high income. A great expansion of seawater desalination is not foreseen for the next future, especially for China despite of its economic growth: huge inter-basins transfers are projected to provide water to the northern and highly populated areas.

The case of India differs from the other Asian countries in the sense that there are more than 200,000 villages with inadequate drinking water, out of which about 50,000 suffer from brackishness problems affecting a population of about 80 million. Approximately one third of these villages are acutely affected by salinity levels above 4000 ppm. Provision of safe water to the villages has been given high priority in recent years, with hundreds of small RO and ED plants (10–30 m$^3$/d) installed in the affected villages.

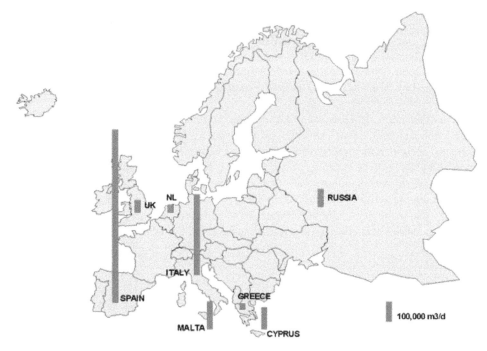

Figure 8.    Role of desalination in Europe (Wangnick, 2001).

### 3.4    *Europe and Mediterranean area*

As it is expected, desalination in Europe (Figure 8) is focused in the Mediterranean countries, where the climatologic conditions and the economic activities (tourism and agriculture)
demand high water volumes. Spain is the dominating country in the scene of desalination,
with almost a capacity of 1.2 Mm³/d at the end of 2001 (Wangnick, 2002b). It is a very
mature technology that represents the 8.3% of the urban consumption, mainly in the Canary
Islands. One thing that does not occur in other places is the use of desalination for agricultural
purposes; in some cases even seawater desalination plants have been constructed to irrigate
high-profitable crops in the Southeast of Spain and the Canary Islands. The RO technology is
dominant because the two old MSF existing units have been replaced, and several innovations
have been proved, especially in the Canary Islands, where the eastern islands (Lanzarote,
Fuerteventura, Gran Canaria) depend strongly on the desalted seawater. The future in Spain
is really promising, for instance the *Spanish National Hydrological Plan* approved by law in
2001 includes a list of desalination plants in the Levante that adds up to 336 Mm³/yr (MIMAM,
2003), and some other new plants in Tenerife and the Balearic Islands. And more recently, the
new *Spanish National Plan*, approved in June 2004, includes new 621 Mm³/yr with 20 new
desalination facilities located in the East and Southeast of the Spanish Peninsula.

By far, the second on that list is Italy, with a total capacity of 400,000 m³/d, most of their
plants located at the south of the country and the island territories. The third and fourth places
are occupied by two Mediterranean islands: Malta and Cyprus. In this case desalted water
is a big percentage of the total water available in those regions. Greece, Turkey, Jordan and
Lebanon are other countries in that areas that also have small RO desalination plants. However,

Table 3. Water supply and demand data for the North Africa states in 2000 (Abufayed *et al.*, 2002).

| Country | Water resources (Mm³/yr) | | | | Water demand (Mm³/yr) |
|---|---|---|---|---|---|
| | Surface | Ground | Desalination | Total | |
| Algeria | 13,500 | 3700 | 200 | 17,400 | 6100 |
| Egypt | 65,500 | 7400 | 100 | 73,000 | 70,500 |
| Libya | 300 | 3400 | 800 | 4500 | 5300 |
| Morocco | 19,500 | 5000 | 100 | 24,600 | 13,100 |
| Tunisia | 2700 | 1800 | 300 | 4800 | 2500 |
| Total North Africa | 101,500 | 21,300 | 1500 | 124,300 | 97,500 |

Israel has planned a major program of seawater desalination (combined with other elements) for the years 2002–2010 to provide new 400 Mm³/yr of fresh water from series of 6 large plants (from a range in size 45–100 Mm³/yr) along the coast of the Mediterranean Sea, to partly balance the water scarcity problems in the country (Shamir, 2003).

Russia only has a reduced capacity of their desalination plants with a total value of about 100,000 m³/d composed by several old MED plants located at the Caspian and Black Sea. In Central Europe, Germany, Austria and Netherlands have several plants to recycle wastewater or produce pure water for industrial processes including power generation. They do not produce drinking water.

Water resources in the North Africa region seem to be limited in time and space, unequally distributed, and remote with respect to centers, suffering from a continuous increase in demand. The geographical situation could be characterized as similar to the Middle East, but seawater water production there is almost negligible with respect to that area, mainly due to the limitation of their financial constraints.

The estimated cumulative installed desalination capacity reached 1 Mm³/d at the end of 1995 (Abufayed *et al.*, 2002). More than 625,000 m³/d are located in Lybia (mostly MSF plants for municipalities), and 192,000 m³/d in Algeria mainly for industrial purposes. Egypt incorporated 95,000 m³/d at the end of 1999, with small RO plants and MSF/VCD plants, and finally Morocco and Tunisia reported respectively 64,000 m³/d and 47,000 m³/d of installed capacity at the end of 1995. In this regard, five North African countries (Morocco, Algeria, Tunisia, Lybia and Egypt) requested in 1989 technical assistance from the *International Agency of Atomic Energy* (IAAE) to study the feasibility of desalination using nuclear power. The future of desalination in this area is also focused on the tourism locations (Tunisia and Egypt) that even include brackish water plants, and the increasing applications of the solar energy in remote and dessert areas for small locations. To sum up, desalination should be the alternative saving solution when the mobilization of non-conventional water resources is impossible or very costly (essentially in coastal zones). Table 3 presents relevant data for the North Africa region corresponding to the year 2000.

## 4 DESALINATION COSTS

The economics of desalination vary from country to country, depending upon the opportunity cost of energy and capital as the main factors of production, and the type of desalination

Table 4.    Investment of seawater desalination plants depending on the type of process and expected plant size (Uche *et al.*, 2002).

| Technology | Capacity range (m³/d) | Specific investment cost (€/m³/d) |
|---|---|---|
| MSF | 10,000–50,000 | 1680–1080 |
| MED-TVC | 5000–20,000 | 1080–800 |
| VC | 1000–5000 | 1500–1020 |
| RO | 10,000–100,000 | 900–550 |

process. In case of dual-purpose plants, in which water and power are two separate products, there are several methodologies that allow quantifying the costs of water and power. Moreover, the progressive privatization of the public sectors in the *Gulf Cooperation* countries (GCC) had provoked the injection of multi-billion investments coming from the private sector, allowing under different contacts the operation and control of new big power and desalination plants or rehabilitated existing ones. In this way, the new independent water and power producers permits the fulfillment of the continuous growing demand for water in countries with financial constraints.

### 4.1    *Investment costs*

Breaking the desalination costs of water, the total costs can be divided in capital costs (up to 40% in worst cases) for interest and depreciation due to the plant investment, and the rest are considered running (or operating) costs. The investment costs for seawater desalination plants have decreased over the years thanks to the still ongoing development of processes but also components and materials, apart from the increasing competence between desalination companies. Specific costs close to 1000 €/m³/d could be obtained for large plants in the Middle East, whereas the same plant for RO process in some other places like the Mediterranean countries could be constructed below 600 €/m³/d, if no additional requirements are demanded. Table 4 shows the expected range for different desalination processes and plant sizes.

The investment costs of the plant can be broken into the following sub-systems:

– The raw water supply system includes the seawater intake structure. It could suppose a high percentage if the specific seawater flow is high and the sea shore is very shallow (up to 30%), but they could be reduced if the water intake is shared with other plants, as for instance the cooling system of the required power plant or hybrid plant. In case of hybrid plants, the feed water for the RO plant could be the cooling water rejected from the power plant (it has been demonstrated that RO units consume less energy when seawater is preheated, or alternatively more permeate is produced consuming the same energy), and therefore the cost is reduced significantly.
– The pre-treatment system is very low for thermal plants and consists of simple equipment to storage and dose some additives (antiscale, antifoam and sodium bisulphite). However, in RO plants costs are considerably higher because the raw seawater must be carefully treated (chlorination, filtration, acidification, inhibition by polyphosphates and dechlorination) before entering the RO racks. Those costs could even increase if membrane techniques are included to reduce some pollutants in raw water. In that case it could reach up to the 20% of the investment costs.

- The desalination system, by far the most expensive system of the plant, differs considerably with the type of process and plant size. For thermal plants the elevated heat exchange area and fine materials required implies that those costs contribute to more than the 70% of the investment. The weight of the membrane costs and the high-pressure pump, as soon as the energy recovery system for RO plants, do not sum up the same contribution that the observed in thermal plants (a maximum of 60%).
- The brine disposal system includes not only the brine discharge for thermal plants, because it exists the cooling water outfall (with identical salt content that raw water) and therefore it is not representative for that plants. However, for RO plants, brine discharge is almost double concentrated than raw water; specific design is required. In case of endemic species nearby the cost, as for instance in the Spanish Levante (*Posidonia Oceanica*), the correction measurements needed to protect the local ecosystems could increase the cost of that system up to 5–7% of the total investment.
- The post-treatment system applied to the produced water is quite simple but it is obliged by the purity of desalted seawaters. Usually, product water is re-hardened by adding carbon dioxide and lime, and finally is disinfected with chlorine. In some cases, caustic soda is also added to increase pH, so compensating the effect of the acid previously dosed. In both cases (thermal or RO plants), this contribution is not significant to the total investment cost of a desalination plant.
- The storage system of the drinking water produced consists of cylindrical tanks with a capacity up to 24–48 hours of water production. Those tanks could contribute less than the 3% of the total investment required.
- The auxiliary equipment (instrumentation, electrical connections, air and oil systems) as soon as the civil works required for a desalination plant does not influence the final costs of the plants in both cases (thermal and RO plants). It could suppose up to the 10% of the total investment of the plant.

## 4.2  *Operating costs*

The long list of components that contains the operating (or running) costs are:

- Energy cost is usually the dominant component (more than 50% of the total operating costs), although its weight depends on the location of the plant. In oil-rich countries, energy prices are much lower than in other areas, but in countries where energy production is expensive, RO is always the best solution (as for example in Spain). Note that energy prices could be increased in the near future. One of the main reasons could be the penalties imposed to power producers in order to fulfill the *Kyoto Protocol* (limited $CO_2$ emissions per country with respect to its production in 1990).
- Personnel costs decrease as the plant size grows, except for very small plants in where no personnel is dedicated to maintain the plant operation or shared with other activities. Large plants are generally controlled by computer systems, and therefore people working in those plants should be highly qualified, in order to properly manage and protect those large investments.
- Chemical costs do not suppose an important item for thermal plants, since only antiscales and antifoams are added to raw water. However, for membrane plants that item increases because of its weakness to hard waters, despite of using MF or UF techniques could reduce that operation cost but increase the corresponding investment. In the same manner, the

costs of chemical cleanings are higher in case of membrane plants, although that measurement is only adopted in the worst cases, after clean washes with brine at counter-current flow.
- Membrane replacement in RO plants, which occurs once each 3–8 years for seawater plants, could suppose an important item if we take into account the percentage of the membranes costs in the investment required. The rest of consumables are almost economically negligible in membrane but also for thermal plants.
- Finally, the annual costs for spare parts can be calculated from 1% to 3% of the total investment, depending on the plant size.

### 4.3   *Total costs*

Summarizing, the total (investment + operating) costs for a typical thermal and membrane plant (RO) are depicted in next figures, remembering that costs could vary from country to country.

Some examples of the total costs obtained in different locations are summarized in the next list:

- The levelized unit cost of water in the Kingdom of Bahrain (a total desalination capacity of 335,000 m³/d) rises from 0.66 to 1.10 US$/m³ during the period 2003–2020, taking into account the investments up to now and an energy cost of 0.95 US$/GJ (Khalaf, 2003).
- The first *Independent Water and Power Project* (IWPP) in Abu Dhabi (Al Taweelah A2) will produce 227,000 m³/d (50 MIGD) in a MSF plant at a cost of 0.70 US$/m³ (Awerbuch, 2002).
- The Shuweihat IWPP plant under construction, based on the BOO (Build-Own-Operate) basis will provide 455,000 m³/d (100 MIGD) of desalted water at a quote of 0.69 US$/m³ (Awerbuch, 2002).
- The Askhelon RO plant, with a capacity of 100 Mm³/yr, is a BOT (Build-Operate-Transfer) contract for 25 years at the lowest costs ever offered with that type of contracts for desalted seawater: 0.527 US$/m³ (Lokiec & Kronenberg, 2003). Regarding the six plants included in the *National Plan* 2002–2010 in Israel, water costs at the plant will range between about 0.50 to 0.55 US$/m³ at the smaller ones (Shamir, 2003). Obviously, it is not included the cost of the conveyance system for getting water to the customers.
- Private companies have now bid less than 0.50 US$/m³ for supplying desalted water to Singapore (Biswas & Tortajada, 2003).

Table 5 resumes some of the recent BOOT (Build-Own-Operate-Transferred) projects.

A recent list of the last large seawater desalination plants (SWDP) under biddings projected in Spain is presented in Table 6.

It shows very clear that technological developments have consistently reduced the costs of desalted water: in the last 20 years the cost of desalination in Spain was around 2.00–2.10 €/m³, and now that cost is four times less.

If a new desalination plant is projected, a simple procedure for calculating the unit product cost for that plant is the *Desalination Economic Evaluation Program* (DEEP) (Ettouney *et al.*, 2002), once main design data of the plant are given. It is based on a hybrid *Microsoft Excel* spreadsheet and *Visual Basic* methodology, and is suitable for analyzing various desalination and energy source options.

Table 5.   Large seawater desalination plants (SWDP) projected or under construction with BOOT contracts (Morris, 2004).

| SWDP | Tampa Bay | Trinidad | Larnaca | Dhekelia | Singapore | Askhelon | Algeirs |
|---|---|---|---|---|---|---|---|
| Capacity (m³/d) | 95,000 | 135,000 | 40,000 | 40,000 | 136,000 | 274,000 | 200,000 |
| Feedwater salinity (ppm) | 26,000 | 38,000 | 40,000 | 40,000 | | 40,000 | 40,000 |
| Contract year | | | 2000 | 1996 | 2002 | 2002 | 2003 |
| Years of contract | 30 | 23 | 10 | 10 | 20 | 25 | 25 |
| 1st year price* | 0.46 | 0.71 | 0.73 | 1.09 | 0.45 | 0.52 | 0.818 |

*A normalized cost of energy (0.04 cents of US$/kWh) is used.

Table 6.   Recent seawater desalination plants (SWDP) in Spain. The reflected values are the selected offers (Uche *et al.*, 2002).

| SWDP | Almería | Cartagena | Carboneras |
|---|---|---|---|
| Capacity (m³/d) | 50,000 | 65,000 | 120,000 |
| Depreciation cost (€/m³) | 0.16 | 0.13 | 0.15 |
| Fixed cost (€/m³) | 0.10 | 0.09 | 0.04 |
| Energy (€/m³) | 0.19 | 0.18 | 0.18 |
| Membrane replacement (€/m³) | 0.03 | 0.02 | 0.02 |
| Maintenance (€/m³) | 0.02 | 0.01 | 0.00 |
| Chemical additives (€/m³) | 0.01 | 0.02 | 0.05 |
| Total cost (€/m³) | 0.51 | 0.45 | 0.44 |

## 4.4   Use of desalted water for agriculture

The basic use of desalination in the past and the near future is the production of drinking water for population. Of course, the use of softening techniques will be increased in order to be consumed in the different industrial processes. But the inclusion of desalination for agricultural purposes, which consumes the 70% of the available water nowadays, is a very promising option that has to be analyzed carefully. Up to now, it has been introduced in really water scarcity areas (Kingdom of Saudi Arabia for irrigating cereals in the inland territories of the Arabian Peninsula), but in this case the search for a policy of autonomy in agricultural resources has been prevailed more than the economic rationality of those projects.

On the other hand, Spain is the leader in using desalination for irrigation. The total volume destined to agriculture was estimated in 212.3 Mm³ (including brackish desalted waters) at the end of 2003 (AEDyR, 2002). The Canary Islands and the Southeast of the country have used since the 1980s desalted water to highly profitable crops produced in greenhouse farms (mainly vegetables and fruits). In this sense, a very interesting study (Albiac *et al.*, in press) has been made to evaluate the economic alternatives to the *Water Transfer* proposed in the *Spanish National Hydrological Plan* for the total volume destined to avoid aquifer overdraft (419 Mm³/yr) and guarantee the supply reliability (142 Mm³/yr). Water management supply and demand alternatives were examined, including banning aquifer overdraft, water pricing, introducing water markets, and augmenting water supply with water from the Ebro River or from seawater desalination. The results show that the best alternative for this paradigmatic example is a combination of measures including water control and aquifer overdraft, water trading between irrigation areas, and water supply expansion through seawater desalination in the coastal counties (from the Alicante to Almería provinces). For instance, the effective demand for desalted seawater (i.e. the water volume that produces positive net income to farmers) in those counties would reach to 387 Mm$^3$, considering a definite cost (e.g. it includes the cost of spreading water into the allotment) of desalted water of 0.52 €/m$^3$ provided by big constructed or projected plants in the area (Carboneras plant, with a design capacity of 42 Mm$^3$/yr that could be doubled). This means that some counties as Campo de Dalías and Campo de Níjar could absorb up to 60 Mm$^3$ (49 Mm$^3$ and 11 Mm$^3$ respectively). Note that the foreseeable costs of transferred water from the Ebro River could reach a range from 0.56 to 1.05 €/m$^3$ in those coastal counties (CIRCE Foundation, 2003), so it is clear that desalination is the more reasonable option in that coastal but with high-profitable crops. The economical model developed in that area also indicates that desalination is not profitable for inland territories and/or less-profitable crops like cereals or some fruit trees.

## 5    ENVIRONMENTAL ISSUES

### 5.1    *Main impacts*

The environmental charges associated to the desalination processes differ from each process and location. Two major classifications should be made, taking into account the geography and the dominant technology involved: the Middle East, in which distillation processes are mainly applied in relatively closed oceans or seas and dessert climate, and the rest of the world with RO as the dominant technology, and big oceans or seas. In both cases, two effects should mainly be studied: the energy consumption of the desalination plant and the brine discharges of plant, as soon as some other minor consequences derived from the chemical treatments and cleaning systems, and the noise provoked in the installations if they are close to population.

In the Middle East zone, and particularly the Arabian Gulf, the long list or large distillation plants (MSF) is provoking a serious environmental impact, in that area where the economic growth of new formed countries is degrading its sustainable development (Al-Gobaisi, 2003). Arabian Gulf has special characteristics that strengthen the problem:

– A reduced area of 100 km long and 300 km wide, and average water depth of 35 m, and an average residence time of 2–5 years.
– An arid sub-tropical climate, very limited annual rainfall and an evaporation/river runoff factor of 10 that explains the gradually increasing salinity (37,000–50,000 ppm).
– A region with political conflicts and the largest oil route in the world (20% of the total production passes through the Strait of Hormuz).

In general, distillation plants discharges brine with a concentration of about 60,000–65,000 ppm starting from raw seawaters of 40,000–45,000 ppm. Therefore salt dilution in the ocean is not very problematic in this area, taking into account the scarce flora of the Gulf soils. However, the continuous flow of saline waters in that closed sea would contribute to the gradually increasing salinity already mentioned. Furthermore, $CO_2$ and $NO_x$ emissions derived from the elevated consumption of those plants, which include thermal but also electrical consumption for pumping seawater, are very significant. Finally, thermal pollution is an

Table 7. Relevant airbone emissions produced by different desalination technologies (Raluy *et al.*, 2003).

| | MSF | MED | RO (4 kWh/m$^3$) | RO (3 kWh/m$^3$) | RO (2 kWh/m$^3$) |
|---|---|---|---|---|---|
| kg $CO_2$/m$^3$ | 23.41 | 18.05 | 2.26 | 1.73 | 1.20 |
| g dust/m$^3$ | 2.04 | 1.02 | 3.10 | 2.37 | 1.56 |
| g $NO_x$/m$^3$ | 28.30 | 21.43 | 5.10 | 3.92 | 2.74 |
| g NMVOC*/m$^3$ | 8.20 | 6.10 | 1.00 | 0.76 | 0.52 |
| g $SO_x$/m$^3$ | 28.1 | 26.31 | 14.7 | 11.86 | 9.08 |

*NMVOC: Non-methane volatile organic compounds.

additional impact associated to distillation plants, because brine is usually discharged from 5°C to 7°C above the raw seawater temperature. In that case marine fauna is also affected in a reduced sphere around the brine discharge pipe (or collector).

The situation in the other countries that incorporate desalination plants is really less dramatic than the Arabic framework. The energy consumption of RO plants is really lower than one consumed by distillation plants (in 5 or 6 times, see Table 7), and therefore the $CO_2$ and $NO_x$ emissions derived from that consumption do not represent a very important contribution. Table 7 presents the airborne emissions provoked by different desalination technologies, revealing the gap between thermal and membrane techniques.

On the other hand, the effects on marine flora provoked by the brine disposal could be important in some places. For instance, in the case of the Mediterranean Sea and especially in the Spanish Levante, endemic specie called *Posidonia Oceanica*, a marine plant leaving off-shore in a range of 5–30 m depth, is strongly protected by the European Union. It has been demonstrated that it is very sensible to brine increments as some recent studies reported (Buceta *et al.*, 2003), and compensatory measures have to be taken (for instance enlarging the brine discharge pipe up to the end of the existing prairies). Consequently, in most cases the environmental load due to the installation of desalination plants can be removed by increasing the cost of the brine disposal system.

Finally, brine disposal has a very problematic solution for brackish waters coming from saline aquifers despite of its lower costs, since in most cases that discharge could even degrade more and more the aquifer quality if it is not correctly designed.

### 5.2 *Comparing with other alternatives*

Anyway, the environmental impact of desalination should be compared with the impact provoked by another alternative supplying the same quantity (and quality if possible) of water. A paradigmatic example (see Serra *et al.*, 2003, for details) could be the comparison of the environmental assessment between the water provided by the *Ebro River Water Transfer* included in the *Spanish National Hydrological Plan* (a maximum of 1050 Mm$^3$/yr) and the same quantity of desalted water obtained by the number of desalination plants that could supply that specified volumes. The environmental assessment tool employed is the *Life Cycle Assessment* (LCA), which is one of the most powerful, recognized and internationally accepted tool to examine the environmental cradle-to-grave consequences of making and using products and services by identifying and quantifying energy and material usage and waste discharges.

Table 8.   Overall LCA scores for different energy consumptions of RO and the *Ebro River Water Transfer* (pay-off period of 25 and 50 years) (Serra *et al.*, 2003).

|  | Unit | RO ($4\,kWh/m^3$) | RO ($3\,kWh/m^3$) | RO ($2\,kWh/m^3$) | Transfer (50 years) | Transfer (25 years) |
|---|---|---|---|---|---|---|
| Eco-indicator 99 | GPts | 2.62 | 2.04 | 1.46 | 1.86 | 2.20 |
| Ecopoints 97 | GPts | 43,400 | 34,200 | 25,100 | 29,900 | 35,900 |
| CML 2 baseline | – | 0.546 | 0.414 | 0.283 | 0.362 | 0.378 |

*Simapro 5.0* is the software used to perform that complex LCA study, which includes the environmental loads associated to the construction and assembly of the installations, their operation and the final disposal once their useful life is finished. Some aspects that have not been taken into account in the LCA of RO versus *Ebro Transfer* are the next:

– Biological effects derived from the brine disposal or the rupture of the river basin unit provoking the loss of endemic species or the invasion of new invasive ones (as for instance the *zebra mussel*), and the environmental impact of the water transfer in the Ebro Delta.
– The different quality of desalted and transferred water, assuming that both waters are valid for human consumption in both cases.
– The secondary water distribution networks from the channel or permeate piping system to the end-users, because of the lack of information about this issue in the *National Hydrological Plan*.
– The uncertainty of the available volumes of the Ebro River in the next future taking into account the historical data revealing faults in the recent years. This problem should be more acute if it is also considered the effect of the climate change and the increasing demands in the conceding basin.

The average energy consumption of the Ebro River water transfer was $2.5\,kWh/m^3$ (calculated from MIMAM, 2003) in the operation phase of the life cycle of the water transfer, whereas the energy consumption of the RO was varied from 4 to $2.5\,kWh/m^3$, considering the decreasing trend of energy consumption in that technology. The Spanish model to produce power is considered for that analysis (51.3% is thermal, 13.3% is hydropower and the other 35.4% is nuclear energy). The overall scores obtained with the analysis are presented in the Table 8. Three different methods have been used. Note that those methods provide a unique number without any physical sense but very useful to compare different alternatives.

It means that with the present state of the art of both technologies, RO is slightly more pollutant than the *Ebro Water Transfer*. The importance of energy can be better understood when the LCA results are presented in percentage from each life cycle stage, as shown in the Table 9.

In both options, the operation stage, in which the energy consumption is the most important factor because the infrastructure and the plants are already built, is clearly the most important phase. Thus, the efforts for reducing the environmental load in both options should be devoted in the address of reducing the energy consumption and/or the usage of environmental-friendly energy production systems. Anyway, the weight of the water transfer construction is quite important with respect to the same phase for RO plants, a factor which is usually not considered

Table 9.   Environmental load, for three different assessment methods, corresponding to each life cycle phase for the RO (4 kWh/m$^3$) and the *Ebro River Water Transfer*, considering the pay-off periods of 25 and 50 years (Serra *et al.*, 2003).

| Process | MSF | Unit | Eco-indicator 99 | Eco-points 97 | CML 2 baseline |
|---|---|---|---|---|---|
| RO | Assembly | % | 6.95 | 10.22 | 2.03 |
| | Membranes | % | 0.77 | 0.28 | 0.06 |
| | Operation | % | 92.28 | 89.5 | 97.91 |
| Water Transfer | Assembly | % | 31.58 | 33.6 | 8.58 |
| (25 years) | Operation | % | 68.41 | 66.4 | 91.42 |
| Water transfer | Assembly | % | 19.09 | 20.32 | 4.50 |
| (50 years) | Operation | % | 80.91 | 79.68 | 95.50 |

for other environmental assessment techniques. Note that the final disposal stage of both alternatives does not suppose any contribution with respect to the other stages.

The results obtained here present a very optimistic future for desalination (especially RO) with respect to conventional hydraulic projects, since:

– Technologic improvements permit a continuous reduction of energy consumption. On the contrary, hydraulic projects are a mature technology.
– The gradual inclusion of renewable energies (see Serra *et al.*, 2003, for different energy production scenarios) favors RO desalination more than conventional water supply techniques.
– By integrating desalination with energy production systems, a very important reduction of the environmental charges of desalination could be obtained (Raluy *et al.*, 2004).

## 5.3   *Using renewable energies*

The use of renewable energies for producing desalted water is not developed yet at a major scale, but it could be a very interesting solution for remote and lowly populated areas, where no other alternatives are available at affordable costs. The use of solar direct heating onto a glass surfaces collecting evaporated seawater in a greenhouse recipient only produces up to 4 L/m$^2$ of glazed cover; and the use of *parabolic through collectors* (PTC) concentrating the solar energy to heat a thermal fluid or produce steam that it is consumed in a thermal distillation process (MSF or MED small plant coupled to the energy storage system) can improve that productivity up to 10 L/m$^2$ (García *et al.*, 1999).

Photovoltaic solar cells connected to small RO units is quite affordable if a storage system and/or batteries are also included in the design project, in order to smooth the solar irradiation along the day. Anyway, the cost of that solution is only recommendable for isolated locations (see Espino *et al.*, 2002). But the most promising renewable energy for connecting RO desalting units is wind, considering than even wind farms would be settled off-shore and therefore its environmental impact would be strongly diminished, and the great power that could be generated in the coastal line provoked by a constant wind regime. The integration of that type of energy with a pumping system that allows the storage of seawater in small dams that could produce energy in a hydroelectric installation, and finally desalted RO in a continuous manner, is a project that will be developed in the island of El Hierro (Canary Islands) and seems to be one of the promising future line for desalination in a low-medium scale.

## 6   FUTURE INNOVATIONS IN DESALINATION

Dealing with the present technologies and their improvement, the future of desalination technology is addressed to the reduction of costs and the minimization of the environmental impacts generated by those technologies.

### 6.1   *Reducing costs*

Several future research lines are opened in this field. The more important ones are remarked here:

- The research oriented to the integration of water and energy systems, is essential to reduce the costs of water and energy in dual purpose plants. Up to now, water and energy production were managed independently, but a combined management and the appropriate tools as for instance *Thermoeconomic Analysis* (Uche, 2000) permits an in-depth knowledge of both plants and their interactions, thus discovering guidelines for the reduction of both costs. Use of residual heats or exhausted flows rejected to the environment is highly recommended in those plants.
- The use of hybrid systems (RO + MSF/MED units, in which RO only works in low-peaks of power demand, usually in winter periods). The concept of aquifer storage could also be combined in those periods, reducing therefore the cost of desalted water (costs are always calculated per year of operation) (Awerbuch, 2002).
- Last idea suggests increasing the plant availability or utilization factor, permitting the cost reduction. The use of experienced designers, consultants and operators, as soon as high qualified personnel is also essential to maintain those high factors. Use of materials resistant to corrosion is another factor improving that availability, especially for thermal plants (Wangnick, 2001).
- In the MSF process, the inclusion of softening membranes (UF/MF) for allowing higher *Top Brine Temperatures* (up to 140°C) without the problems of scaling and precipitations, and therefore increasing the process productivity (e.g. higher Performance Ratio) (Ejjeh, 2001). Another way of reducing water costs in thermal techniques is augmenting the *Heat Transfer Coefficients* (HTC) of evaporators/condensers, and therefore the required surface area could be diminished. If surface area is maintained, seawater production can be increased considerably on the other way (Wangnick, 2001). Alternatively, as MED plants work at low temperatures (up to 70°C), the inclusion of aluminums or even plastic materials will reduce considerably the investment cost of evaporators.
- For membrane processes, one of the prominent lines is the search for a more simplified scheme in the pretreatment system, mainly due to the enormous space required for the conventional treatment – flocculation, sedimentation and filtration, followed by diverse additives including chlorination and dechlorination. And the principal equipment of RO plants (membranes) is yet susceptible for major improvements (Fariñas, 2001): flow through membranes could be increased (or alternatively the pressure required to produce the same flow could be reduced), and/or reduce the specific costs of membranes (e.g. the cost per unit of surface in spiral membranes). Remember that the pressure required to produce a significant flow across the membrane is more than two times the osmotic pressure required for a typical seawater.

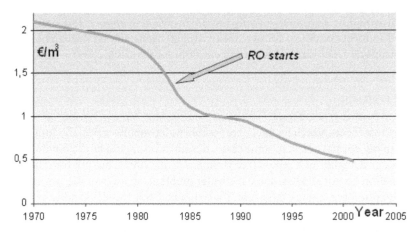

Figure 9.   Water cost obtained by seawater desalination in Spain along the time (AEDyR, 2002).

– The inclusion of the energy recovery systems for large RO plants (for instance the pressure exchanger supplied by RO kinetic) substituting Pelton turbines could reduce up to $1\,kWh/m^3$ the specific energy consumption, reaching a total consumption (intake pumping + RO process + permeate pumping to storage system) of less than $3\,kWh/m^3$ nowadays. In this way, the advantage obtained by these techniques is yet close to its limit, having an efficiency of about 95%.

   All those measurements could reduce the costs of desalted water on the next future but not as recent decades, as we can see in the historical trend for the water cost desalted in Spain (Figure 9). This is mainly due to the compensation produced by the expected increasing prices of energy, taking into account for the political situation and the environmental charges that will be associated to energy.

## 6.2   Reducing the environmental impacts

To reduce the environmental impacts of desalination plants, starting from the present situation, some guidelines are suggested here:

– First of all, it is essential to reduce the energy consumption for all technologies, and integrate them with energy production systems, particularly with those driven by renewable energies.
– Impede the corrosion of materials.
– Use sound-insulation materials to reduce the impact of noise.
– Reduce the addition of chemicals and use the less aggressive ones to the environment.
– Avoid (if possible) the use of chlorine in desalination processes by investigating new affordable disinfection methods, in order to prevent the formation of trihalomethanes (Wangnick, 2001).
– Reduce the thermal contamination by immersing cooling exchangers.
– Increase the investment required for enlarging the brine discharge pipes, also for inland territories to collect saline waters from brackish plants.

Finally, it would be desirable to apply the *Environmental Impact Assessment* (EIA, usually required in most countries) for new facilities, and especially strategic EIA that consider cumulative impacts form various sources in one region (Schiffler, 2004).


# 7   CONCLUSIONS

Desalination represents nowadays a reliable potable water supply, especially for providing safe fresh water to big coastal cities. In some cases, desalination should be the dominant source for balancing water demands of a region (see Bremere *et al.*, 2001), but not for other ones: up to now, desalination do not allow to solve the water problems in all over the world. The reasons for that assumption are, among others, briefly explained and classified here:

## 7.1   *The Integrated Water Resources Management (IWRM) solution*

– Desalination should always be the last solution to avoid water availability problems: water *demand* strategies as efficient irrigation methods, reduction of leakages in pipes, use of water saving devices in households, introduction water markets in drought periods, and the intensive use of reused water should be strengthened previously. A paradigmatic example of that inadequate projection is Malta island (Schiffler, 2004): when leakage losses were reduced, desalination plants had overcapacity.
– Desalination should be the selected water *supply* alternative if provokes less economic and environmental charges to conventional alternatives for supplying water (other solutions may involve inter-basin transfers with a huge budget and uncertainties derived from the climatology).

## 7.2   *Costs*

– Currently, only 3% of global drinking water is supplied through desalination, but it is concentrated in developed countries and in the Arab Gulf countries (remember that 2700 million people live on less than 2 US$/d) (Schiffler, 2004).
– Costs have fallen to affordable levels for many communities (up to 0.50 €/m$^3$), but in the future they will continue to fall but not as fast as previously, since no major breakthroughs are envisaged in the immediate future.
– The use of desalted water for agriculture should be studied carefully. Usually, the economic rationality of this solution is not justified, and the use of subsidies is quite extended. Moreover, in water scarce areas desalination for urban supply should not be applied to free up water for irrigation.

## 7.3   *Environmental impact*

– Desalination has local impacts but they can be mitigated, usually at low or moderate costs, which obviously must be charged to the users. And in most cases, they can be outweighed by the benefits (for instance avoiding the use of alternative conventional resources harmful with the environment).
– However, it has a serious global impact derived from its energy consumption. Remember that the thermodynamic limit for desalting seawater is about 0.8 kWh/m$^3$, i.e. the minimum

realistic consumption for the complete plant would be estimated in 1.5 kWh/m$^3$, equivalent to a groundwater pumped out from a well at about 400 m depth. Therefore, seawater desalination will not be the definite solution for water scarce areas till renewable energies would be massively introduced (maybe in the next 20–30 years), taking into account that 1200 million people lack potable water and 3000 million lack appropriate sanitation conditions.

## REFERENCES

Abufayed, A.A.; Elghuel, M.K.A. & Rashed, M. (2002). Desalination: a Viable Supplemental Source of Water for the Arid States of North Africa. *Desalination*, 152: 75–82.

AEDyR (Spanish Association for Desalination and Water Reuse) (2002). Publicity brochure of desalination in Spain. Presented at the *IDA 2002 Conference*. Manama, Bahrain.

Albiac, J.; Hanemann, M.; Calatrava, J. & Uche, J. (in press). Evaluating Alternatives to the Spanish Nacional Hydrological Plan. *Water Resources Development*, special issue.

Al-Fraij, K.M.; Al-Adwani, A.A. & Al Romh, M.K. (2003). The Future of Seawater Desalination in Kuwait. *IDA 2003 Conference,* Paradise Island, Bahamas.

Al-Gobaisi, D.M.K. (2003). Sustainability of Desalination Systems – an Essential Consideration for the Future. *Integrated power and desalination plants*. EOLSS Publishers Ltd., London, UK.

Al-Weshah, R. (2002). The role of UNESCO in Sustainable Water Resources Management in the Arab World. *Desalination*, 152: 1–13.

Andrews, W.T. & Verbeek, V. (2002). Regional Report on Desalination – Latin America & Caribbean. *IDA 2002 Conference*. Manama, Bahrain.

Awerbuch, L. (2002). Vision for Desalination – Challenges and Opportunities. *IDA 2002 Conference.* Manama, Bahrain.

Birkett, J. (2002). Desalination in North America. A Regional Review. *IDA 2002 Conference*. Manama, Bahrain.

Biswas. A.K. & Tortajada, C. (2003). Assessment of Spanish National Hydrological Plan. Reports of International Experts required by the Government of Aragón.

Bremere, I.; Kennedy, M.; Stikker, A. & Schippers, J. (2001). How Water Scarcity will Effect the Growth in the Desalination Market in the Coming 25 years. *Desalination*, 138: 7–15.

Buceta, J.L.; Gacia, E.; Mas, J.; Romero, J.; Ruiz, J.; Ruiz-Mateo, A. & Sánchez, J.L. (2003). Estudio de los incrementos de salinidad sobre la fanerógama marina Posidonia Oceánica y su ecosistema, con el fin de prever y minimizar los impactos que pudieran causar los vertidos de agua de rechazo de plantas desaladoras. Final report and recommendations. [Pers. comm.]

CIRCE Foundation (2003). Alegaciones al Proyecto de Transferencias aprobado en el artículo 13 de la Ley 10/2001 y su Estudio de Impacto Ambiental. [Pers. comm.]

Ejjeh, G. (2001). Desalination, a Reliable Source of New Water. *International Conference Spanish Hydrological Plan and Sustainable Water Management*. Zaragoza, Spain.

Espino, T.; Peñate, B. & Piernavieja, G. (2002). Proyecto Dessol: experiencia piloto de desalación de agua de mar por ósmosis inversa alimentada con energía solar fotovoltaica. Una opción de abastecimiento para zonas costeras. *AEDyR 3rd Annual Congress*. Málaga, Spain.

Ettouney, H.M.; El-Dessouky, H.T.; Faibish, R. & Gowin, P.J. (2002). Evaluating the Economics of Desalination. *Cep magazine* (www.cepmagazine.org).

Fariñas, M. (2001). Novedades tecnológicas en la desalación por ósmosis inversa. *International Conference Spanish Hydrological Plan and Sustainable Water Management*. Zaragoza, Spain.

Freshwater Action Network (2002). NGO Guide to the Water-Dome, the World Summit on Sustainable Development and International Water Policy. *World Summit on Sustainable Development*. Johannesburg, South Africa.

García, L.; Palmero, A.I.; & Gómez, C. (1999). Application of Direct Steam Generation into a Solar Parabolic Trough Collector to Multieffect Distillation. *Desalination*, 125: 139–145.

Goto, T. (2002). Water Problems in East and South Asia. *IDA 2002 Conference*. Manama, Bahrain.

Hsu, S.K. (2001). Seawater Desalination as an Alternative Water Supply in Taiwan. *IDA International Conference.* Singapore, March 2001, Session VII.

Khalaf, A. (2003). The Effective Unit Cost of Electricity and Water in the Kingdom of Bahrain. *IDA Conference.* Paradise Island, Bahamas.

Lokiec, F. & Kronenberg, G. (2003). South Israel 100 million/yr Seawater Desalination Facility: Build, Operate and Transfer (BOT) project. *Desalination*, 156: 29–37.

MIMAM (Ministerio de Medio Ambiente) (2003). *Proyecto de Transferencias autorizadas en el artículo 13 de la Ley 10/2001 del Plan Hidrológico Nacional y su Evaluación de Impacto Ambiental.* Spanish Ministry of Environment. Madrid, Spain.

Morris, R. (2004). Technological Trends in Desalination and Their Impact on Costs and the Environment, *World Bank Water Week.* Washington, D.C., USA.

Raluy, G.; Serra, L. & Uche, J. (2003). Life Cycle Assessment of MSF, MED and RO Desalination Technologies. *Second Conference on Sustainable Development of Energy, Water and Environment Systems,* Dubrovnik, Croatia.

Raluy, G.; Serra, L.; Uche, J. & Valero, A. (2004). Life Cycle Assessment of Desalination Technologies integrated with Energy Production Systems. *Desalination*, 167: 445–458.

Schiffler, M. (2004). Desalination: Recent Trends and the Role of the World Bank. [Pers.comm.]

Serra, L.; Raluy, G.; Uche, J. & Valero, A. (2003). *Environmental Impact of Water Production Technologies. Life Cycle Assessment of Ebro River Water Transfer versus the Reverse Osmosis Desalination.* Report financed by the Government of Aragón. [Spanish version available at www.trasvasebro.com].

Shamir, U. (2003). Review and Evaluation of Certain Aspects of the Spanish National Hydrological Plan. Reports of International Experts required by the Government of Aragón.

Shiklomanov, V. (1998). *World Water Resources. A New Appraisal and Assessment for the 21st century.* UNESCO, Paris, France.

Uche, J. (2000). *Thermoeconomic Analysis and Optimization of a Dual-Purpose Power and Desalination plant.* Ph.D. Thesis. Department of Mechanical Engineering, University of Zaragoza, Spain.

Uche, J.; Valero, A. & Serra, L. (2002). *La desalación y reutilización como recursos alternativos.* Administrative document printed by the Government of Aragón.

Uche, J.; Serra, L.; Herrero, L.A.; Valero, A.; Turégano, J.A. & Torres, C. (2003). Software for the analysis of water and energy systems. *Desalination*, 156: 367–378.

Wangnick, K. (2000). *2000 IDA Worldwide Plants Inventory.* Report no. 16. Wangnick Consulting.

Wangnick, K. (2001). A Global Overview of Water Desalination Technology and the Perspectives. *International Conference Spanish Hydrological Plan and Sustainable Water Management. Zaragoza, Spain.*

Wangnick, K. (2002a). *2002 IDA Worldwide Plants Inventory.* Report no. 17. Wangnick Consulting.

Wangnick, K. (2002b). Regional Review for Europe – The Region Leads in Seawater Desalination. *IDA 2002 Conference.* Manama, Bahrain.

World Water Council (2000). *World Water Vision, Making Water Everybody's Business.* Earthscan, London, UK.

CHAPTER 19

# The potential for desalination technologies in meeting the water crisis: *comments*

E. Custodio
*Technical University of Catalonia (UPC), Barcelona, Spain*

ABSTRACT:  These notes add some comments to the chapter by Uche *et al.* (this volume) on desalinizated water use. The comments do not refer to the desalinization techniques themselves. Raw water for desalination is often coastal seawater, but also brackish and saline groundwaters, both in coastal and inland emplacements. Treated sewage water, and also brackish water from other sources, such as return irrigation flows, are susceptible of being desalinized. Desalinization technology and costs have improved dramatically in the last few decades with respect to specific energy consumption, the processes themselves, the materials, and the membranes when they are used. But in spite of a current, fairly well known, and proven technology, information on water costs remains uncertain for people and even for policy-makers due to diverse intervening economic parameters, and how they are modified to yield to social pressure and political goals. The relative low cost of water storage implies that under normal circumstances the most economic way of operating a desalinization plant is continuously at nominal capacity. Return flows from a desalinization plant must be safely disposed of and this adds significantly to water cost. Desalinization is an energy and capital intensive process, as are also long water transfers and tertiary treatment of waste water. Produced water cost therefore is intrinsically expensive, more for seawater than for brackish water. It becomes a new water resource in areas and for uses that can support high prices and have water resources at hand to be desalinized. However, this is not a resource for poor areas or for water intensive processes except if heavily subsidized. Under normal circumstances desalinization is just one more alternative for water supply to be considered among others.

Keywords:  *Desalinization, sea water, brackish water, groundwater, environment*

## 1  INTRODUCTION

The chapter prepared by J. Uche, A. Valero and L. Serra (this volume) on *the potential for desalination technologies in meeting the water crisis* prove to be very comprehensive covering the different aspects of desalinization and its role in water resources. The emphasis in these comments will be on water resources and some related environmental aspects and not on the technical aspects.

In this chapter, the term desalinization shall be used with preference to desalination. In fact the *Webster's Ninth New Collegiate Dictionary* (1983) says: "Desalination: *see desalt*; Desalinization: *see desalt*; Desalt: *to remove salt from . . .* ". It is clear that these are synonymous terms. The preferential use of desalinization tries to point out that the focus is in making the water less saline and fresh instead of taking out salts, although some processes accomplish the latter. Some experts on this topic argue that desalination is the right term when raw water is sea water, and desalinization is preferable when raw water is brackish water. In the opinion of the author this is too artificial and does not introduce any significant language improvement.

Desalinization is any process whose objective is to reduce the saline content of water that is naturally rich in dissolved salts or has been salinized due to some natural and/or anthropic processes.

Saline and brackish waters contain an excess of salts that prevent or make their use inconvenient for the intended purposes (potable, urban, industrial, rural or agricultural). The reduction of some constituents, such as inorganic carbon, $Ca^{2+}$, $Mg^{2+}$, $Fe^{2+}$, can be accomplished at low cost by means of conventional physico-chemical treatments. But the removal of $Cl^-$, $Na^+$ and $SO_4^{2-}$, which are generally the main constituents of brackish and saline waters, need desalinization treatments due to their high solubility and chemical stability. The reduction of other constituents such as $NO_3^-$ from anthropically polluted water may also be a complementary objective for desalinization. The reduction of organic contaminants and of bacteria and viruses are a consequence of desalinization but not a main objective. Processes aimed at these objectives, such as micro-, ultra- and nano-filtration through membranes, although they share similar technologies with membrane desalinization, they are not properly such.

Desalinization technologies are well known and have been around since the late 1950s and some of them now in use were developed in the 1970s (Custodio & Llamas, 1976: section 23). During this time some major improvements have been made such as less energy consumption, more suited materials, enhanced head transfer, and much better membranes. Sufficient experience has been also gained to allow the making of reliable projects and valid cost estimations. Some of the distillation processes (MSF, MED, VC) and membrane processes (RO, ED) are currently well established and developed, as commented by Uche *et al.* (this volume).

Other processes such as freezing and solvent extraction have not progressed beyond their pilot plant stage, and perhaps will not be developed any further since no economic advantages are expected.

Figure 1 presents a simple scheme of the desalinization processes with the main items and stages that will be considered and discussed later on. It is quite self-explanatory.

It seems quite important to mention that fresh water and potable water availability is not clearly related with water availability except in particular situations, such as arid and semiarid areas, where the only solution is to cut down on water demand for human activities or import water from other areas. In this case, the reuse of water may be significant.

**DESALINIZATION**

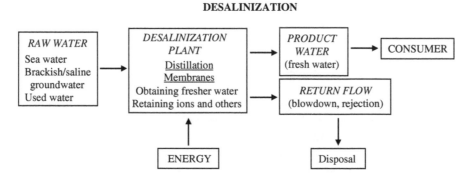

Figure 1.   Simplified scheme of desalinization processes with the main items and stages.

Oftentimes, available water is too saline or it has some solutes in excess or it is inorganically or organically contaminated. This of course will reduce usable water resources and may even create an understandable concern or the feeling of an oncoming water crisis. Nevertheless, usable water can be produced using existing technology from any kind of water (brackish, saline, contaminated), even though a large area for technical improvement still exists. But this implies, however, that energy is available to do it and it can be done at an affordable price in a way that does not further harm the environment. Jointly with waste disposal, this is a major challenge, and a reason to not give much weight to Shiklomanov's (1998) figure of 10,000 km³/yr for total renewable freshwater resources, or to the threshold limits of necessary available water per person, as pointed out by Llamas & Custodio (2003). Also, the demand for water does not necessarily coincide with water needs (Merrett, 2004).

One major challenge is how can human populations have access to fresh and potable water within their economic capability when a large fraction of them are poor, technologically deprived and energy deficient. In these cases desalinization is of little use.

## 2  ROLE OF DESALINIZATION

A main result of the present situation is that technological progress and cost reductions have allowed desalinization to be considered as an alternative for solving a given set of freshwater demand problems. It becomes a real alternative when freshwater is an expensive local commodity due to the cost of making it available or due to market competition when it is scarce. The concurrence is to respect other water sources that are also capital intensive and energy consumption intensive, as occurs with long water transfers or the reclamation of sewage water, and even deep aquifer development in extreme situations.

Real circumstances are quite variable, and normal situations may be changed into exceptions when an area has a high water demand or is arid. These situations do not permit simple rules to be established or detailed economic analyses to be requested in order to help in the decision-making process. It may so happen that there is not an optimal solution, but rather a set of possibilities where in some of them desalinization may play a role when brackish and/or saline water is or can be made available. This includes used water and return flows from agriculture, mining or industrial usage when too brackish for direct use.

The abundance of brackish and saline water in the ground, both in costal and inland situations, even if large parts are non-renewable reserves, opens a complementary desalinization role for groundwater development. One primary problem that remains is what to do with the return flow (blowdown or rejection). This will be considered later on.

As mentioned in the Introduction, will also be commented later on, and is well stressed in the papers by Uche *et al.* (this volume) and Semiat (2000), desalinization is inherently expensive, even with foreseeable technological improvements, unless energy is available at very low cost, which is unlikely except in heavily subsidized situations.

## 3  DESALINIZATION GOALS

Desalinization, as a real alternative or complement to other freshwater sources, has three different aspects.

(a) Desalinization of sea water in coastal areas (including continental saline surface water bodies). Raw water is either sea water directly uptaken from the sea or groundwater that is directly recharged into the ground from the sea to avoid turbidity, solid particles and biological problems. Usage in this situation usually takes place near the coast. Inland transportation may imply an important added cost, especially if water demand is in high elevation areas or far away. Every 100 m elevation or head loss adds about 0.4 kWh/m$^3$. The blowdown brines (return flow) can be disposed of into the near-by sea (see later on).

(b) Desalinization of brackish and saline groundwater, either near the coast or at a distance, in continental emplacements. Desalinization of brackish water by means of membranes has a relative low cost and a reduced energy consumption. Its feasibility will depend on available brackish and saline water reserves or the rate at which they are produced in the case of seawater intrusion. Feasibility will also depend on the existence of difficult-to-pretreat inconvenient solutes, any possible side-effects of groundwater exploitation on existing freshwater resources, and on land subsidence. The possibility of obtaining raw water and its cost will also be another feasibility factor. Every 100 m elevation above the dynamic groundwater level adds about 0.5 kWh/m$^3$ to the desalinization cost for the common energy efficiency of submerged pumps. One of the main problems in inland areas is return flow disposal, which may vary from saline water to a brine. This may add a significant cost increase and may even make desalination unfeasible. If there is not a deep saline aquifer to inject them into, return flows must be carried out to the coast or to evaporation ponds by means of brine-ducts. Brine-ducts are prone to failure and breakdown, with the added risk of polluting continental water resources. Abstraction of saline water from a coastal or continental aquifer may help in the protection of existing freshwater resources. This effect must be taken into consideration in water balances, and it may affect freshwater availability if wells produce brackish water through the mixing of freshwater with more saline water.

(c) Improving the quality of treated sewage and used water by removing a part of the salts as well as dissolved organic matter, bacteria and viruses at the same time through the use of appropriate membrane technologies. The cost will depend on raw water salinity and the degree pre-treatment to be applied, and on the final quality to be obtained. Product water may be for direct use (irrigation, industry, municipal, even household) or for further treatment and storage by means of artificial recharge into the ground.

## 4   ENERGY FOR DESALINIZATION

One of the main drawbacks for desalinization is the high energy consumption and the associated environmental problems produced. Energy production for desalination adds to the increased impact on the Earth's thermal balance through the release into the atmosphere of greenhouse gases and particles. But long water transfers may also be energy consumption intensive (up to some kWh/m$^3$).

Current reverse osmosis processes with return flow pressure recovery may consume 3.0 to 4.0 kWh/m$^3$ of product water for raw water from the sea, and proportionally less for lower salinity raw water. These values approach the thermodynamic minimum energy for conversion. From here on, only relatively small improvements in energy consumption can be expected for membrane technology. There is still some more room for distillation processes. MED scores are also currently good for sea water desalinization in large plants, although they cannot compete with membrane processes for brackish water.

To the cost of in-plant product water, the cost of elevation and distribution to the supply centres, which is far from being negligible, must be added.

The price of energy will vary from region to region and may include economic charges, taxes and subsidies of many kinds. This makes real circumstances quite variable. The future trend is probably on the increase. In many cases the relatively cheap hydropower potential is already fully used, fossil fuels are being depleted and renewable sources of energy are expensive (or heavily subsidized) and represent only a relatively small contribution to total energy demand. Variable renewable energy sources, such as wind and solar power, are poorly suited to the capital expensive desalinization plants if used directly.

Nuclear power is a real alternative, although not a cheap one and with limited resources, but it suffers from a wide and sometimes irrational social rejection. This has been a serious brake for solving energy problems in many countries, and as a consequence adds to the energy-related problems of desalinization. From the point of view of global environmental impact on the climate, nuclear energy is potentially much cleaner than fossil fuel energy. The future for relatively clean and affordable energy prices still relies on nuclear energy. But the gap between the present situation and what could be a significant contribution for the future is widening. The need is for a bold decision now and a sustained maximum effort during a series of years. For improved nuclear fission energy production using breeder reactors, too much time has been lost on moratoria. Any significant impact on energy availability may come too late. Nuclear fusion reactors have a more important potential, but just to be practical this technology will probably need more than 25 years to achieve a significant impact on energy availability.

## 5   DATA ON DESALINIZATION COST

When comparing desalinization with other freshwater supply sources in areas where water prices are sufficiently high, very diverse figures are found in both technical and mass media reports. These differences not only misinform citizens but they also mislead anyone having any sort of decision-making capacity. Assuming that figures are not expressly modified or falsified, which does occur sometimes to reach some pre-established result or support some pre-established solution, the different results released for public information are quite widespread. In addition to technical issues, the basis for economic calculations may also clearly differ. Some of the main factors for this are the interest on money and the discount rate to be applied, the duration of the pay-off period and the financing conditions, which may conspicuously change from case to case. This is often due to differing public and private investment viewpoints grounded on social interest and enterprise benefits. Public sponsoring tends to apply lower interest and discount rates, longer pay-off periods and improved financing conditions through hidden guarantees and diverse kinds of subsidies. Crucial also is knowing where the water cost is calculated: at the desalination plant, or at distribution storage reservoirs, or at the mains, or where consumed. Water quality is another factor to be considered, since pre-treatments are expensive, a low salinity product will allow increasing water availability by mixing with other brackish waters or waters with excessive contents of some constituents.

The manner in which taxes, fiscal exemptions, subsidies and special prices are applied also play a significant role on the resulting costs and the economic parameters used to compare projects and alternatives. Publicly sponsored activities often take advantage of lower taxes, low interest loans, subsidies on energy, land purchases or construction, increases in the price of other commodities (e.g. electricity and motor fuels), in order to compensate for water cost if considered too high from a social perspective. Sometimes brine-ducts are built as a common interest investment (e.g. paying others).

These and other similar considerations mean that comparisons can only be carried out fairly when economic circumstances are similar or respond to well-defined protocols.

A continuous source of debate about these different circumstances has taken place in the Eastern Canary Islands, Spain, where desalinization plants of very different ages and sizes are financed, operated and maintained from a wide combination of public and private initiatives. There and in other areas of Eastern Spain water prices charged to customers or used for accounting purposes may differ by more than a factor of two. For a large reverse osmosis plant about to be commissioned in Eastern Spain, free-market product water cost at the plant is calculated at about $0.9 €/m^3$ but will be sold at $0.4 €/m^3$, the difference being compensated by subsidies and other economic advantages. Currently, in the midst of a controversy between a long transfer canal and desalinization, the media is showing cost informations that varies between $0.3 €/m^3$ and $1.2 €/m^3$, and are seldom complemented with further data, such as the raw water origin and salinity (brackish or marine). Only experts have some information.

Other factors that are not easily obtainable are the applied repayment rate and time of facilities, interest rates, the point where water is made available, cost sharing with other uses (energy, water production), return-flow disposal and associated storage, transportation facilities, energy costs and subsidies on water, energy and personnel. Results are often presented in a rather obscure form to the public that prevents desalinization development from properly solving real water issues, or it results too in favour. Notwithstanding, poor knowledge, and to some extent irrational political pressure, have been responsible for these shortcomings (some plants have been installed as a political gift and they have never been exploited since there were no provision for operation expenses) and suboptimal use of integrated water resources, especially with poor attention to groundwater.

Current trends to desalinization of tertiary treated sewage water are gaining ground, but some of the experimented defects have not been adequately corrected. Sometimes expectancy is not fully supported by well-proven technologies. However, research is still active and many difficulties should soon be resolved.

## 6   OPERATION CONDITIONS OF DESALINIZATION PLANTS

Since desalinization is a capital-intensive process, as are surface water developments and long-distance water transfers, operation at nominal capacity should be on a continuous basis, thus keeping the number of stops for maintenance and repairs to a minimum. One important difference regarding energy production facilities, which are also capital-intensive processes, is the storage of energy to compensate supply variability. There is a high cost attached to energy storage. Thus, lower capital and high energy cost facilities can be used for regulation. Water storage, on the other hand, is generally much easier and cheaper, which means that continuous operation of desalinization plants is feasible if enough water storage capacity is added. Surface water dams in many cases are storage facilities whose electricity production units are relatively cheap but long-transfer pipelines and canals however are so capital-intensive, that projects must generally look for full continuous use, even if in many cases they are also energy intensive.

The most difficult water storage conditions appear to be in densely populated coastal areas due to urban pressure and the scarcity of adequate sites. The capacity for a one month supply volume may contribute less than 0.03 €/m³ to water cost, except in very extreme situations. This means that the continuous operation of desalinization facilities is both advisable and feasible, and that the storage facilities required will normally have no important effect on the economic competition among desalinization facilities or other alternatives for freshwater supply.

The conclusion is less clear for a dual purpose facility (water and electricity), depending on the conditions for electricity production. This may be just one more cause for the difficulty in managing these types of facilities.


7   ENVIRONMENTAL ASPECTS

There are some environmental aspects to be taken into consideration related to desalinization.

Some water managers and environmentalists are concerned about the negative results of abstracting brackish or saline groundwater to feed desalinization plants, arguing the risk this may have on fresh groundwater resources in the exploited aquifer. In most cases these concerns are not real (Custodio & Bruggeman, 1987). If very simple precautions were taken, then what would be produced would be a protection of freshwater resources against further salinization. This is one of the *hydromyths* (Custodio, 2005) that must be overcome. When brackish water is abstracted, only a fraction of the freshwater resources from the aquifer system may be affected, but this flow must be considered in freshwater balances. A combination of freshwater abstraction and brackish water development for desalinization is under study in the Netherlands (Kooiman *et al.*, 2005).

A much more serious concern is the disposal of return flows, which are often brines. For coastal desalinization plants the disposal site is generally the sea. This slightly warm and denser-than-seawater brine not only is a threat to the raw water supply for the plant if some of it gets back into the feed system, but it is also a hazard for marine life, especially bottom species and sensitive *posidonia prairies*, which are important habitats in some Mediterranean sea submerged platforms. Control can be obtained through longer outfall pipes (up to 5 km and more) and increased diffusion, thus swelling the cost of desalinization. It is assumed here that hazardous chemicals are being disposed of safely through a separate system, or the effluent is being treated before disposal. For example, in electrodyalysis brines high free chlorine concentrations come from the process.

For inland plants using brackish or saline groundwater brine disposal may become a major problem, risking aquifer saline pollution and land degradation. In fact when environmental authorities fail to carry out controls, this is a real problem. It is even more acute for small reverse osmosis plants, where the brine is disposed of in some cases out in the *back-yard* or in old wells, or injected into the ground without a sound practice and a careful study of future hazards.

Brine disposal requires brine-ducts to collect and convey the return waters to the disposal site. This is expensive and prone to leakage and failures as well as requiring good surveillance and maintenance. Disposal may be carried out in the sea, with the aforementioned restrictions, if the coast is not too far away. Otherwise, brines must be evaporated in natural or forced-evaporation ponds, and the precipitated salts must be contained to prevent future dissolution. If there are certain favourable circumstances and it can be carried out safely, after some treatment brines may also be injected into deep saline aquifers. This, however, is not a common situation. This all adds up to desalinization costs.

Disposal is easier when there is a relatively small return flow of sewage or used water from desalinization plants, but the problem is still worthy of consideration, especially if there is an accumulation of some harmful substances, meaning that resulting sludges and cartridges must be disposed in a security landfill. The waste problem here is not only for the desalinization plant itself but also for the pre-treatment stages.

The environmental concerns of desalinization adding to global problems through the generation of green-house gases and particles are real for the direct or indirect consumption of fossil fuels. But other water supply alternatives may also be energy intensive such as the power required for long water transportation or the prior treatment of used water.

Furthermore, energy spent on the construction of large water facilities (canals, pipes, drainages, earth displacement, etc.) and the production of associated construction materials (cement, iron, etc.) should also be taken into consideration in energy balances.

## 8    SOCIAL IMPACT OF DESALINIZATION

No doubt desalinization is rapidly expanding and will continue to expand solving real water supply problems. Nevertheless, the cost of the water produced is limited to the demands of those that can afford its price. This means relatively rich towns, tourist areas, non water-intensive industries and special cash crops. Poor citizens and farmers will be left out of the benefits unless important subsidies are applied or unless the water made available is only for subsistence, or for strict potable uses, say a few litres per day per person. Any remaining water needs may have to be supplied with less quality water (brackish, with some inconvenient and non-toxic constituent, reused). For this purpose small reverse osmosis plants have been installed in small urban communities, for instance in some inland villages in Southeast Spain, in some convenient public emplacement for local inhabitants having easy access to it. Brine disposal may often become a neglected problem, and a future nightmare.

## 9    CONCLUSIONS

The excellent chapter by Uche et al. (this volume) is complemented here with comments to point out that desalinization is currently a significantly well-developed technology that has achieved important cost reductions. Even though improvements can still be expected, they are not expected to be dramatic. Therefore, the water produced by it will remain costly and energy intensive, even though other alternative water supply resources may be as equally costly and energy intensive.

Desalinization plants must be operated continuously, which means that enough storage volume of water has to be provided. Artificial recharge of aquifers is an interesting alternative for this type of storage in coastal areas.

Desalinization produces brines that must be safely disposed of, adding significantly to the water the cost.

Brackish groundwater is an important and often cheaper resource for desalinization; however, the duration of resources and the brine disposal must be considered carefully.

Desalinization is a proven technology and its processes are well known. This means that produced water costs can be calculated with some degree of confidence once the economical parameters have been defined. These economical parameters, however, may vary from case to case and may have been modified through social and political circumstances and pressure. This is often done in a non-explicit manner resulting in cost scenarios that cannot be easily compared. The result is misinformation and difficulties in making a good selection among competing alternatives. Citizens and even policy-makers lack often reliable information. This situation fuel obscure political manoeuvrings and the rise of a *water fundamentalism* that unnecessarily complicate sound water management and the fulfilling reasonable social goals.

ACKNOWLEDGEMENTS

The author thanks the Marcelino Botín Foundation and particularly Dr. M. Ramón Llamas for his invitation to prepare these comments. Mr. Argimiro Huerga and the General Director's secretariat, of the Geological Survey of Spain, have take care of the finishing of this chapter. The ideas expressed in these comments are the author's and not necessarily that of the organization he is linked with. With this paper I would like to pay a tribute to my friend and universal developer of Ecology as a science, Dr. Ramón Margalef, who passed away in Spring 2004. Humans, and especially those dealing with water owe him and this confidence in God's role, many guidelines and basic knowledge.

REFERENCES

Custodio, E. (2005). *Myths about seawater intrusion in coastal aquifers.* In: Groundwater and Saline Intrusion (18 SWIM, Cartagena 2004). Hidrogeologia y Aguas Subterráneas 15, Instituto Geológico y Minero de España, Madrid: 599–608.

Custodio, E. & Bruggeman, G.A. (1987). Groundwater problems in coastal areas. *Studies and Reports in Hydrology*, 45: 1–576. UNESCO, Paris, France.

Custodio, E. & Llamas, M.R. (1976). *Hidrología subterránea* (Groundwater hydrology). 2 vol.: 1–2350. Ediciones Omega, Barcelona, Spain.

Kooiman, J.W.; Stuyfzand, P.J.; Maas, C. & Kappelhof, J.W.N.M. (2005). *Pumping brackish groundwater to prepare drinking water and keep salinizing wells fresh: a feasibility study.* In: Groundwater and Saline Intrusion (18 SWIM, Cartagena 2004). Hidrogeologia y Aguas Subterráneas 15, Instituto Geológico y Minero de España, Madrid: 625–635.

Llamas, M.R. & Custodio, E. (2003). Intensive use of groundwater: a new situation which demands proactive action. In: M.R. Llamas & E. Custodio (eds.), *Intensive Use of Groundwater: Challenges and Opportunities*. Balkema: 13–31.

Merrett, S. (2004). The demand for water: four interpretations. *Water International*, 29(1): 27–29.

Semiat, R. (2000). Desalination: present and future. *Water International*, 25(1): 54–65.

Shiklomanov, V. (1998). *World water resources: a new appraisal assessment for the 21st century.* UNESCO, Paris, France.

# Author index